CULTURES OF CONTAMINATION: LEGACIES OF POLLUTION IN RUSSIA AND THE U.S.

RESEARCH IN SOCIAL PROBLEMS AND PUBLIC POLICY

Series Editors: William R. Freudenburg and
Ted I. K. Youn

Recent Volumes:

Volume 9: Environmental Risks: Perception, Evaluation and Management – Edited by G. Böhm, J. Nerb, T. McDaniels and H. Spada, 2001

Volume 10: The Environmental State under Pressure – Edited by Arthur P. J. Mol and Frederick H. Buttel, 2002

Volume 11: Terrorism and Disaster: New Threats, New Ideas –Edited by Lee Clarke, 2003

Volume 12: The Organizational Response to Persons with Mental Illness Involved with the Criminal Justice System – Edited by Stephanie W. Hartwell, 2005

Volume 13: Long-Term Management of Contaminated Sites – Edited by Thomas M. Leschine, 2006

RESEARCH IN SOCIAL PROBLEMS AND PUBLIC POLICY
VOLUME 14

CULTURES OF CONTAMINATION: LEGACIES OF POLLUTION IN RUSSIA AND THE U.S.

EDITED BY

MICHAEL R. EDELSTEIN
School of Social Science and Human Services, Ramapo College of New Jersey, Mahwah, NJ, USA

MARIA TYSIACHNIOUK
Center for Independent Social Research, St. Petersburg, Russia

LYUDMILA V. SMIRNOVA
Division of Education, Mount St. Mary College, Newburgh, NY, USA

Emerald

United Kingdom – North America – Japan
India – Malaysia – China

Emerald Group Publishing Limited
Howard House, Wagon Lane, Bingley BD16 1WA, UK

Reprints and permission service
Contact: booksandseries@emeraldinsight.com

British Library Cataloguing in Publication Data
A catalogue record for this book is available from the British Library

ISBN: 978-0-7623-1371-6
ISSN: 0196-1152 (Series)

Printed and bound by CPI Group (UK) Ltd, Croydon, CR0 4YY

Awarded in recognition of Emerald's production department's adherence to quality systems and processes when preparing scholarly journals for print

INVESTOR IN PEOPLE

CONTENTS

PART 4: PERSPECTIVES

ACKNOWLEDGMENTS

Many deserve thanks for this volume. The Center for Independent Social Research in St. Petersburg has proven to be an important partner for our projects, a gracious host during our time there and very supportive in every way. Elsewhere in our travels we were assisted by Volgograd State Pedagogical University, and wish to particularly acknowledge Rector Valerie Danielchuk and Dean Tatiana Andrushenko and also the staff of the Ecological Gymnasium. The deepest thanks to Natalia Mironova and the staff of the Movement for a Nuclear Free Russia and all of the participants in their grassroots network, all activists from the Chelyabinsk Oblast; Olga Tsepilova for her tour organization and to the mayor and many corporate heads in Kirishi; and the citizen activists and welcoming villagers in Dzerzhinsk, where Alla Bolatova and Dima Vorobiov arranged a magnificent interview and tour brilliantly led by Dimitri Levashov of Sierra Club. Thanks also to Boris Filatov for his ever gracious and informative discussions, Alexander Sherstyuk for his assistance in meeting with oblast officials in Volgograd, to Stanislav and Oleg Glazechev for their warm welcome in Moscow, and to Polina Agakhaniants for the Lions Bridge.

Ramapo College of New Jersey provided program support, hosted the grant, and was the locus for much of the project activity. Ramapo must be particularly lauded for its international focus, its support of interdisciplinary projects, its willingness to host Maria Tysiachniouk's Fullbright study, and for its long-term and pioneering partnership with Volgograd State Pedagogical University, which, among many positive things, originally brought Ludmila Smirnova to the United States. During the Russian team visit, Rutgers University, Penn State University, the New Jersey Meadowlands Commission, Black Rock Forest Preserve, Orange Environment, Inc., GAG (Group Against Gas), and many others were gracious hosts. Jean Ann McGrane, now City Manager of Newburgh, NY, Manna Jo Greene of Clearwater and Kyle Rabin, then of the Riverkeeper, presented superb programs on the brownfield redevelopment of Kingston, NY and on Indian Point Nuclear Power Plant; Adeline and Murray Levine and Janice Hastrup and the EPA Superfund office at Love Canal under the direction of environmental scientist Michael Basile and Love Canal activist Luella Kenney collectively offered a powerful Love Canal case visit; Catherine McVay Hughes and George Thurston conducted an insiders tour of the World Trade Center neighborhood; Bill Hallman organized a successful mini-conference; Michele Garcia of the Ironbound Community Center provided an important environmental justice tour of the Ironbound section; Steve Couch gave an unmatchable tour of Centralia, Pennsylvania; EPA's Pat Sceppi and her colleagues and Ramapo's Sue Scher gave an unforgettable tour of the Federal Creosote Superfund site, and Judith Johnsrud, Abe Wandersman, and Madelyn Hoffman gave wonderful guest presentations. And last, but definitely not least, George and Prudence Thurston and their family opened their home as a bed and breakfast during the entire site visit for most of the visiting Russians and hosted work and social events for the entire group.

Jonathon Reisman, as Administrative Assistant, proved to be an invaluable ingredient in our mix. This extraordinary young mathematician took a course with Masha at Ramapo when she came for a Fullbright scholarship in 1999. He became so excited with her work that he spent the next two years as her research assistant and collaborator in Russia, becoming fluent in Russian and eventually undertaking his own funded research. In our travels in the U.S. and Russia, Jonathon was always with us doing whatever was needed for the project's success, including countless hours spent translating. He could hold his own with anyone in the exchange, adding countless observations and additions to our discourse and learning. We hoped he would become a social scientist, but he ran off to Medical School instead. The fact is that Jonathon can do anything and be great at it.

This project was originally conceived with the support of the Ramapo College Foundation, specifically Vice President Cathy Davey. Her colleagues Ron Kase and especially Claudia Esker supported the grant writing effort and also grant administration. Claudia was a constant and invaluable guiding light. The irascible and exacting Russ Rheinhart was accountant extraordinaire.

Bill Freudenberg, co-editor of the *Journal of Research in Social Problems and Public Policy* with Ted Yuon, has been a long time colleague and friend. He has long supported, encouraged, and enabled the lead editor in many of his best opportunities, and it was not surprising that he stood behind this one as well.

It has also been a pleasure to work with the staff at Elsevier Publishing, London, who proved very tolerant of our breaking many rules and deadlines while always supporting the project to move forward. Since most of these staff were united by the name Scott – Carol Scott, Joanna Scott, and Scott Bentley – for a time I dubbed myself Michael Scott Edelstein. Thanks to them and also to Chris Tancock, Paul Penman of Elsevier, who somehow were not named Scott, the production team of Macmillan India Limited and Joel Edelstein for graphic support on the maps.

And the core thanks for their continuing support for this project rests with the Trust for Mutual Understanding, a foundation that facilitates many wonderful projects that break down barriers, create collaborative opportunities, and unite east and west in pursuit of the arts and the environment. Masha, Lyudmila and I are forever grateful.

On behalf of Masha (Maria), Lyudmila and myself,
Michael R. Edelstein

LIST OF CONTRIBUTORS

Tatyana J. Andrushchenko	State Pedagogical University, Volgograd, Russia
Robert Binford	Metropolitan Detention Center, Los Angeles, Los Angeles, CA, USA
Alla Bolotova	Center for Independent Social Research, St. Petersburg, Russia
Nikolay G. Britanov	Research Institute of Hygiene, Toxicology and Occupational Pathology (RIHTOP), Volgograd, Russia
Stephen R. Couch	The Schuylkill Campus, The Pennsylvania State University, Schuylkill Haven, PA, USA
Michael R. Edelstein	School of Social Science and Human Services, Ramapo College of New Jersey, Mahwah, NJ, USA
Boris N. Filatov	Research Institute of Hygiene, Toxicology and Occupational Pathology (RIHTOP), Volgograd, Russia
Margaret Gibbs	School of Psychology, Fairleigh Dickenson University, Teaneck, NJ, USA
Oleg S. Glazachev	P. K. Anokhin Research Institute of Normal Physiology RAMS, Moscow, Russia
William K. Hallman	Food Policy Institute, Rutgers University, New Brunswick, NJ, USA

Janice L. Hastrup	Department of Psychology, State University of New York at Buffalo, Amherst, NY, USA
Madelyn Hoffman	New Jersey Peace Action 89 Walnut St. Montclair, NJ, USA
Catherine McVay Hughes	Community Board One, New York, NY, USA
Valentina V. Klauchek	Research Institute of Hygiene, Toxicology and Occupational Pathology (RIHTOP), Volgograd, Russia
Sergei V. Klauchek	Volgograd State Medical University, Volgograd, Russia
Ivan Kulyasov	Centre for Independent Social Research, St. Petersburg, Russia
Antonina Kulyasova	Centre for Independent Social Research, St. Petersburg, Russia
Nadezhda Kutepova	NGO The Planet of Hopes, Ozersk, Chelyabisnk region, Russia
Adeline G. Levine	Department of Sociology, SUNY Buffalo, NY, USA
Natalya Makarevich	Department of Psychology, Volgograd State Pedagogical University, Volgograd, Russia
Nicholas D. Martyniak	Bloustein School of Planning & Public Policy, Rutgers University, New Brunswick, NJ, USA
Anne E. Mercuri	The Schuylkill Campus, The Pennsylvania State University, Schuylkill Haven, PA, USA
Natalia Mironova	Movement for Nuclear Safety, Kaslinskaya, Chelyabinsk, Russia
Jonathan Reisman	Robert Wood Johnson Medical School in New Jersey, Piscataway, NJ, USA

Lyudmila V. Smirnova	Division of Education, Mount St. Mary College, Newburgh, NY, USA
Sherilyn N. Thomas	Psychology Services Center, University at Buffalo (SUNY), Amherst, NY, USA
Olga Tsepilova	Sociological Institute RAN, St. Petersburg, Russia
Maria Tysiachniouk	Center for Independent Social Research, St. Petersburg, Russia
Abraham H. Wandersman	Department of Psychology, University of South Carolina, Columbia, SC, USA
Irina A. Zykova	Research Institute of Radiation Hygiene, St. Petersburg, Russia

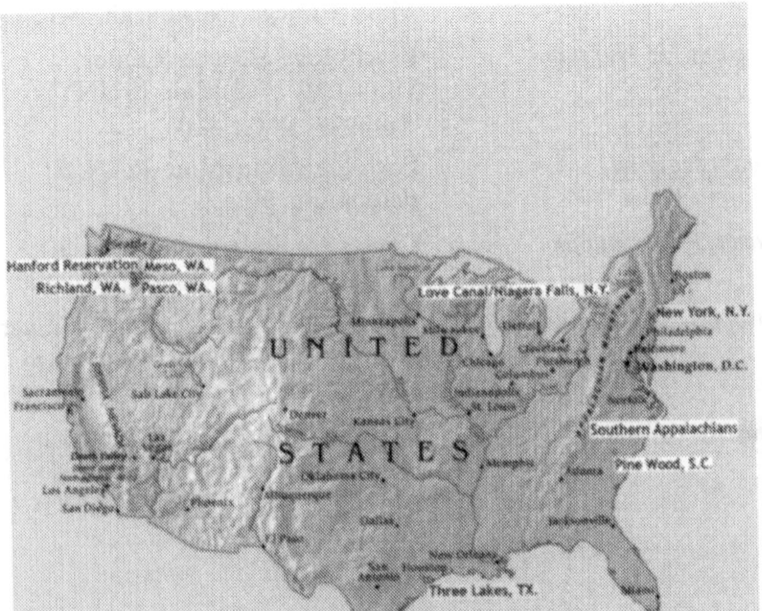

US Map from "North America" Plate, CIA World Factbook 2006. https://www.cia.gov/cia/publications/factbook.index.html. Manhattan Map from USGS Topo Dataset, NASA World Wind Project. http//worldwind.arc.nasa.gov

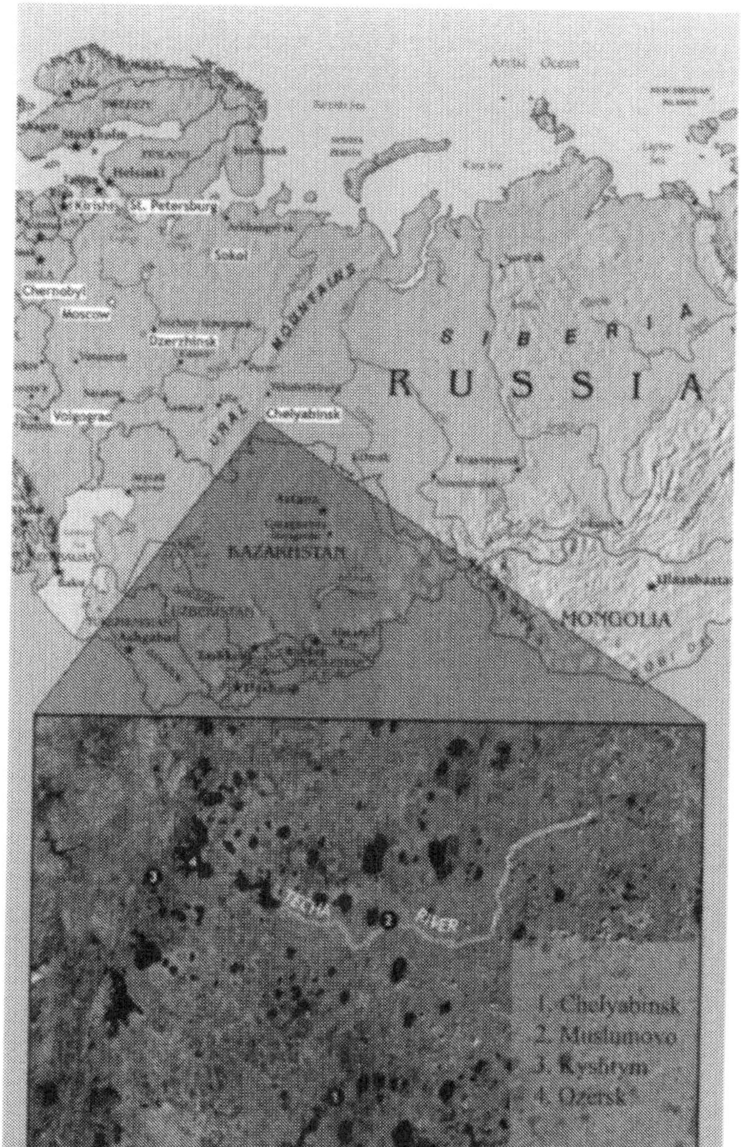

Russia Map from "Central Asia" Plate, CIA World Factbook 2006. https://www.cia.gov/cia/publications/factbook.index.html Chelyabinsk Plate from USGS Geocover 2000 Dataset, NASA World Wind Project.http://worldwind.arc.nasa.gov

SUSTAINABILITY AND THE NEED TO DEAL WITH THE CONTAMINATED LEGACY: A COMPARISON OF RUSSIA AND THE U.S.

Michael R. Edelstein

INTRODUCTION: SUSTAINABILITY AND THE LEGACY OF CONTAMINATION

Sustainability is a catch term for the different way of life required to counter the ills of the modern era. It encompasses new social and economic as well as ecological relationships. Sustainability theory is, by its nature, hopeful in envisioning an alternative corrective course of action. This volume deals with "the negative legacy problem" that confounds this optimism because we have so profoundly contaminated and altered the earth in lasting ways. Any effort to create a sustainable future will have to deal with this legacy. It is a huge and profound burden faced unevenly by people and non-humans today and that we have left for future generations (see Edelstein, 2006).

In my previous work (Edelstein, 2004), I have concluded that environmental pollution is an artifact of a "contaminating culture." In this volume

Cultures of Contamination: Legacies of Pollution in Russia and the U.S.
Research in Social Problems and Public Policy, Volume 14, 1–10

ISSN: 0196-1152/doi:10.1016/S0196-1152(06)14021-1

we examine a fascinating but sad historical fact: the two dominant and dominating societies of the 20th century, Russia[1] and the U.S., shared a disregard for the health of earth, people and bio-diverse species through the parallel pursuit of industrialization and militarization. This commonality occurred despite the divide between east and west, communism and capitalism, collectivism and individualism, frugality and over consumption and Cyrillic and Latin culture. The real issue of significance all along was less the iron in the curtain of separation between these societies during the Cold War period than their unifying toxic cultures. This common denominator poses a new curtain of constraint to the prospects for future sustainability. Even if we agree that sustainability is the desired alternative paradigm to modernity, the way to get from here to there remains the challenge. Our legacy of contamination is a clear barrier to this transition.

In this volume, we ponder the cultures that generated the bulk of this global toxic legacy. We seek insight into its creation, information on its impacts and an understanding of differences in how it is being addressed. Hopefully our illumination of the problem will contribute to a better understanding of the constraining conditions and will help guide us in meeting these twin challenges created by the legacy of contamination: how to address what we already have done and how to stop doing more of it.[2]

A SHORT ECO-HISTORICAL ANALYSIS

For the reader who may be unfamiliar with key historical referents that appear throughout this volume, allow me to offer a one paragraph "eco-history". The "Cold War" and "Iron Curtain" formulations that separated Russia and the U.S. during the period after the Second World War locked them into a parallel universe of technological and, to some extent, social development. The foundation of the Cold War was nuclear parity and the unhappy idea of mutual assured nuclear annihilation. Both sides were assumed to have the ability to maintain nuclear arsenals, with all attendant environmental consequences. At the same time, the ability to conduct chemical and biological warfare was not ignored. Beyond the military arena, both sides pushed civilian nuclear power and the use of chemicals in non-nuclear arenas, especially food production. In the west, this reliance upon petrochemicals became the backbone of a "synthetic paradigm" that underscored growth in consumerism and its underside, waste, in every facet of the society. If the driving force in the west was overconsumption driven by the market place, in the east it was state controlled production. To maintain their respective positions in the world, the U.S. was frequently forced to be anything but democratic in its choices. Russia was anything but "communitarian" in its.

In the 1980s, USSR President Mikhael Gorbechev, aware that the Cold War was a lose-lose proposition, began to focus on democratic reforms needed in Russia to correct social and environmental harm and to allow for a change in global dynamics. Gorbechev was interested in promoting the potential for political and social reform (i.e., Perestroika) through a period of openness (i.e., glasnost) that could not be tolerated under the closed requirements of the Cold War. Spurred by the fiasco of Chernobyl, during glasnost, much of the past ecocide in Russia came to light, including the disasters at the Mayak nuclear facility that receive considerable attention in this volume. And Russian society reverberated with reform and experimentation. With the subsequent fall of the union between socialist republics, Russia initially rushed toward the U.S. model. But when its rapid development brought corruption and near economic collapse, the country shifted into a regressive phase of its transition. Russia abounds in unachieved potential. It is still unclear what it is transitioning toward (see Yanitsky, 1996).

The unilateral fall of the Berlin Wall and opening to the west that followed was seen by the U.S., within the Cold War frame of mind, as surrender. In this view, the West won, proving the superiority of its pro-growth, market economic and synthetic paradigm (loosely and incorrectly defined as "democracy"). While Russia was thrown into transition, as the "winner," the U.S. became resistant to change precisely at the point when the need for movement toward sustainability was becoming paramount. As a resulting regression, Cold War victory helped abrogate the trends of the 1960s and 1970s, America's own period of perestroika and glasnost, characterized by a new awareness of the costs of the synthetic society apparent during a period of ecological concern, public participation and openness.

These dynamics, of course, have now been overshadowed by further contradictory global developments – globalism and world trade on one hand and global terrorism on the other. And the United States is facing a transitional period itself given that it has remained totally dependent upon the modern petrochemical "synthetic paradigm," its pollution levels climbing despite an industry exported abroad and its economic standing in the postmodern world threatened by the comparative advantage found in formerly "unmodernized countries" to choose to more flexibly grow in alternative sustainable directions. And global climate change has emerged as the metachallenge for all.

THE PROJECT

In Spring 2002, Ramapo College of New Jersey received a grant from the Trust for Mutual Understanding to explore the comparative issues of contamination as a constraint to sustainability in Russia and the United States through a project of exchange and joint investigation. The project grew from two previous positive collaborations between myself and Maria Tysiachniouk of the Center for Social Research in St. Petersburg. A wonderful and surprising side benefit of these projects was that a third collaborator, Ludmila Smirnova, former Professor and Dean at Volgograd State Pedagogical University, is now my wife as well as colleague.

This project grew from my own work on the psycho-social impacts of environmental contamination and from my interest in sustainability as a social movement. Masha (Maria) shared these interests. What better way to explore the topic than to convene an exchange between Russian and American scholars and activists to explore differences and similarities in the contamination legacies of our respective countries and how they were being addressed?

We succeeded in creating a fascinating series of exchanges, allowing for first hand investigation and site visits in Russia and America. The level of depth in these exchanges and investigations should be emphasized. We were able to attract outstanding social science professionals from both countries. We were then able to take these professionals to some of the most important contaminated sites in each country, creating encounters with local professionals knowledgeable about the specific cases. A series of conferences, roundtables and workshops was held that afforded ample opportunity for idea and research sharing, opportunities for training, ways to improve policy and practice and means to better respond to ecological disaster and its victims. The combination of these quality site visits with ample opportunities for participants to share, discuss and debate their work resulted in a level of high professional excitement and new learning. This volume was proposed as a way to capture and share some of this learning.

The American Team

The American team that traveled to Russia in summer 2002 included experts in psychology (Margaret Gibbs of Fairleigh Dickinson University, William Hallman of Rutgers University and myself), environmental information (Susan Maret), education (Lyudmila Smirnova of Mount St. Mary College and Volgograd State Pedagogical University) and public health (George Thurston of New York University). Two students joined the group – Jonathan Reisman, who served as project assistant and assistant translator, and Sara Tice.

Additional American colleagues participating during the visit of the Russian team were sociologists Stephen Couch of Penn State University and Adeline Gordon Levine of the State University of New York at Buffalo; psychologists Janet Hastrup of the State University of New York at Buffalo, Michelle Garcia of the Ironbound Community Center, Newark, N.J., Abraham Wandersman of the University of South Carolina, Patricia Seppi of the U.S. EPA and Murray Levine of the State University of New York at Buffalo (also a lawyer); environmental lawyer Jean Ann McGrane, now City Manager of the City of Newburgh, New York; social worker Susan Scher of Ramapo College of New Jersey; and activist scholars Madelyn Hoffman of the Grass Roots Environmental Organization of New Jersey and anti-nuclear activist Judith Johnsrud.

The Russian team traveling to the U.S. was led by environmental sociologist and biologist Maria Tysiachniouk of the Center for Independent Social Research in St. Petersburg, Russia and her colleagues from the center, sociologists Alla Bolotova and Dima Vorobiov; sociologist Olga Tsepilova of the Institute for Sociology of the Russian Academy of Sciences, St. Petersburg branch; physicist, activist and sociologist Natalia Mironova, Director of the Nuclear Safety Movement, Chelyabinsk; and psychologist Tatiana Andrushenko of Volgograd State Pedagogical University.

Additional Russian colleagues participating in Russia were economist Antonina Kouliasova of the Center for Independent Social Research, St. Petersburg, Russia; sociologist Nadezhda Kutepova, an activist from Ozersk; and medical doctor and public health specialist Boris Filatov of the Research Institute of Hygiene, Toxicology and Occupational Pathology (RIHTOP), Federal Directorate of Medical, Biological and Extreme Problems, Russian Ministry of Health and Social Development, Volgograd, Russia.

Exchange One: Americans to Russia, Summer 2002

In summer 2002, the American team carried out three phases of study travel. The first involved a visit to Volgograd in late May 2002, visiting with a network of people involved in addressing environmental contamination issues, including government officials, scientists and educators. We were told by the top environmental leadership of Volgograd oblast that there were no contamination issues to address; spent a day with Dr. Boris Filatov, Director of the Southern Center for Hygiene, an environmental health institute located in the former chemical weapons facility. Filatov is one of Russia's primary experts on contamination and remediation, as well as the health effects of environmental exposures, principally dioxin; he is a leader of efforts to reduce the stockpiles of chemical weapons and clean up weapons facilities. We also visited the Ecological Gymnasium, a remarkable alternative school, in the hyper-contaminated Krasno-armeysk district of Volgograd.

In early August, members of the American and Russian teams traveled by train several days from Moscow to the former closed city of Chelyabinsk in the Ural Mountains. There we joined Natalia Mironova and her staff at the Headquarters of the Movement for a Nuclear Free Russia on a trip into the countryside to an old Soviet-era rest camp where we spent several days conferencing with some 20 local environmental activists from the Chelyabinsk region in an open door amphitheater sitting above beautiful Lake Turgoyak.

Much can be learned just from reading the list of organizations represented at the conference: "Aigul" (a women's association which works with victims of radioactive exposure), "Chelyabinsk Public Fund Ecology" (publishes *Open Position*), "Environmental Enlightenment Organization Techa" (formed by Mayak nuclear facility victims; monitors the safety of food), "the Movement for Nuclear Security" (which had recently won a lawsuit with the plant Mayak resulting in one of the first instances of victim compensation in Russia), "White Mice" (a group of "nuclear hostages" of the village Muslyumovo, radiation victims of Mayak releases who work to unite people living on contaminated territories), "Youth Organization Consciousness of Rights" (litigated unsuccessfully to stop discharge of radioactive waste into surface waters by the Mayak facility) and "Youth Movement Ecofront" (formed in 1999 to stop import of radioactive wastes and launched the Ecological Student Inspection in Chelyabinsk). The stay at the rest camp was itself fascinating, with hundreds of Russian families vacationing much as Russians have done since the camp was originally built in the 1920s. We even had an apparent spy along for the ride. The team gave up a chance to visit Muslumovo, some fearing exposure to plutonium, some not wanting to exploit toxic victims and others suspicious that information gained during "toxic tourism" was contrived. We interacted with residents instead either at the camp or in Chelyabinsk.

Mironova accompanied the team as we headed back toward St. Petersburg by rail, detouring on the way to visit with activists and tour chemical plants and affected neighborhoods in Russia's leading chemical city, Dhzerzinsk.

In a small village of traditional farmhouses under the shadow of a long row of huge chemical smoke stacks, Bill Hallman distributed presents to the children sent by his own kids, resulting in a quick bond with the local residents.

The team returned by the night train to St. Petersburg on August 7, next to accompany Olga Tsepilova to her longitudinal study site, the industrial city of Kirishi, to the north. Kirishi was the epicenter of the new Russian environmental movement that emerged during Perestroika in response to the city's major concentration of contaminating industries. Tsepilova orchestrated a program of presentations by civic leaders and activists, culminating in a meeting with the mayor, himself a former environmental leader. Before leaving, the team visited a Russian version of beating swords into plowshares: a polluting industrial plant retooled into a vodka factory. The Russian and American teams joined in a toast drawn directly from the production line.

Back in St. Petersburg, we were joined by Nadezhda Kutepova from Ozersk during a three-day conference at the Center for Independent Social Research. Here each team member gave one hour presentation on his or her work with an hour for group discussion. Various other professionals dropped in and the event proved to be an uncommon opportunity for deep sharing among peers.

Exchange Two: Russians to the U.S., Fall 2002

The visiting Russian team arrived in early November. On the 2nd, NGO Orange Environment hosted a program at the Black Rock Forest Preserve to discuss the movement to close the Indian Point Nuclear Power Plant with Manna Jo Greene and Kyle Rabin of the "Indian Point Safe Environment Coalition." Later, at the home of Margaret Gibbs, an amazing exchange transpired between leading Russian and American anti-nuclear activists, Natalia Mironova and Judith Johnsurd, comparing their nations' respective anti-nuclear movements.

Sunday found the team in Kingston, New York, eating a picnic lunch at the restored Kingston waterfront, a successful Brownfields program and discussing the coalition formed to promote cleanup of PCBs from the Hudson River over dinner with environmental attorney Jean Ann McGrane.

On Monday, November 4, the team toured the Centralia, Pennsylvania underground mine fire site with Penn State Sociologist Steve Couch, co-author of a classic book about the community. We lunched at the

Hazleton Senior Citizens' Center as guests of a local grassroots group, GAG (Group Against Gas), publicizing the problems faced by residents of a mile-square area contaminated with petroleum products, mainly benzene. After a tour of the neighborhood and a reception in one of the activists' home, the team ended the day in a seminar with Couch based upon his contamination work.

Tuesday found the team in New Brunswick, N.J. for a workshop with leading environmental social scientists from Rutgers, New Jersey Institute of Technology and other nearby colleges organized and chaired by team member William Hallman, followed by a public presentation at Rutgers by the team. On Wednesday, an Environmental Justice tour of the Ironbound section of Newark was led by Michele Garcia of the Ironbound Community Center, and Thursday, team members gave a number of presentations at Ramapo College on "Sustainability and Contamination." Friday brought the team to Manhattan for an intensive tour of the World Trade Center site and neighborhood with NYU's George Thurston and Catherine McVay Hughes.

On Sunday the team traveled to Buffalo, New York for a two-night visit hosted by Adeline and Murray Levine and Janice Hastrup of the University of Buffalo. Highlights of the visit included Jan Hastrup's discussion of the University of Buffalo Hodgkins Disease Cluster, an intensive tour of Love Canal and neighboring hazardous waste sites directed by activist Luella Kenney, Murray Levine's description of the national grass roots environmental movement from his vantage point as the long term head of the board for the Center for Health, Environment and Justice (the national group headed by celebrated toxic activist Lois Marie Gibbs), Addie Levine's slide show and discussion of Love Canal and a briefing by the EPA Superfund office at Love Canal under the direction of environmental scientist Michael Basile.

The following Wednesday found the team in an intensive tour of the Manville, New Jersey Superfund site facilitated by Ramapo Social Worker Susan Scher and EPA psychologist Pat Seppi, including discussions with affected residents. The New Jersey Meadowlands Commission hosted a luncheon at their Environmental Center and Headquarters, on land reclaimed from an old landfill, featured New Jersey Environmental Activist Madelyn Hoffman discussing how contamination issues are addressed in the U.S. and the American grassroots movements.

In these two trips, we had covered much ground and seen a good glimpse of our respective contaminated legacies.

THIS VOLUME

I have organized the volume as follows. Section 1 focuses on the psycho-social dynamics of chemical contamination. The area of psycho-social impact study has advanced in the U.S. over the past 25 years, but is relatively new in Russia. Each of the six chapters presented here not only offers a case study, but further offers insight into the dynamics of the response to contamination. The chapters in Section 1 consider such questions as: What causes people to ignore or mobilize over a threat? What is the nature of contamination impact and how likely is recovery from it? What are the methodological choices for studying the impacts of contamination?

Section 2 deals with nuclear accidents and the closed cities and closed society needed to support particularly military nuclear operations. Here six out of seven chapters address Russia, its closed cities and its hidden history of accidents. Section 3 addresses mitigations for contamination impacts, dominated by Russian cases that offer a sharp contrast to U.S. practice. Section 4 offers perspective by examining issues of disclosure by examining distorted risk communication in the aftermath of the World Trade Center disaster, exploring environmental altruism in the two societies and presenting a final comparison of the social responses to environmental challenges in Russia and the U.S. based upon the parable of the Lion's bridge.

NOTES

1. For readability, in this volume, we generally refer to both the former U.S.S.R. and the modern Russian Federation as Russia or, occasionally, as the FUSSR/RF (Former USSR/ Russian Federation), unless greater precision is required to differentiate the U.S.S.R. and the RF.

2. And, the problem is not only what we have already done. As the former Soviet Union reshapes itself and other more populous "undeveloped" nations modernize, they add to the cumulative insult. Unless we can learn to live differently, we will continue to add to this negative legacy. Much depends on the ability of new development to follow a sustainable path. Less wedded to the synthetic paradigm than the U.S., the developing world may have an easier transition to sustainability (Sachs, 2006).

REFERENCES

Edelstein, M. R. (2004). *Contaminated communities: Coping with residential toxic exposure.* London: Westview Press.

Edelstein, M. R. (2006). The realities of sustainable practice: Constraints and opportunities in Orange County, New York. In: T. Schroyer & T. Golodik (Eds), *Creating a sustainable world: Past experiences, future struggles* (pp. 297–310). New York: Apex Press.

Sachs, W. (2006). Fairness in a fragile world: A memo on sustainable development. In: T. Schroyer & T. Golodik (Eds), *Creating a sustaining world: Past experiences, future struggles* (pp. 31–58). New York: Apex Press.

Yanitsky, O. (1996). *The ecological movement in Russia (Ekologicheskoe dvizhenie v Rossii)*. Moscow: RAN.

PART 1: THE DYNAMICS OF A CONTAMINATED COMMUNITY

SECTION 1: INTRODUCTION

What are the dynamics of a contaminated community – in Russia or the U.S.? These six diverse chapters chart what I have called "Environmental Turbulence," the disrupted complacency of everyday life caused by accepting information about one's exposure to environmental contamination as fact. The Theory of Environmental Turbulence (Edelstein 2004) posits that normal social and institutional networks regularly fail to help toxic victims address their needs, forcing victims to band together in common response.

The section opens with Adeline Gordon Levine's *The Love Canal: Social Science Research in a Community in Crisis.* This is an autobiographical statement by a seminal lead researcher about a seminal lead event. Here Levine recalls the Love Canal disaster, the first instance in the U.S. of a formally defined contaminated community with its newly defined category of "toxic victims." When I went to Love Canal in 1979, Levine was already writing her classic book on the disaster (Levine, 1982). She was correcting the page proofs for her book when, later, in 1981, fresh from my own research in nearby Jackson Township, I spent a day sitting on the New Jersey Shore comparing notes with her. Our conversation was fueled by the pure adrenaline of trying to understand a newly defined type of human event.

Some of that adrenaline is captured by Levine in this account of her pathbreaking efforts to document an unfolding toxic event. Levine discusses residents' exercise of their first amendment rights in forming a powerful grass roots organization able to influence government. This organizing model from Love Canal was subsequently replicated across the U.S., in no small part through the influence of a master organizer, Lois Marie Gibbs, who was transformed by the events at Love Canal from a housewife into a major public figure (see also Hoffman, this volume). Levine offers wonderful insights into the mind of a sociologist encountering a novel event. She recounts how she was drawn to study Love Canal, the issues she confronted as a social scientist encountering toxic victims in the field and her choice of a mode of qualitative research appropriate to the situation. She considers the fine line between research and advocacy, so familiar to all who work in this field.

In the second chapter, Olga Tsepilova's *Forging Change in a Contaminated Russian City: a Longitudinal view of Kirishi,* the author's observation of modern Russia might as well apply to the U.S.:

> We created "industrial monsters" that rapidly consumed natural and human health alike.

Like Levine, Tsepilova is a sociologist concerned with how groups form in the face of crisis. Her focus is the town of Kirishi, north of St. Petersburg, where she has chronicled the relationship between industrial pollution and civic action since Perestroika. Polluting industries in Kirishi had operated for a long time without public opposition. Tsepilova's longitudinal surveys suggest that for activism to develop required the twin triggers of perceived threat and government inaction. Kirishi's history as an industrialized city stayed in the background while one particular enterprise became "the threat." When a vitamin plant, BXZ opened in 1982 and, subsequently suffered an accidental release, existing organizations served as springboards for the development of an environmental movement whose massive demonstrations became the epicenter of a short lived green revolution throughout Russia. With a turnover in city government and the leading environmental expert and activist installed as Mayor and the closed BXZ plant converted to vodka production, the problems that had motivated citizen mobilization appeared to have been addressed; activism became once again less important. Other polluting enterprises continued operations unquestioned.

This theme of selective mobilization is further explored in Chapter 3, *Sokol: Seeing the Forests for the Trees.* Antonina Kulyasova and Ivan Kulyasov ask the provocative question of whether residents of a mill town are even aware of pollution given its ubiquity. Sokol has long been a major pulp and paper center. One requires no expertise to detect the pollution there. It assaults the senses. All are aware of it, but the pollution is such a constant background feature that no one but the experts pay it heed. Based on their longitudinal interview study, the authors identify a typology of concerns over pollution; common to each category is that all respondents ignore the pulp and paper mills that historically have been the primary contamination threat. However, as in Kirishi, when a new industry opened that produced a highly noxious and visible pollutant affecting all residents, an environmental movement formed in Sokol to specifically oppose the plant. Kulyasova and Kulyasov explain this difference by telling us that, in Sokol, the pulp and paper plant is the "father of the town." To oppose the pulp industry, perhaps even to admit that it is polluting, is like striking out at one's family, to threaten jobs and the economy and to cause deep social discord. In short, the authors suggest, to address the high levels of environmental risk would necessitate taking on high levels of social risk; the two have an inverse relationship.[1] Nevertheless, Sokol mirrors all of Russia, with its outmoded industrial plants and no funds for pollution control, where economic downturn is the only environmental saving grace, slowing production or shutting down plants producing serious reductions in pollution levels otherwise unattainable.

In Kirishi and Sokol, citizen opposition was mobilized by new polluting industries. In the U.S, citizens more frequently mobilize to oppose proposed facilities perceived as threatening or stigmatized. "Siting" conflicts are of particular interest because, in the U.S., a rational decision making process has been imposed over the permitting of hazardous facilities, giving voice to applicants, experts, public interests and citizen concerns (Edelstein, 2004). In Chapter 4, *The Case of the Pinewood Landfill: The Politics of Risk, Rationality and the Disposal of Hazardous Waste*, authors Nicholas D. Martyniak, William K. Hallman and Abraham H. Wandersman confront the rationality, rationales and rationalizations that underlie "siting" a hazardous waste landfill, concluding that the advocate experts were wrong and the opposing citizens correct in their predictions for the landfill's performance. Neither environmental impacts nor injustices stopped the dumping, leading the authors to further conclude that the entire knowledge system underlying the process of environmental decision making is flawed.[2]

The chapter also illustrates an underlying dilemma of post World War II consumerism in the U.S. by questioning the adage: "wastes must go somewhere." Inevitably, this formulation victimizes the residents of "somewhere." Their resistance will only serve to dislodge the problem "elsewhere" to those less powerful. Such victimization of people and environment is external to what is valued in modern growth economics, even if we assign value to solving the resulting problems. The rationality of hazardous facility siting thus overlooks the fact that, for those on the receiving end, facility siting is always irrational. Unless there is a sustainable societal shift toward eliminating wastes through changes in production, recycling and reuse (i.e., waste exchange), as well as changes in wasteful consumerism and its role in driving the economy, there is no way to get off this treadmill. As Russia (and others) considers following the U.S. path toward consumerism, this is a useful lesson to consider.

Exemplifying these externalities are the psychological costs to toxic victims, explored by Hastrup, Thomas and Edelstein in their case study of a southern Appalachian community contaminated with industrial wastes

(Chapter 5, *Fear of Cancer in a Rural Appalachian Community Following Notification of an Environmental Hazard).* Unlike Pinewood, this location was never deliberatively chosen as the site for a hazardous waste facility. Rather, a farmer paid to haul wastes from a local industry took them to his property, where they were either dumped or burned, contaminating surrounding hills and hollow, as well as the water table. The authors approached the case from divergent perspectives, achieving convergent findings. Hastrup and Thomas conducted a quantitative analysis of indicators of psychopathology, comparing the "hazard" group to a comparison community not close to a source of contamination. Their principle interest was fear of cancer. In contrast, Edelstein's qualitative study was focused on a full examination of psycho-social impacts for extended families living in this culturally unique setting. The convergent approaches demonstrate that residents living with the contamination hazard not only feared cancer as a result, but fear of cancer helped to stigmatize the local environment and undermine the prevalent family-based subsistence lifestyle. As a result, residents of the hollow are caught in a bind: do they break up their extended families and their bonds to the land and move; or do they bring up another generation in a contaminated environment (see Edelstein and Tysiachniouk, this volume, for a chapter on Russians in a similar bind)?

This raises the issue of whether people can recover from such psycho-social impacts. In Chapter 6, "*Toxic Water and the Anthill Effect: The Development of a Subculture of Distress in a Once Contaminated Community,*" Stephen R. Couch and Anne E. Mercuri conclude that recovery is a difficult prospect. Their detailed study for litigation, triangulating qualitative and quantitative measures of impact, describes Three Lakes, a Texas community where groundwater was contaminated by Exxon Corporation. Residents did not learn of the groundwater problem until after it had already been mitigated. However, instead of considering the problem fixed and over, they viewed their exposure to pollution as an injustice and a threat to their health. The resulting psycho-social impacts interfered at every level of individual and community life in Three Lakes. Impacts were surprisingly persistent over time (6–8 years), prolonged by a "subculture of distress" caused by the victims' anger, community conflict, lawyers and litigation and the activities of emergent and permanent grass roots organizations. The authors conclude that, rather than expecting victims to recover from chronic toxic disasters, one must instead understand how they are transformed by the experience.

NOTES

1. For a discussion of related U.S. literature, see Edelstein (2004, pp. 116-117).
2. Their blaming of the decision process provides a stark contrast to the common tendency to blame facility opponents as NIMBYs ("Not In My Back Yarders" acting to protect their private interests; see Edelstein, 2004) or for exhibiting Chemophobia (see Filatov et al., this volume).

REFERENCES

Edelstein, M. R. (2004). *Contaminated communities: Coping with residential toxic exposure.* Boulder: Westview Press.

Levine, A. G. (1982). *Love Canal: Science, politics and people.* Boston: Lexington Press.

Michael R. Edelstein
Editor

THE LOVE CANAL: SOCIAL SCIENCE RESEARCH IN A COMMUNITY IN CRISIS

Adeline G. Levine

INTRODUCTION

This paper is about some of my experiences as a sociologist, doing research at the Love Canal. The Love Canal remains an important story, because it aroused the consciousness of the world to the human and social consequences of environmental pollution. As an indication of continuing interest in that area, the 25th anniversary ceremonies at the site, early in August 2003, were well attended and publicized nationally by the mass media.

I will outline the story of Love Canal first. Then, I will describe how I became involved in research at Love Canal, the kind of research I did and why, and then will discuss some of the methodological and ethical issues I encountered while working there.

THE LOVE CANAL STORY

The Love Canal story started over 100 years ago. The phrase, "Love Canal," has become widely known, a metaphor meaning a community polluted by man-made toxins. The name comes from a man named William Love who

Cultures of Contamination: Legacies of Pollution in Russia and the U.S.
Research in Social Problems and Public Policy, Volume 14, 19–29

ISSN: 0196-1152/doi:10.1016/S0196-1152(06)14001-6

built the canal in the mid-1890s. His business plan to use the power of Niagara Falls failed and the canal itself, some 3,000 feet (900+ meters) long, became a recreational waterway.

Fifty years later, in the 1940s, the Hooker Chemical Company used the canal as a disposal site (or dumpsite) for more than 20,000 tons of waste from the production of over 200 different chemicals. They then covered the canal with clay, earth, and grass, so it looked like a big field. In the early 1950s, the area was "sold" to the city of Niagara Falls, New York, for one dollar. A public elementary school was built on top of the covered-over canal. A neighborhood of neat, modest homes grew up around the school – a very pleasant neighborhood.

Most people did not even know about the concealed canal and its contents. But in the 1970, after several years of unusually high amounts of rain and snow, the old covered-over canal was found to be leaking chemicals into the back yards and homes near the covered dumpsite.

On August 2, 1978, the Commissioner of Health for the state of New York officially announced that the leaked materials seriously threatened the health of those families.

That announcement was publicized around the world. One cause of controversy was that no one knew then – nor do they know even now – precisely what materials were buried there, how far they had migrated under the ground and through the air, nor exactly the impact on human health and the environment, of long-term exposure to a mix of chemicals.

However, despite the lack of complete information, decisions had to be made by government officials and by the people who lived in the neighborhood, about what to do in this crisis situation.[1]

First, some background to clarify some of the actions taken by the government and the people. The United States has a system of overlapping governmental structures. The federal government and the state governments carry out many similar functions and also some different ones.

In this case, the federal government started immediately to try to prevent further leakage from the canal and to repair some of the damage to the physical environment. The massive engineering project continued for years and although less active now, constant monitoring of the area is still ongoing.

Our state governments bear the chief responsibility for caring for the health and welfare of their citizens. In this case, the New York state governor established a task force, with representatives from nine state agencies to carry out the decisions of the governor and his advisers.

The state task force did many things, but their major assignments concerned health and homes. The health department's responsibility was to conduct studies of the people in the whole neighborhood. However, few health studies resulted and they proved of limited use in policy decision making. To this day no one knows with scientific certainty the health impact of the Love Canal chemicals. However, controversy over the studies added to the confusion and anxieties of the people who lived in the Love Canal neighborhood.

The second decision was about homes. At the very beginning, before it was possible to make even a good estimate of the damage, the Governor announced that the state of New York would purchase the 239 privately owned homes closest to the canal and would also help people move away. This decision angered other residents who lived in the same neighborhood, but whose homes were not to be purchased. They were frightened about the unknown hazards to their families' health and saw no way they could ever sell their homes in a private real estate market, and move without major assistance from the government. The plight of these residents captured national and international attention.

The Love Canal case provides a good illustration of the relationship between the citizens and government of the United States. In the United States, the first amendment to the federal constitution guarantees the rights of citizens to assemble, and to petition the government. Within days of the first public announcement that the area was a hazard to health, hundreds of families established the Love Canal Homeowners Association to seek government help. The organization represented most of the families in the two-year struggle to force the government to buy more homes, in nearby streets.

The right to criticize and to petition the government, and protections for a free press (that is, not under government censorship or control) are also guaranteed under our constitution. In fact, the mass media played a very important role, publicizing the Love Canal story and keeping it before the public for years.

In 1980, the government, for political and public relations reasons, announced that they would buy about 1,000 more homes. Hundreds in the most polluted sites were torn down and buried under the ground where they stood, and so was the school that had been built on top of the covered canal. Seven to eight hundred homes remained standing.

Between 1980 and the present time, the neighborhood has been reestablished. It took years of work to contain the leachates in the old canal, to clean drains, storm sewers, and small creeks, and also to maintain the homes left in the neighborhood. All these efforts were made through joint work by federal, state, and city government bureaucracies.

At present, hundreds of homes, cleaned up and repaired, have been sold to new homeowners at prices about 20 percent lower than for comparable homes in the city. Today, the area of the old canal looks like a large, clean field surrounded by a fence, with trees and bushes tastefully planted around the area. In the surrounding streets, children once again are playing in their own yards, on the quiet streets and in the new playground.

When I attended the 25th anniversary ceremonies in August 2003, the neighborhood reminded me of my first visit in August 1978, when the impression of the pleasant homes and streets and playing children contrasted sharply with the chilling knowledge that a huge pollution source lay invisible under the ground.

As a result of the Love Canal events, existing environmental protection laws have been strengthened, and there are new laws in our country to prevent and clean up pollution. Lawsuits were brought to court by the government and by affected people, against the privately owned corporate polluter to recover the costs of dealing with the problems. These were won or settled out of court. Furthermore, the environmental movement at the grass-root level has been revitalized all over the United States, chiefly inspired by the Love Canal story.

Life changed for the various people involved at Love Canal, whether they lived, worked, or studied there. My own life changed because I studied and wrote about the Love Canal and I want to turn to that work now.

RESEARCH AT LOVE CANAL

Why and how do we do research? First, *why*? Motivated by curiosity I visited the Love Canal for the first time nine days after it was declared a health hazard, and four days after the Governor had announced that the state would buy the first two rings of houses around the old disposal site. Every newspaper, every radio station, every television channel reported the story, day and night. I wanted to see for myself what was happening.

The situation was very chaotic at that time. On that first visit I saw people moving out of their well-kept homes, met a pregnant woman convinced that she carried a monster; spoke to another woman afraid that her daughter would be unable to bear children; saw worried men and women lined up to get information from newly established government offices; talked to some of the government workers trying to bring some order out of the chaos; and met the young woman (Lois Gibbs) who was suddenly thrust into prominence as the leader of a brand new citizens' organization. After eight hours, I came home, determined to do research at Love Canal.

Second, *how*? Research requires resources. So, I did what American academics do: I tried to find a grant, money to support research, from a government or private funding agency. I could see that the Love Canal situation was changing from hour to hour; I knew that I had to start working immediately.

To save time, I quickly contacted many funding sources, described the situation and asked whether they would even consider an application from me for research support. I learned right away that no one would fund my research! "Why should a sociologist be concerned about an environmental problem?" I was asked. I was told "We only fund research using well-established paper and pencil tests"; "we provide funds only to test theories"; "we have a 26 page form to fill out and you will hear from us in 6 months"; and so on, and so on.

I had to choose between spending time looking for funding, or doing the research, using assets I had in my university position. As a tenured professor, I was expected to devote one half of my work to research of my own choosing, and I had access to various university resources.

Within days, I recruited five very enthusiastic graduate students to work with me in a field research course. Afterwards, I wondered how I could have been so lucky as to NOT obtain grants, lucky because the people who lived near Love Canal would not have trusted me. In their eyes, I would have been profiting from their problems.

We started the field research seminar at the end of August, with the beginning of the university school year. From the very first, we decided that we would not use standard instruments. In this new situation not only did we not know the answers people might provide us, we did not even really know the questions to ask. Our research was to be exploratory and we would learn as we did the work.

We anticipated various methodological problems and as the dynamic story unfolded we dealt with others. We solved some and simply had to live with some. I will focus on two sets of issues, first that of collecting good data and second that of the research team's activities, including the issue of advocacy in a situation where we were chiefly using a participant observation method of research.

COLLECTING DATA

We quickly learned that there were three major groups or sets of "actors" at Love Canal:

1. The residents living in the area who were frightened about their health and especially worried about their children's health. The majority of the residents owned their homes; the minority rented apartments in a government-subsidized housing project. The homes were private property and the people anticipated tremendous financial losses as the market value of their homes plummeted. They wanted all the help possible from the government.
2. The second group was comprised of the representatives of the state agencies organized into a temporary task force. They felt they were doing the best job they could, given the scope of the problems and the amount of resources they had to work with. (The federal government was also involved, but focusing chiefly on the canal containment, they were not in daily contact with the residents.)
3. Third was the chemical company, which had used the canal for some 20 years as a dumpsite. This large corporation was in the private sector, as are most of our industries. It was a major employer in the community and wanted to protect its reputation. The company was concerned about the lawsuits it anticipated and wanted to minimize the entire problem and the company's part in creating it.

In the setting of uncertain information about the migration of chemicals and the damage to health and environment, the three groups struggled about the best way to solve the immediate problems.

They were also involved in a long-term struggle for what was to become the widely accepted history of Love Canal. The accepted version of reality might affect court decisions, legal settlements, government regulation of the chemical industry and disposal practices, the careers of professional government workers, the market value of the residents' homes, the structure of their neighborhood, their sense of the state of their health, and their feelings of comfort, security, and emotional well-being.

Each of the three groups not only had different kinds of information but also interpreted differently the information available to everyone. Depending on the consequences they perceived, the groups varied in their members' willingness to share with researchers their knowledge and beliefs about the events. They were not chiefly concerned with our desire as social scientists to do research. Rather, they viewed researchers as interpreters; disseminators of information which could help or harm them; and possibly as moral judges, not simply as neutral collectors of facts.

Our training as social scientists leads us to believe that we not only *should* be but also *can* be neutral and objective in our methods, and we do try to be. However, in a community in crisis, we too are an "interest group," working with information and helping to create a valuable commodity, that is, the long-term final history and interpretation of the events. If we understand this, then we may better understand why our access to information in these cases is not and may never be complete.

The residents were very eager to share their information and opinions with us. The government people provided reports and papers freely but were cautious in expressing personal opinions. The chemical company confined information, using a public relations office which issued carefully worded statements absolving the company and reminding everyone that the company was a major local employer.

Our research group depended on other sources and ways to collect and check information. As participant-observers,[2] we spent at least one full day a week – usually more – in the field, observing, visiting, and attending meetings. We kept extensive, detailed descriptions of our numerous visits to the area. We collected newspaper articles, announcements, reports, letters, and other materials prepared by the government, or by the residents. We took many photographs and we kept records of our seminar discussions. Some materials were obtained from the state health department through the state's Freedom of Information Act.

In addition to our constant conversations with everyone we encountered, we conducted formal interviews with families, and with members of the state task force. Also, I interviewed 20 key individuals.

In sum, we collected a vast amount of materials. This is very common in field research, which necessarily includes elements of historical research and journalism as well as social science. Those research materials are now in library archives, available to scholars.

To understand that phenomena appear different from different viewpoints, the graduate students and I each focused on different groups of people: the state task force, the major citizens' group, the smaller citizen groups, the mass media, and the corporation. In the field setting, we did not interact with each other, or go from one group to another. In our weekly class discussions we acted as informal controls for each other, checking facts and providing the viewpoints of the different groups we were each with in the field. We also invited key people to address our group.

As a sociologist I was always concerned about presenting a biased picture. I found the structure of presenting and advocating opposing views to be very helpful. We called our process an "adversarial discussion method" and I recommend it to researchers in similar settings.

INTERVIEWING AS A SOCIAL PROCESS

We developed interview schedules, to help us learn the problems people faced and their perceptions about them. In the autumn of 1978, we interviewed 102 people from 61 families in their homes and 63 people from 38 of these families again, a few months later. The semi-structured, open-ended interviews allowed people to talk as much as they wanted to and some interviews went on for several hours. More than 30 state task force members were interviewed as were 20 other key people.

Interviewing is a social process. Some of us felt hesitant about creating pain by probing in sensitive areas and we feared that our own emotions would prevent us from probing enough to understand the views and feelings of people who were under great stress.

During my fifth or sixth interview, when a woman described her very sick children, I suddenly felt I was going to cry, and I changed the subject. Deciding to use my reaction as a learning opportunity, I invited a clinical psychologist to listen to my taped interview in our research class. That session helped us all to understand that our personal feelings could intrude on the interview process.

Interesting problems arose for us in the course of the interviews. Sometimes we did not immediately understand the importance of what we were being told – for instance, people's concerns about their pets, or their gardens. One man said he would like to take a seriously illegal action to ease his frustration. Sometimes spouses quarreled during the interview about whether or not to move immediately. The very fact that we were interested in them made a few interview respondents aware that their situation was seen by others as serious.

The student who observed and interviewed the state task force found it an emotionally draining experience. She was tolerated, not welcomed in their headquarters. To do her task, she simply stayed around and depended on people's unwillingness to ask her to leave. Her interviews yielded information that was very consistent from person to person, obviously an official position.

It was helpful to use our class sessions to openly discuss our own and our subjects' reactions to the interviewing process. I learned that the emotional aspects of working in such highly charged settings should be taken into account, to avoid "interviewer burnout" where the interviewer might take self-protective measures, such as changing the topic at crucial moments.

ACTIVITIES AND ADVOCACY

We soon realized that we had to find something to do while observing. Using this method of research, the researcher must relate to the very people being observed; you are not inconspicuous, and you cannot just stand around and take notes without making people nervous and suspicious. Also, we wanted to do something for the people who were giving us time and information.

It was most difficult for the student observing the state task force going about their salaried jobs. She could find little to do during the time she spent in their offices. Once, someone asked her if she was a spy. Treating them as fellow professionals, we invited a few of the state task force members on a few occasions to meet with our seminar and to participate in a public forum.

It was far easier to find activities with the citizens' group, all volunteers who welcomed helpers. Many times I made the coffee and ran errands. My major activity was to clip items about Love Canal from the newspapers stacked in a huge pile in the office. This gave me a useful task and a place, literally, to be: in a chair at a table, wielding scissors. I consulted in setting up a speakers' bureau and helped one speaker prepare for her first public appearance. As "the professor" I was sometimes asked for the interpretation of technical reports. While the subject matter was usually out of my area of competence I could explain the scientific approach, or at least provide a sympathetic ear.

The very fact that I kept returning to spend time with the largest citizens group provided the core workers with some sense of outside sympathy for their cause. They have told me that my interest was important for their morale during the long days when nothing seemed to be happening to help them.

I thought carefully about my own and my students' activities. The ones I have mentioned seemed appropriate to me in my role of participant-observer. However, beyond simple helpfulness lies the problem of outright advocacy, of "taking sides." We made sure that everyone knew our roles and our interests as researchers, not activists. But, when the stakes are high, the situation controversial, the researchers may be asked to publicly show their loyalties. The researcher who refuses may be cut off from crucial sources. The researcher who advocates too much may severely bias the data and lose credibility as a scholar. The researcher who does not help and speak up for victims may violate personal and common standards of decent, ethical behavior.

I am sure that any researcher in a community in crisis will have to consider this issue. Let me describe some of my own decisions.

I refused an invitation to become a member of the governing board of the largest citizens group. I attended but did not participate in demonstrations, by carrying signs for example. But to show my ambivalence, during a month-long demonstration – starting daily at dawn, during a bitterly cold December – I did not picket but I made gallons of hot chocolate for the shivering residents.

I started lecturing extensively about Love Canal within three months of starting the research project. I presented all viewpoints as best I could; I always felt that it was important to get the story out to the part of the public I could reach.

As time went on, while I remained objective as far as possible, I lost my feelings of neutrality. The sides in this struggle clearly had vastly different interests, resources and power: the affected residents desperately worried about their families, government leaders eager to show that they had solved the problems without wasting taxpayers' money, and a multi-billion dollar corporation intent on minimizing the consequences of their business practices. My sympathies lay with the residents.

Because a field researcher is the main "instrument" of research, the problem of selective bias must be of concern, always. Being aware, constantly, of the problem helps make one cautious in using findings and in drawing conclusions. The adversarial discussion I described was very helpful in limiting the inferences to be drawn from data.

It is not possible to have "control groups" in these settings, so completely unlike an experimental model of research. But I have often heard from researchers engaged in other communities in crisis caused by man-made pollution. These researchers encounter the same general issues. Their reports have given me a validity check for my own conclusions, a form of replication if you will.

In my published writings and presentations, I set myself a standard of having solid documentation for every statement of fact. I have tried to distinguish clearly between such statements and expressions of my own opinions and conclusions.

To conclude, in this brief paper, I have described but a few of the issues. I touched on the sponsorship of research when I talked about grants. There are other problems, such as what to do when the mass media becomes interested in your research project, or asks for interviews. Or, what to do if problems arise in personal relations within the research team or between researchers and subjects. Any of these may affect the research outcomes.

Our research methods courses, our social science concepts, and our professional ethics offer us a good deal of guidance. But we do face some new challenges when we conduct field research in communities in crisis.

NOTES

1. Please note that I am merely sketching a part of a complex story, which is far more fully described in my book, *Love Canal: Science, Politics and People* (Levine, 1982).

2. Most helpful to me in using the participant-observation method was Willliam Foote Whyte's book, *Street Corner Society: The Social Structure of an Italian Slum* (Whyte, 1955). Originally published in 1943, the second edition in 1955 provides a valuable 85-page appendix describing his methodological approach.

REFERENCES

Levine, A. (1982). *Love Canal: Science, politics and people*. Lexington, MA: D.C. Heath, Lexington Books.

Whyte, W. F. (1955). *Street corner society: The social structure of an Italian slum*. Chicago, IL: University of Chicago Press.

FORGING CHANGE IN A CONTAMINATED RUSSIAN CITY: A LONGITUDINAL VIEW OF KIRISHI

Olga Tsepilova

INTRODUCTION

The 20th century was characterized by increased risk to both natural ecosystems and humanity due to the combined effect of fast industrial development and the increasing scarcity of natural resources. The former Soviet Union was one of the most polluted regions on earth. Within the country, it became necessary to delineate zones of recognized ecological disaster. This condition resulted from a policy that placed industrial development as the highest priority and the well-being of people and nature as the lowest. We created "industrial monsters" that rapidly consumed natural and human health alike. Beyond such well-known ecological disasters as the Aral Sea and the Bashkirian and Chernobyl disasters, other severe ecological problems screamed out across the nation for attention. No place was immune from such tragedy, even the relatively small town of Kirishi, Leningradskaya oblast, an hour's train ride north from St. Petersburg.

Cultures of Contamination: Legacies of Pollution in Russia and the U.S.
Research in Social Problems and Public Policy, Volume 14, 31–46

ISSN: 0196-1152/doi:10.1016/S0196-1152(06)14002-8

The Kirishi case is of interest, not merely for the scope of its problems, but for the energetic local response these problems evoked. From 1987 through 1991, a local NGO lead a powerful ecological movement in Kirishi, providing perhaps the most successful and inspiring example of Soviet era grass roots response in a zone of ecological disaster. That this occurred at the point of transition in Soviet society was no accident. The Kirishi movement both reflected changes nationwide, but this much publicized local response also had a heuristic influence on the national course of events. There was no doubt that it deserved an in-depth sociological investigation.

A longitudinal study of Kirishi was started in 1989 and continues today. The study, much as the events in Kirishi, played a heuristic role for the entire nation, modeling new possibilities at a crucial juncture. Not only did these developments help to bring about the end of the Soviet Union, but they offered a framework for holistic understanding of the development and functioning of social movements in contemporary Russian society.

In this article, I discuss the emergence of the Kirishi social movement, its prerequisites for emergence, the role played by public opinion, its social-ecological and political consequences, and its implications for social movement theory.

METHODS OF INVESTIGATION

A number of convergent research methods were employed in the Kirishi sociological study, including comparative surveys of the population combined with in-depth interviews with representatives of the Kirishi city administration, representatives of industry, leaders of the environmental movement, and representatives of the media.[1] In addition, between 1989 and 1997, periodic content analyses were made of the local newspaper *Kirishski Fakel.* Participant observation has also been employed over the course of the study, the author directly participating in direct action events, sessions of the Kirishi City Council, and in the meetings of the city municipality. Finally, throughout the study period, analysis was made of data on social-economic and medical statistics and of a wide variety of other relevant documents, including election campaign materials, election results, programs, and by-laws from active local citizen groups and organizations, selective protocols of City Council sessions between 1985 and 1993, and relevant decrees and orders from 1985 to the present.

THEORETICAL FOUNDATION

This study was guided by contemporary theories of collective behavior, relative deprivation, resource mobilization, and political opportunity structure. Because deprivation played a key role at the peak of the ecological disaster and in the resulting sense of crisis in Kirishi, the earliest stages were particularly influenced by the theories of collective behavior and relative deprivation, which dominated American social movement research in the 1950s and 1960s (Gurr, 1968; Smelser, 1963).

Resource mobilization theory[2] became useful for later describing the evolution of the environmental movement in Kirishi and comparing it to movements in regions and societies not enmeshed in responding to a disaster (see Tsepilova, 2002). Its emphasis on resources and mobilizing abilities of organizations as essential factors for the emergence of social movements made it an ideal framework for analyzing empirical material about successful Soviet social movements.

The final stages of this research drew upon the theory of political opportunity structure to understand the institutionalization of the Kirishi environmental movement in the formation of local environmental policy. This institutionalization can be understood in four distinct ways:

- as the literal conversion of "the movement" into "the Institute" (Offe, 1990), where social practices became regular, routine, and long-lasting (Abercromby et al., 1984; Berger & Luckmann, 1967).
- as increased organization, professionalization, and widening of activities (Mitchell, Mertig, & Dunlap, 1992).
- as the inclusion of the movement or its members in institutionalized activity, acceptance by the movement of the demands of the society at large and/or acceptance by society of the grievances of the movement and their inclusion in official policy (Dunlap & Mertig, 1992).
- as integration of the grievances of the movement into the programs and agenda of legal and legitimate parties and political organizations (Putnam, 1996).

Collectively, these perspectives describe the integration of a protest movement into public life. This institutionalization of an environmental movement is the sum of both organized and spontaneous actions involving a range of social agents who promote environmental policy on the basis of conscious environmental values. Such agents include governmental officials from environmental protection agencies, scientists and ecologists from the governmental, academic and non-governmental sectors, environmental organizations, and activists.

PREREQUISITES OF AN ENVIRONMENTAL MOVEMENT IN KIRISHI

The ecological disaster that enveloped Kirishi in 1987 served as the catalyst for the emergence of an environmental movement there. The ecological situation further worsened and became associated with a perceived extreme increase in disease rates after the December 1974 opening and subsequent expansion of a large biochemical plant, BXZ.

Public perception reflected the objective environmental situation. In 1982, environmental conditions in Kirishi had worsened due to the opening of a facility for the hydrolyzed production of yeast and protein–vitamin concentrate called BVK at the BXZ facility. As public worry over the plant's impact grew, specialists and experts came to share this concern. At first, there was little corroborating information to support these concerns.

Under local procedures adopted in 1982, Kirishi was run by commissions, representing different ministers and agencies that worked behind closed doors. As the quality of the city's environment continued to erode, it became more difficult to hide information about the conditions. By the mid-1980s, party and soviet bodies began a limited informational campaign at meetings conducted in workplaces and through the local media, particularly the newspaper *Kirishski Fakel*. The public learned that the complex environmental situation was being addressed by building, reconstructing, and modernizing treatment plants and production facilities.

The only significant and authoritative source of information about the BZX situation was the Kirishi regional hospital, whose main doctor, V. P. Esinovsky, strongly opposed production of the protein–vitamin concentrate. Dr. Esinovsky and his colleagues began their own health studies of the impacts of this chemical's production in the late 1970s.

The overall paucity of official information forced citizens to find alternative informational sources. Given the extent of public concern over health threats, information began "leaking" through multiple channels, including local party and soviet bodies, heads of major city enterprises, and workers at the biochemical plant. People also began to share news across social networks, feeding rumors, and fabricated stories. Panic spread across the city. By 1987, social tension in Kirishi had reached the boiling point. This heightened level of concern was transformed into a social movement through the "social capital" provided by two soviet-style non-political civic associations that appeared in the 1970s, the "Society of Book Lovers" and the "Natural History Society."

By the mid-1980s, the Natural History Society's 40 members drew both from the local intelligentsia and veterans of World War II. The larger Society of Book Lovers had 120 members at the beginning of 1987. Its core consisted of technical intelligentsia drawn mainly from the construction bureau of the "advanced biological machinery industry." Meeting several times a month, the book lovers held literature parties, exhibitions, and book fairs. By the mid-1980s, these groups were preoccupied with the environmental situation in Kirishi.

As the two groups drew closer together, they aroused the suspicions of the KGB. In the beginning of 1987, some of the activists were invited to visit the "1st department" for a conversation addressing the "wrong thematic direction" evident in the work of non-politically oriented public organizations. Rather than give up, however, the activists sought a way to legalize their environmental protection activities.

Another source of social capital for the development of environmental initiatives came from workers' collectives at major local industrial enterprises, including the oil refinery and the hydroelectric station.[3] These collectives began to discuss the unfavorable environmental situation by the early 1980s. By the mid-1980s, their words became strident, reflecting "nakal" (inflamed passions), and the collectives became decidedly more activist.

The principal source of protest came from the Kirishi regional hospital workers' collective, whose chief doctor continued to informally supply the city with information on the harmful impact of the vitamin concentrate on the health of the population. This collective was the first to demand that production of the product cease.

By the mid-1980s, therefore, Kirishi had all the necessary ingredients for a mass social movement. The environmental crisis determined the direction of this movement, creating and maintaining a high protest potential among the concerned citizenry. The framework of existing social networks provided the requisite "social capital" for these first civic initiatives. Only one last major condition was required for a movement to form. This was a national-scale development, Perestroika. President Gorbachev's policy of freedom of speech, along with his political, social, and economic changes, offered new resources for the development of social movements. It only remained to see how a particular movement would mobilize these resources.

The burning match that ignited the situation in Kirishi was an accident at the Biochemical Plant in January 1987. When the treatment plant failed in the hydrolysis production facility, the environmental situation turned disastrous. Within a short period, the incidence of allergic reactions had skyrocketed, as did the rate of child mortality. Local unrest was evident across the city. Graffiti appeared on the walls of buildings opposing production of vitamin concentrate. In April, police dispersed protesters gathered outside the Communist Party Headquarters and the Executive Building decrying the lack of official action. Additional troops of militia and army were rumored to be on the ready. The public was ordered to remain calm at the May 1st celebrations. Nonetheless, masses of people came with posters demanding that production of the vitamin concentrate be stopped and the facility shut down. The first rows of demonstrators were parents who had recently lost a child. They marched forward holding up portraits of their dead child framed in black. Their presence helped dissuade officials from calling in police.[4]

DEVELOPMENT OF THE ENVIRONMENTAL MOVEMENT

The political situation had changed. Under the pressure of public opinion, on May 17, 1987, Kirishi city authorities were forced to register their first local non-governmental organization, called the "6th section of VOOP" (or VOOP 6). VOOP is the acronym in Russian for the "All Russia Society for Nature Protection," which had at that time five official sections. The sixth section was formed under the slogan "Movement against the vitamin concentrate plant." From 1987 to 1991, VOOP 6 was an active organization with approximately 50 involved members. While the organization was hierarchically structured, it had strong horizontal connections to powerful support groups in all major city enterprises (excluding the vitamin plant), as well as other organizations and enterprises. The group could thus act with authority and organization while maintaining broad interaction and cooperation. In my opinion, this mixed vertical–horizontal structure allowed the organization to effectively solve even the most complicated questions. Activists from "VOOP 6" successfully raised funds from both homes and workplaces to support the movement against the Biochemical Vitamin Plant.

"VOOP 6" mounted an effective media policy that spread its message across media outlets with local, regional, and national audiences, attracting many into the movement. Continuous publicity through local media allowed VOOP 6 to maintain a high level of involvement of local citizens over a protracted period of time. Attention to the regional and national media exported the problem from a local concern into an issue with a much broader audience. Residents of a small regional industrial town turned out to be the center of attention of the whole country. Passage of an August 1990 law on print media broadened the possibility for using mass media as the key resource of the new movement. From this point on, the Kirishi environmental movement entered a new stage of institutionalization. The new possibilities allowed by mass media were exploited to their fullest extent in the formation of an effective political force at the local "parliamentary" level.

Kirishi Greens, flush with power from their emerging broad-based support, adhered to a hard line in discussions with the official power structure. At the same time, unlike the majority of Greens in the early era of perestroika, they never rejected dialogue with those in power. Although not a deliberate strategy, their openness supplied them with additional resources that contributed to the success of the movement.[5]

A wide variety of actions were undertaken by the Kirishi environmentalists, including holding town hall meetings, citywide demonstrations, petitioning drives and organizing a mass-media information campaign, and preparation for a citywide strike. One of the most valuable resources of the group for realizing its stated goals was the nonpolitical use of political methods. From its inception, VOOP 6 exemplified democratic tendencies, both politically and ideologically. It embraced a wide range of social questions stretching beyond pure and narrow ecological concerns. In this way, the organization was predestined to a long life and long-term influence.

Across Russia at the time, a liberalization of voting regulations and creation of local democratic deputy councils offered new possibilities and resources for local groups, such as participation in the preparation and passage of various resolutions, including those in the sphere of nature conservation. This new tool radically changed the tactics of groups like VOOP 6. For the first time, they gained a real chance to influence politics and, in so doing, to realize their goals. In March 1990, 12 of the 20 members of VOOP 6 who participated in local elections were elected deputies. Not only did the Green deputies comprise 10% of the City Council but also they received enough support from the overall body to form a powerful Green faction.[6]

Up through the dissolution of City Council in 1994, the Green faction played a major role in resolving political, socioeconomic, and ecological problems in the city and contributed to the nonpartisan formulation of new citywide ecological policies. Between 1990 and 1994, the Greens made environmental quality a serious political issue in Kirishi. Nearly a third of deputy's inquiries at the time addressed environmental concerns. New laws were passed for environmental protection that barred siting any harmful new production plants in the city and rigorously regulated the environmental consequences of preexisting industry.

The development of the Kirishi ecological movement represents one of the most successful social movements of the early period of perestroika in the entire former Soviet Union. This success was not accidental. Strong ecological organization rested on an existing firm base of "social capital" and a motivating condition of ecological crisis. The organization employed a wide array of tactics involving both political and nonpolitical participation to realize its goals, most notably the use of mass media and direct participation in local parliamentary affairs. The movement flexibly and effectively used both the new resources resulting from structural changes in society as well as resources made available by the old political system.

PUBLIC OPINION ABOUT ECOLOGICAL PROBLEMS

It is interesting to examine the results of the surveys taken during this period of change. In 1989, citizens of Kirishi considered the unfavorable ecological situation to be the most pressing problem faced by the city. This was evident on the survey taken that year. An open question asked which general problems of the city demand special attention of city government. Out of 952 surveyed, 525 people answered the question. Around 80% of respondents questioned the city government's handling of ecological problems, far more than evidenced concern about social-economic development. In contrast, by the 1993 and 1997 surveys, ecological problems had been forced to the periphery of people's attention by the morass of acute social-economic problems (see Table 1).

In 1993, the ecological situation in the city ranked eighth behind unemployment and deteriorating material conditions for the underserved, including their inability to access for-pay services in the spheres of health and education, the lack of social equality, rising crime rates, and access to goods and services. In 1997, ecological issues ranked seventh behind social security and job creation, rising social problems associated with drunkenness and crime, housing problems, and aggravated social inequities.

Although the ecological situation was rated similarly in these years, in 1993 it was cited as the primary motive for people wanting to move from the city and in 1997, when nearly every resident expressed a wish to move away, ecological alarm was significantly supplanted as the reason by the difficulties residents encountered finding work and housing.

Table 1. Significance of City Problems (Kirishi, 1993, 1997).

Question to the Respondent	Time of the Survey			
"What Problems do you Consider the Most Significant in the Life of your Town?" (Choice Available between 2–3 Problems)	1993 ($N = 911$)		1997 ($N = 1018$)	
	%	Ranking	%	Ranking
Issue of employment/unemployment	38	2	60	1
Issue of social welfare	39	1	45	2
Increase in alcohol consumption	27	6	31	3
Increase in crime rate	35	4	28	4
Housing	18	7	27	5
Increase in social inequality	35	3	21	6
Unfavorable environmental situation	18	8	15	7
Decrease in spiritual and cultural values	11	10	11	8–9
Not enough cultural recreational activities	12	9	11	8–9
Foreign citizens living in the city	1	11	5	10
Not enough goods and services	28	5	3	11

The low relative importance of the ecological problem does not mean that the citizens of Kirishi stopped supporting environmental protection as one of the assigned priorities of city government. As seen in Table 2, the majority of respondents were unable to make a forced choice between economic well-being and a safe environment. Nevertheless, for those making this choice, twice as many selected material prosperity over an ecologically clean environment. This bias in favor of resolving socioeconomic problems is understandable given the state of social and economic crisis in Russia at the time. It is, in fact, remarkable that economic goals did not totally dominate environmental concerns.

In 1997, 15% of the city's population continued to name the unfavorable ecological situation as one of the most acute problems of city life despite the competing weight of aggravated social and economic problems. How was it that some residents of Kirishi continued to consider environmental protection a priority? Analysis of the survey data suggested that neither sociodemographic variables nor the experience of changed material conditions explain these rankings.

Rather, the urgency of ecological problems appears to depend on such factors as personal observation of problems, adverse impact on the person's health and their rating of government's management of the problems. Table 3 implies that people most likely to rate the ecological problem as one of the most acute problems of Kirishi and who also consider the

Table 2. Choice Between Social-Economic and Environmental Goals[a](Kirishi, 1997).

Index of "Social-Economic/Environmental Priorities"	$N = 1018$ (%)
Resolution of economic and social problems	38
Lack of certain distinct position	41
Environmental protection	21

[a]The index is made up of answers to the following question on the survey: "Recently there are suggestions to close industrial plants that pollute the city environment, but implementation of this suggestions will cause problems: former workers will have to change their qualification, possible decrease in salaries, difficulties with new employment and many others. Can you please determine your position on this issue by evaluating the following statements on a 5-point scale where 5 is "agree" and 1 is "disagree":

1. The economic situation in the country will not allow government to spend significant finances on ecological improvements. Currently, the priorities should be economic and social issues.
2. Industries that pollute the environment should be closed and they should be restructured into environmentally safe enterprises, despite existing economic and social problems, which are unavoidable in this kind of situation." Interaction between variables 1 and 2: tau-$b = -0.33$.

environment to be in poor condition are unsatisfied by the results of government environmental policies. They also tend to be experiencing the consequences of pollution on their own health or on the health of their close acquaintances.

SOCIOECONOMIC AND POLITICAL CONSEQUENCES OF THE SOCIETAL ECOLOGICAL MOVEMENT

The change of priorities by residents of Kirishi in the perception of city problems took place, in the author's opinion, not only due to significant worsening of socioeconomic conditions of the population but also because the environmental policies of the government had positive consequences. The change in local ecological policies, in turn, significantly aided the success of the mass ecological movement in Kirishi. The results of the surveys in 1993 and 1997, which are rankings by respondents of the action of civil servants in the local government in their resolution of environmental problems, are displayed in Table 4.

We see the significant change in the attitudes toward the environment in 1997 compared with 1989. In 1989, 70% of those who responded to the

Table 3. Factors Determining the Relative Significance of Environmental Problems (Kirishi, 1997).

Factors	Environmental Condition Indicated Within Acute Problems					
	%	*N*	Tau-*b*	%	*N*	Tau-*b*
Evaluation of environmental condition in 1997 Russia			−0.21 ($p<.00$)			−0.16 ($p<.00$)
Bad	31	(226)		1	(150)	
Satisfactory	11	(642)		30	(425)	
Good	9	(96)		43	(65)	
Evaluation of influence of environmental condition on health			0.11 ($p<.01$)			0.11 ($p<.01$)
Not dangerous	8	(132)		37	(93)	
Possibly dangerous	14	(540)		27	(357)	
Feeling not well	20	(190)		10	(119)	
Sick	22	(86)		10	(65)	
Evaluation of governmental environmental policy			−0.05 ($p<.16$)			−0.01 ($p<.51$)
No effort to improve the situation	22	(87)		10	(61)	
Make some effort, but no results	16	(268)		29	(180)	
They managed to do certain things	16	(340)		21	(225)	
They have done a lot	5	(36)		28	(25)	

Table 4. Evaluation of the Activities of Local Authorities in Solving Environmental Problems in Kirishi by Citizens of Kirishi (Kirishi, 1993, 1997).

Question to the Respondent	Time of the Survey	
"Do Local Authorities Try to Solve Environmental Problems in Kirishi? If Yes, to what Extent they are Successful?"	1993 ($N = 911$)	1997 ($N = 1017$)
Do not make any efforts to solve environmental problems	14	9
They are trying to solve environmental problems but without significant results	34	26
They managed to achieve certain things in solving environmental problems	21	33
They managed to achieve a lot in solving environmental problems	2	4
It is difficult to say	29	28

survey gave a negative evaluation of government's performance in solving environmental problems. In contrast, the 1997 survey suggests that the implementation of environmental policy by local authorities was viewed as having been successful. Similarly, improved evaluation of the authorities' environmental performance is also evident in comparisons of 1993 and 1997 survey results. As a further indication of this positive shift of opinion about government environmental policy, only a tenth of respondents felt that authorities were doing nothing to improve the local environment and a quarter found the environmental policy to be ineffective. This positive evaluation of environmental policy in 1997 is notable because government was rated more highly in this area than it was for its social-economic, employment, and crime policies.

In sum, the urgent priority given environmental issues in 1989 was accompanied by a negative assessment of local governmental policy. As we compare responses in 1993 and 1997, the dynamics are reversed. Government is credited with solving environmental problems, making them less a priority.

It is important to stress the significance of the period in Kirishi when the environmental movement moved into the government structure. This period served to institutionalize Green practices, many of which were formerly opposed by authorities. Environmental management activities for all branches of government become routine managerial practices, part of everyday life. These activities proved effective and were highly evaluated by citizens of the city.

What then are the consequences of the mass environmental movement in Kirishi for the contemporary social-economic and environmental situation in the city?[7] According to Safronov (1999), the level of protest readiness in Kirishi citizens remains very high. Only 12% of his sample of residents said that they would not participate in any protest actions; another 5% restricted their activism to a readiness to sign petitions to government. In contrast, 45% of the citizens said they were ready to go to the streets, to participate in authorized demonstrations and civic disobedience actions. The prospect of unauthorized strikes, picketing, unsanctioned demonstrations, and other actions was acceptable to 38% of the adult population. A tenth of the sample reported a willingness to participate in the whole repertoire of collective action if this is necessary (see Safronov, 1999, p. 56).

Additional support for a potential high level of protest in Kirishi are contained in a comparative analysis of Kirishi protest level with data on St. Petersburg and five Western democracies (see Table 5).

It is possible to suggest that high level of protest readiness in contemporary Kirishi is determined by the events that took place at the end of 1980s and the early 1990s, when a successful mass environmental movement in the city proved that civic action can influence political decisions. It is not by chance that the highest readiness to protest is found among veteran "activists of the 1980s."[8] Having successfully participated in actions against local polluters, they gained confidence that their cause was worthy and that they could achieve positive results for their city.

CONCLUDING REMARKS

Despite the successful efforts of the city environmental movement in Kirishi, the 1989 governmental decision to close and restructure BXZ was not fully implemented. Although discharges of BVK into the atmosphere were not curtailed, actual production of BVK was not stopped until 1993 because of economic reasons. BXZ was converted to a vodka factory. Its trademark vodka, called Tigoda, is widely sold in markets in St. Petersburg and throughout the Leningrad region.

Despite the earlier efforts, Kirishi district continues to be the most environmentally unsafe part of the entire Lenigradskaya region. Today its economy is dependent upon the oil refinery plant, an enterprise of high ecological risk. In 2001 it was estimated that the average citizen of Kirishi district is exposed to nine times the pollution of the average statistical citizen of Leningrad region. Moreover, the environmental situation in Kirishi is not

Table 5. Readiness to Participate in Protests: Comparison with Western Democracies[a].

Actions	Data (%)	
	Possible Participation	Absolute Readiness
Signing petitions		
Netherlands	88	75
UK	75	53
USA	91	78
Western Germany	84	71
Austria	75	65
St. Petersburg	58	30
Kirishi	71	43
Authorized demonstrations		
Netherlands	68	46
UK	55	31
USA	68	39
Western Germany	66	42
Austria	44	27
St. Petersburg	58	29
Kirishi	64	43
Unauthorized strikes		
Netherlands	33	16
UK	27	12
USA	28	7
Western Germany	23	7
Austria	15	4
St. Petersburg	17	5
Kirishi	24	9
Combination of actions		
Netherlands	29	14
UK	23	8
USA	18	3
Western Germany	27	10
Austria	20	6
St. Petersburg	8	2
Kirishi	20	6

[a]The source on Western democracies: Barnes et al. (1979). Time of the survey: 1974. Number of surveyed: Netherlands, 1201; UK, 1483; US, 1719; Western Germany, 2307; Austria, 1584. The questions in St. Petersburg and Kirishi were identical. Characteristics of the surveys: St. Petersburg, 1994, $N = 581$; Kirishi, 1997, $N = 1018$. The index "possible participation" = answers "have done" or "no doubt-would do" or "might do." The index "absolute readiness" = answers "have done" or "definitely would do." Actions: (1) signing petitions, (2) participating in authorized demonstrations, (3) joining wildcat strikes, (4) blocking traffic.

expected to improve. The city remains in need of a strong environmental policy and constant environmental monitoring and control.

The first steps toward formation of effective environmental policy in Kirishi were achieved to a large extent because of protest strategy. However, it is important to develop conventional strategies of decision making. Given its activist history, Kirishi offers a unique prospect in modern Russia for successfully developing forms of public participation in its political, economic, and environmental decisions. In this way, residents of a contaminated community can become a model for all.

NOTES

1. Surveys were conducted in 1989, $N = 950$; 1993, $N = 911$; and 1997, $N = 1018$. A representative sample of the adult population of Kirishi was measured allowing for comparisons of response according to gender, age, and the level of education. Interviews were conducted during the years of the public surveys, and also in 1991, 1995, and 1999.

2. Resource mobilization theory was developed in the United States in the 1970s (see Gamson, 1975; Obersholl, 1977; Tilly, 1984; Zald & McCarthy, 1987).

3. At that time in the industrial zone of the city were two industrial giants: the oil refinery plant (NPZ) and a hydroelectric station.

4. Staff of BXZ stayed on the other side of barricades. Despite the danger of BVK, most people working at the enterprise tried to protect the plant from the movement's pressure.

5. For example, after difficult negotiations with authorities, already by Summer 1987, VOOP 6 had received official permission to set up an information desk in the center of the city to inform citizens regarding environmental conditions.

6. As a consequence, the social environmental movement in Kirishi never had to function as a mass movement before it became institutionalized. VOOP 6 lived on through December 1991, serving to coordinate and disseminate information between the Green fraction of City Council and citizens.

7. The data received by V. Safronov (1999) on the protest potential in Kirishi are useful for our analysis.

8. During the analysis of empirical data of the 1997 survey, the index named "activist 1980s" was constructed from five questions. It united participants of actions against environmental pollution (signing petitions, participation in meetings and demonstrations, material support of the environmental movement in the city), which continued to believe that these activities were of continued value to the city.

REFERENCES

Abercromby, H., Hill, C., & Turner, B. (1984). *Dictionary of sociology*. Harmondsworth, NY: Penguin Books.

Barnes, S. H., Kaase, M., & Allerbeck, K. R. (1979). *Political action: Mass participation in five western democracies*. Thousand Oaks, CA: Sage.

Berger, P. L., & Luckmann, T. (1967). *The social construction of reality: A treatise in the sociology of knowledge*. Garden City, NY: Doubleday & Company, Inc.

Dunlap, R. E., & Mertig, A. G. (1992). The evolution of the U.S. environmental movement from 1970 to 1990: An overview. In: R. E. Dunlap & A. G. Mertig (Eds), *American environmentalism: The U.S. environmental movement 1970–1990* (pp. 1–10). Washington, DC: Taylor and Francis.

Gamson, W. (1975). *The strategy of social protest*. Homewood, IL: Dorsey.

Gurr, T. A. (1968). Causal model of civil strife: A comparative analysis using new indices. *American Political Science Review*, *62*(4), 1104–1124.

Mitchell, R. C., Mertig, A. G., & Dunlap, R. E. (1992). Twenty years of environmental mobilization: Trends among National Environmental Organizations. In: R. E. Dunlap & A. G. Mertig (Eds), *American environmentalism: The U.S. environmental movement, 1970–1990* (pp. 11–26). Washington, DC: Taylor and Francis.

Obersholl, A. (1977). *The decline of the 1960's social movements. Research in social movements, conflict and change*. Greenwich, CT: JAI.

Offe, C. (1990). The institutional self-transformation of movement politics: A tentative stage model. In: R. J. Dalton & M. Kuechler (Eds), *Challenging the political order: New social and political movements in Western democracies* (pp. 232–250). Cambridge: Polity Press.

Putnam, R. D. (1996). *Making democracy work: Civic traditions in modern Italy*. Moscow: Ad Marginem (Russian edition).

Safronov, V. (1999). *Problems in establishing liberal democracy: Values, grievances and protest potential in small industrial city in Russia*. Final Report on the project granted by the Russian State Scientific Foundation, No. 99-03-00205a. Manuscript. <http://socio.ru/public/safronov/opr.zip>

Smelser, N. (1963). *Theory of collective behavior*. New York: Free Press.

Tilly, C. (1984). Social movements and national politics. In: C. Bright & S. Harding (Eds), *Statemaking and social movements: Essays in history and theory* (pp. 297–317). Ann Arbor, MI: University of Michigan Press.

Tsepilova, O. (2002). Social movements in the zone of ecological disaster: History, development and social consequences in the case of Kirishi. *Journal of Sociology and Social Anthropology*, *1*, 70–100.

Zald, M., & McCarty, J. (Eds). (1987). *Social movements in an organizational society*. New Brunswick, NJ: Transaction Books.

SOKOL: SEEING THE FORESTS FOR THE TREES

Antonina Kulyasova and Ivan Kulyasov

INTRODUCTION

This chapter analyses conditions under which residents of a small Russian town accept the concepts "pollution" and "ecological risk." The town in question is Sokol in the Vologda oblast of the Russian Federation, where there are two pulp and paper mills and other forest industries. Sokol is a typical small town with a population of about 40,000. The pulp and paper mills are locally run. The issues surrounding Sokol's pulp and paper mills generally present a typical Russian picture (Kuliasova & Kuliasov, 2002a, 2002b) with one major exception. Industrialization in Sokol goes back more than a century and thus reflects the broader history of the 20th century.

The authors conducted qualitative sociological research in Sokol from 2000 to 2003. This article is based upon 15 expert key informant interviews with representatives of various sectors of society in the cities of Vologda and Sokol, as well as 62 local resident interviews taken in Sokol and its immediate suburbs and reflecting differences in gender, age, social status, and employment.

In pulp and paper mill communities, residents need not rely upon technical equipment to record the extent of pollution of air, water, and soil. They can directly detect such pollution on a daily basis by means of their sense organs. However, given the industrial history of the community, signs

Cultures of Contamination: Legacies of Pollution in Russia and the U.S.
Research in Social Problems and Public Policy, Volume 14, 47–72

ISSN: 0196-1152/doi:10.1016/S0196-1152(06)14003-X

of pollution are so normal that we wanted to learn if they are actually perceived by people, and if so, whether they are defined under the concepts of "pollution" and "ecological risk." Related questions of importance were the impact on perception of pollution due to the source of contamination being the waste products of a Central Construction Bureau and whether residents connect their state of health with ecological risk.

During the period in which industrial production developed in the USSR and Russia, ecological pollution was not publicly discussed. The concept of "ecology" and "ecological risk" was virtually absent from people's daily discourse. At that time the priority in cities was the development of the economy, the growth of industrial production and the construction and broadening of manufacturing. In the 1980s, the situation started to change and the concepts of "ecology," "ecological pollution," and "ecological risk" became widely used. At the end of the 1980s and beginning of the 1990s, with glasnost and perestroika, there began an active struggle for a clean environment in Russia. These years were characterized by a mass ecological movement and protest campaigns. However, later in the 1990s, Russia entered a period of economic decline, and once again ecological problems faded into the background.

The beginning of the 21st century in Russia has been characterized by ongoing restructuring of the state agencies concerned with natural conservation and a weakening of ecological legislation, which have in turn led to a weakening of state ecological control (Kuliasova & Kuliasov, 2002a, 2002b).

THE DESIGN OF THE CONCEPT OF "POLLUTION" AS A REFLECTION OF ECOLOGICAL RISK

The modern concept of risk has been explored in the writings of several Western social theorists, most notably Anthony Giddens, Ulrich Beck, and S. Lash (Beck, Giddens, & Lash, 1994; Giddens, 1999). According to their analysis, the development of science and technology has not only raised the level of social risk and risk in nature, but has also limited our ability to meaningfully evaluate the consequences. In a condition they term "reflexive modernization," the concerns of modern people increasingly encompass the very consequences of modernization itself. Future alternatives are interpreted so as to reflect present realities and thus become all the less predictable in their consequences.

In conceptualizing ours as "the risk society," Beck writes about the ubiquity and invisibility of risk. The industrial destruction of life's ecological and natural foundations has released a social and political dynamic without historical analogy, the development of which compels us to rethink the independent relationship of nature and society. Nature can no longer be understood without society or society without nature. A side effect of the socialization of nature has been the socialization of nature's destruction. The damage suffered by nature and the destruction of natural habitats are now becoming medical, social, and economic threats to humankind globally. It is specifically this conversion of threats to nature into threats to social, economic, and political systems that represents a challenge to the present and future (Beck, 1998, 2000).

Oleg Yanitsky (1998, 1999) has applied the theories of the risk society and reflexive modernization to Russia's conditions of an economy transitioning between communism and capitalism. He argues that such transitional societies experience "the paradox of modernization," simultaneously advancing into high modernization and regressing backward into a phase of "de-modernization." Such paradoxical development is the fundamental pattern in creating a modern, Western technological society. It entails intensified risk production on both ends. Accordingly, Yanitsky has characterized Russia as a society of general, widespread risk. This risk is magnified culturally by the absence of institutionalized reflection on risk in both the professional sphere and in scientific knowledge, as well as an absence of any regular analysis of the social and natural costs of personal activity. Meanwhile, there is a need to spend an ever-increasing proportion of society's material and intellectual resources on the creation of procedures capable of regulating the level of risk built into the process of social manufacturing. It follows that the acceptance of risk comes to be viewed as an inescapable condition of human existence. To Yanitsky, such risk acceptance must be viewed as a form of pathology.

Collectively, these risk theorists give special attention to the growing isolation of experts from the public. This isolation is of concern because experts are in a position to provide institutionalized calculation of and reflection on risk and guide decision making accordingly. As experts have lost contact with the values and wisdom of the larger society, expert knowledge itself has acquired a specialized and fragmented character. As a result, abstract systems that are global in their influence become fully accessible to an increasingly small minority of experts. Crossing so many political and social boundaries, moreover, experts are even isolated from their peers and find that they are unable to coordinate and compare their evaluations and recommendations.

This isolation of expertise has important consequences for understanding how risk is perceived. The concepts "pollution" and "ecological risk" as a whole are understood differently by experts and by local residents lacking expert knowledge. Edelstein has observed that risk is an abstraction to experts who can view it in relative terms. The American public, in contrast, responds to contamination when it is a personal threat, not an abstraction, and they consider it to be an absolute rather than relative issue Edelstein (2004). Perhaps the only condition that puts both lay people and experts on the same footing is when pollution can be directly perceived by the senses.

Using the case of Sokol, we seek to explore the extent to which risk-reflection, in the form of the construction of the concepts of "pollution" and "ecological risk," is characteristic of the residents of a typical Russian city with locally run industries.

The Setting

The town of Sokol is situated in the Vologda Region, in the northwest of Russia. Sokol and its countryside are very beautiful, the scenery dominated by both coniferous and pine forest. Sokol is situated on the banks of the river Sukhona, a stream of striking beauty. The Sukhona is a navigable river belonging to the White Sea watershed and is the largest river in the Vologda Region. The Sukhona flows from Kubensk Lake, one of the largest lakes in the Vologda Region, and then runs into the Northern Dvina and on to the White Sea. The entire Sukhona Region has been populated since time immemorial, as evidenced by ancient, picturesque villages and the cities of Totma and Velikii Ustiug, whose histories go back almost 1,000 years. Because Sokol is situated in the upper reaches of the Sukhona, its manufacturing activities affect the river as a whole. Due to ice floes, in Spring the river flows in the opposite direction. Thus, its source, Kubensk Lake, is also impacted by industrial activities in Sokol.

THE ECO-HISTORY OF SOKOL

The history of Sokol is very closely connected with the history of its pulp and paper mills. As the largest forest industry enterprise in the Vologda Region, the Sokol pulp and paper mill is vital to the economy of Sokol and the whole region. The history of pulp and paper production in Sokol dates back to the late 19th century. At that time, there were about 20 small villages along the banks of the Sukhona, surrounded by forest, swamps, and flood plains. In 1896, a forest industrialist and timber merchant named A.Yu. Surkov cofounded the "Sokol Northern Pulp and Writing Paper Production Company," headquartered in Arkhangelsk. By the next year, Belgian experts were overseeing construction of a pulp mill on the bank of the Sukhona. The location offered excellent transport connections: the road between Moscow, Vologda and Arkhangelsk ran nearby, as did rail connections from Vologda to Yaroslavl and Arkhangelsk. The first paper machine was established in 1899 and the second one in 1903. Workers came from the 17 villages surrounding the mill. The mill gradually grew, and by 1912 it had 1,030 workers. By that time workers and specialists were already coming to the combine from all over Russia and abroad. Very early in the 20th century, hostels, workers' barracks, offices and houses for the engineers were built around the mill. At the turn of the century, other industries also began to develop rapidly in this area of the upper reaches of the Sukhona. In 1911, businessmen from St. Petersburg began to construct a second pulp mill, the Pechatkinsk pulp mill (today the Sukhona Pulp and Paper mill). A sawmill opened next door.

Over time, the settlements around the Pechatkinsk and Sokol mills grew together to form the town of Sokol, officially recognized in 1932 as the administrative, cultural and industrial center of the Sokol district (Loshchilov, 1999). From the beginning, a portion of the population acknowledged the ecological dangers of Sokol's industrial combines. Thus, opposition from peasants who refused to sell their lands forced the second combine to be relocated from its original proposed site in the upper reaches of the Sukhona. These peasants took their advice from "knowing people" (i.e., experts) who convinced them that the future production plant was too dangerous.

In the pre-revolutionary period prior to 1917, working and living conditions at the combine were difficult. Work occurred all day long in two shifts of 12 hours each. There was not enough accommodation for the workers, and therefore temporary barracks and apartments for specialists were constructed. After the revolution, the integrated combine continued to develop to meet the paper needs of the young Republic. The town accordingly expanded rapidly as well. By the 1930s, it brimmed with peasants relocating from villages throughout the entire district. The attraction to the town also reflected an escape from changes occurring in the rural Soviet Union. In the villages at that time collective farms were being built, animals and the means of production were collectivized, and the government demanded high taxes for natural products. As a result there was famine.

Factory workers were given a food ration. Because so many workers came to escape famine and collectivization, it is unlikely that people thought about the risks and dangers that awaited them in the town.

The ecological problems of this period are indicative of fast growing industrial towns. A rapid expansion of production resulted in discharge of substantial factory wastes directly into the Sukhona. Meanwhile, the growing number of houses discharged sewage directly to the river. A crisis ensued in which many lacked potable water. While the integrated combine provided running water and some wealthier residents had private wells, most transported water for drinking directly from the river. As a result of the lack of clean drinking water, enteric infections, dysentery, and abdominal typhoid spread, killing many. An interviewee recalled this period:

> … before 1934 we lived in a village. My father died that year. Mother could not feed six children in the collective farm, and we moved to the town because they gave out food rations there. Mother got herself work at the factory as a cleaner. After a few months two of my sisters and my brother died in hospital of dysentery. (female, 84 years old)

During the 1950s the level of waterborne infectious diseases in Sokol was the highest of all the Vologda Region districts (Loshchilov, 1999). However, this negative legacy for environmental health was overshadowed by the positive contributions of Sokol for industrial growth, as reflected in the books and newspapers chronicling Sokol's development from the 1930s. By 1936, the Sokol mill was a leading forest industry plant, second in Soviet production (Loshchilov, 1999). In the 1930s, a thermal power station was built close to the Sokol Pulp and Paper Mill and produced enough surplus power to supply the regional city of Vologda. Soviet propaganda emphasized these successes of the 1930s and 1940s, recalled by many middle-aged interviewees with pride in their combine, their town and their contribution to the development of industry.

Further development of the integrated combine occurred from the 1950s to the 1970s. Modern papermaking machinery was introduced. At the same time, chlorine bleach washing started to be used at the Sukhona Central Construction Bureau. In 1976, the Sokol and Sukhonsk mills joined to form one combine. Labor was divided equally. The Sokol mill specialized in producing different sorts of unbleached paper, fiberboard, and technical spirit. The Sukhonsk mill specialized mainly in producing bleached paper and fiberboard. In the early 1980s the annual production of the two mills totaled 132,000 tons of paper and 171,700 tons of pulp. This industrial output and modernization was matched with a rapid increase in pollution levels, forcing the crisis of air and water pollution to become acute.

This industrial expansion continued into the 1980s. Of most importance was the opening of a new hydrolytic factory for producing yeast fodder. This industry made the ecological situation in the town intolerable. As a result, the period of the late 1970s and early 1980s marks a turning point toward serious ecological improvement.

During the post-Soviet privatization process, the Sokol and Sukhonsk enterprises became independent companies once again. At that time the entire forest industry sector in Russia was in crisis. This situation caused trouble in Sokol as well, and the combined production volumes of the two mills were reduced by 8,000–10,000 tons per year during the 1990s (Krasnyy Sever, February 2, 1999). But in 1998 the Moscow investment group "fox" bought the Sokol mill. It started to develop once again, and new, modern paper production lines were opened.

This resurgence of the industry lacked the same social benefit as had prior economic development. Early in the 21st century, the Sokol pulp and paper mills had ceased to be locally operated. As a result, the mills abdicated various social responsibilities that they had historically undertaken. The town's water intake station, its cleaning facilities, kindergarten, technical institute, and other social objectives were no longer provided by the industry. Similarly, the mills stopped building housing for the residents of the town. And then downsizing hit; the number of workers was reduced at the enterprise to less than 1,000.

ENVIRONMENTAL PERFORMANCE IN SOKOL DURING 1970–1990

The topics of ecological problems and pollution received widespread discussion in the 1970s in the Soviet Union, much as they did in the US. These concepts began to be constructed at the level of social discourse. This trend fit the worldwide goal of that time to realize the conservation of nature. In the USSR, this period of awareness corresponded to heightened pollution, reflecting the primitive state of the majority of old industries, lacking as they were in pollution controls of any types. If anything was to be done about these circumstances, it was clear that the central state had to take the lead. Accordingly, in the 1970s, the government of the USSR accepted a raft of resolutions on the construction of pollution control facilities in many cities and enterprises. The impacts for Sokol were typical of the nation as a whole.

Technological Fixes

The first environmental improvements were carried out in the 1970s, when the Sokol mill built a wastewater treatment plant that also served other enterprises and households in the town. This plant was not part of any local initiative. Rather, it was responsive to a government decree from the Ministry for the Pulp and Paper Industry issued specifically for this particular pulp and paper combine. Illustrated was how an attempt to solve local water pollution problems was initiated at the highest level of the Soviet Union. The workers of Sokol and Sukhonsk mills participated in the construction of the purification plant. The construction was rapid and was carried out in very difficult conditions on marshy grounds. The last part of the purification plant began operating in 1980 (Loshchilov, 1999).

The purification plant has had a great influence on the Sokol mill and the development of the town of Sokol during the last 20 years. Of greatest importance, it has also improved the ecology of the river Sukhona. Before the construction of the purification plant, local systems with cesspools and septic tanks were used and a large amount of wastewater was not cleaned at all. The drains from the Sokol mill flowed into the river Sukhona and the sewage from the Sukhonsk mill was released into the river Pelshma, which flows into the Sukhona downstream from Sokol. The absence of any purification plant, water cleaning station or sewerage system constrained house building in certain areas and the development of the town overall. In many interviews interviewees spoke of how, before the construction of the purification plant, masses of cellulose fibers would flow along the river. In addition, acid was being continually poured into the river. One of the interviewees said of this that:

> I didn't work on the combine full-time, but we were part of a student construction group, and in the summer we worked on repairing the combine. I myself saw then how tanks of acid were simply thrown straight into the river. (male, 47 years old)

The second investment in the environment also concerned water purification. In 1990 an integrated wastewater purification system was inaugurated at the Sokol plant to recover lost material. In the first application, 35% of fiber waste was recovered and returned to the production process. This fiber was reused as raw material for 720,000 square meters of soft wood fiberboard in 1990. Since then the process has become even more efficient and about 90% of fiber waste returns to the production process. This system has dramatically reduced the consumption of fresh water at the mill. It is interesting that this ecological improvement did not result from pollution reduction efforts. Rather, it resulted from an initiative announced in the mid-1980s by Mikhail Gorbachev to promote economic efficiency through resource conservation.

The third environmental improvement involved the local thermal power station. As has been the rule for single-industry towns in the Soviet Union and Russia, the Sokol mill took responsibility for developing and operating municipal services and utilities. For example, the integrated combine built much of the housing in Sokol, as well as recreation and sports facilities. And it also actively participated in constructing the town's infrastructure. The district heating system was and is one of the most important public services provided by the mill.[1] The power station was originally built in the 1930s and renovated in 1978 in order to provide heating and hot water to the residents of the town center. The integrated Sokol and Sukhona combine has since provided heating for the majority of houses in the town.

Pollution benefits accrued from ongoing improvements. In the beginning of the 1990s, the thermal power station was rebuilt. At the time, 24% of total emissions in the air were produced by boiler houses, including those produced by the Sokol and Sukhona pulp and paper mills (17% and 7%, correspondingly). The reduction of emissions has been promoted by the conversion of the boiler houses to gas (instead of using coal and mazut). This improvement was noted by practically all our interviewees:

> Earlier in the town there was so much soot that it was impossible to hang washing outside since it would be covered in soot. Between the double-windows there would always be a black coating, and now there isn't any longer. (female, 58 years old)

After the boiler houses' conversion to gas, all industries in the town improved. As one interviewee said:

> The smoke that comes out of the chimneys is now white, not black, without any soot. (male, 19 years old, 2002)

Thus we see that in the Soviet era there were attempts to solve problems related to the local environmental conditions in Sokol. From a certain perspective these could be categorized as early attempts toward ecological modernization and the reduction of environmental risks. In the 1970s there was a more general tendency in Soviet industry to build water purification systems in industrial localities. This was a reflection of a new social and environmental protection policy by the Soviet state in the 1970s (Ziegler, 1987). This modernization was connected to a new discourse about environment and social health in Soviet state institutions (Kelley, Stunkel, & Wescott, 1976). These technological improvements and attempts to make environmental investments were very significant for improving environmental conditions in industrial localities. However, as in many other cases, these attempts were not sufficient to clean the surroundings of the industrial plant in Sokol. Today, the purification system is only partially operative, and it cannot ensure the cleaning of the polluted sewage waters. The ecology of the rivers Pelshma and Sukhona has suffered greatly from this problem.

Social Fixes

The other measure for the environmental study in Sokol was residents' reflections on pollution. As was noted, a hydrolytic factory for the production of yeast fodder was built in Sokol in 1986. This enterprise became the symbol of ecological problems in Sokol because of its noxious impacts. Yeast was discharged into pond sedimentation tanks. During warm weather, the albumen protein would start to decay, producing, in the words of local residents interviewed, an unbearable, putrid smell.

> On this side of the town there used to be a terrible stench. Because they made some sort of yeast for cattle feed there. I saw a dark blue, stinking, foaming liquid flowing straight from the hydrolytic plant into the Sukhona – it was simply impossible even to go near it. (male, 48 years old)

These visible effects prompted widespread discussions about the ecology of the town. A protest movement formed. It was against the background of the battle with the hydrolytic plant that the ecological discourse of the people developed, as they actively constructed their own understandings of "pollution," "ecological problems" and "ecological risk." Mass protests by the population of Sokol took place in the form of a large number of both spontaneous and organized meetings and demonstrations, critical newspaper articles and radio broadcasts, and telephone calls and letters to those responsible for the ecology in the town, as well as to the head of the local self-government organization. As a result of the protests, the factory was closed at the beginning of the 1990s.

> And thus we came to the conclusion that, in the already complex ecological situation that we had, it was impossible for us to support yet another new project which would have aggravated our situation with respect to ecology. This is why the authorities applied every effort for this not to be the case for us. It was proposed, that the orientation of the production plant to be changed to woodworking. (member of the Sokol administration responsible for the social–economic programs in health and ecology).

This event with the hydrolytic factory is very interesting. Despite the long-standing pollution of water and air, residents of Sokol had previously given little attention to the ever-present pollution and environmental risk. What was different now? First, the mobilization of protest strongly reflected the milieu. This was a time in the USSR, at the end of the 1980s and start of the 1990s, characterized by the democratization of society and a mass ecological movement that took the form of protests across all of Russia.

A second factor, in our view, was more important. The hydrolytic factory was a new source of highly visible pollution. It directly affected the sense organs with its strong and disgusting smell. An interesting consequences of this noxious invasion was that everyone in the town was placed on a par – whether town administration, specialists or simple residents. However, authorities brought their power to the task.

A third factor also helped. The hydrolytic factory was new. It represented a novel phenomenon to which people had not yet had time to adjust. It was not connected to the hallowed past, in which people took pride. And it was not vital to the town's functioning as were the pulp and paper mills. All these reasons made this factory vulnerable to being viewed as an ecological risk in ways that the pulp and paper industry was comparatively buffered from.

Contemporary Environmental Problems in Sokol as Defined by Experts

Specialists have identified the quality of air and water to be the main ecological problems in the town of Sokol. Official statistics for the sanitary zone of the Sokol mill, which encompasses most of the town, report levels of sulfur dioxide in the air eight times higher than the maximum level allowed by federal law and standards. One of the neighborhoods located near the Sokol mill was abandoned in 1970 because unfavorable winds carried toxic compounds to it. There was a smell of acid in the town for years, but during the early 1990s the situation improved because of the decline in production. Recently the air quality has been getting worse again. Today ecological problems are severe and affect residents' health. According to an official of the environmental administration, among the residents of Sokol there are, for example, increased levels of allergic diseases, diseases of the respiratory tract, and cancer.

It is very difficult to reduce air pollution at a Russian industrial facility. An expert at the Sokol Sanitary and Epidemiological Inspectorate (SEI) explained:

> We work by usually imposing fines in emergency conditions. We cannot act as far as the air in the atmosphere is concerned, because what are needed are radical measures, such

as the construction of a new acidity regulating unit. Only in these circumstances can we achieve any positive results. (female, 49 years old)

In most cases, industries chose to pay modest fines rather than do the costly needed upgrades. For this reason, the only major force for pollution reduction in Russia has been the fact that economic crisis has produced a downturn in industrial production. Closed or largely quiet factories produce less pollution.

Thus, the principal improvements in Sokol occurred in 2003 when the Sukhona mill was virtually closed. According to specialists of the Vologda Regional Centre of Hydrometeorology, pollutant emissions from stationary sources were reduced 4.7 times (in 1990 they totaled 34,700 tons, while in 2001 this figure was 7,400). The level of sulfur dioxide emitted in the pulping process also decreased. However, the location of boiler houses in the immediate proximity to living quarters makes it difficult to ensure that air quality corresponds to standards (Sokolskaya Pravda, March 10, 2002). The aforementioned interviewee confirms this point in her interview:

In fact, the town was built in such a way that almost the entire residential zone fell within the buffer area of the pulp-and-paper and wood industrial enterprises. So it is meaningless to speak of any 'safe distance.' People live here and polluted emissions affect them negatively. That is why the ecological situation here is very poor. As far as I know, we hold second position in this respect after Tcherepovets [the biggest town in the Vologda region, with even larger paper mills and chemical processing plants]. (female, 49 years old)

The other big problem is the quality of the water. In addition to the two pulp and paper mills, there are many other enterprises located along the river Sukhona, including ceramic handicraft and woodworking enterprises. Other smaller streams in Sokol that flow into river Sukhona are also heavily polluted by the solid and liquid effluents of settlements and industries. There are three main reasons for this pollution. Firstly, the capacity of the purification system is insufficient. The purification plant of the town and mill was designed to process 29,000 cubic meters of wastewater per day, but in practice the effluents reach 35,000–36,000 cubic meters. Secondly, there are sewers from municipal services that flow directly into the river. Thirdly, accidents are quite commonplace, and industrial waste also flows directly into the rivers. For example, in 1999 entire tanks of acid leaked directly into the river (Sokol'skaya Pravda, July 18, 1999). Lastly, there are many small rivers in the town. All of them are polluted to a great extent, particularly the Makhrenga, which flows in the neighborhood of the Sokol Pulp and Paper Mill. All wastewaters also accumulate in the small river Pelshma, which flows near the town of Kadnikov, outside Sokol.

According to the statistics of the Vologda Regional Centre of Hydrometeorology, the main water pollutants are as follows: copper, iron, zinc, manganese, oil products, phenols, nitrites, and nitrates. The water a kilometer downstream, which is where there is waste discharge, is rated 7 on the pollution classification scale; 5 falls within the range "extremely polluted water" (Sokolskaya Pravda, October 23, 2001).

The reduction in water draw off and waste discharge seen from 1990 to 1998 was connected with a setback in production. Since 1999 these indices have been gradually increasing. The main polluter of water is Sokol's pulp and paper mills, which produce 41,500 tons of contaminants. The remaining enterprises in the region together produce 41,800 tons. Treatment plants function inefficiently (Sokolskaya Pravda, July 10, 2002).

According to the State SEI, the overflow of contaminants in sewers at output is as follows: the amount of phenol exceeds permissible norms 62–450 times, formaldehyde exceeds permissible norms 2–36 times; methanol exceeds permissible norms 18–164 times; and lignosulfonic acid and oil products exceed permissible norms 17–34 and 8–36 times, respectively (Vologodskaya Nedelya, March 21, 2002).

We must note that the situation with drinking water is also critical for residents of Sokol who live in houses with running water. An employee of the Sokol Sanitary and Epidemiological Inspectorate (SEI), who is in charge of community facilities in Sokol and the Sokol Region, described the causes of this problem as follows:

> In the town of Sokol the quality of water is bad because our river has a reverse current. In these conditions, adjacent waste discharges affect water intake. That is, water at inlets contains harmful impurities typical of discharges. I have in mind our water inlet, which we use for the supply of drinking water. (female, 49 years old)

It is notable that the official state agency statistics clearly show the very high levels of air and water pollution in the town. Agency reports of pollution are covered regularly by newspapers and thus have the potential to make some impression on the public. Nevertheless, there is little documentation of any public awareness on these issues. And, given the frequency of pollution reports, people may have become numb to the statistics. They may know that there is pollution in the abstract, but at the same time, they may not consider how polluting substances affect their lives and health. While there are many experts reporting the pollution results, none seek to explain their significance to the public.

Nevertheless, experts have made a direct connection between pollution in the town and a high level of morbidity among Sokol residents with regard to certain diseases. According to the statistics of SEI, the Sokol Region in the Vologodskaya oblast ranks first in the average annual rate of congenital malformations. Sokol suffered 1,922 cases per 100,000 population while the corresponding average regional level is 823 cases (State Report on the Sanitary and Epidemiological Situation in Vologda, 2000, 2001). Cited is the opinion of an official at the Sokol administration responsible for social programs, including health and ecology issues:

> We are among those towns where cancer morbidity is the highest in the oblast. Even higher than in Tcherepovets and Vologda. This is also linked, to a great extent, with the appearance of other diseases caused by the ecological problems and directly connected to the industrialized nature of the area. There were times when the maximum permissible concentration [of toxins] in the air was exceeded in more than 50 indices. (Vice Mayor of Sokol)

Likewise, an employee of the Sokol SEI responsible for community facilities in Sokol and the Sokol Region stressed the connection between the mortality rate and ecological problems.

> It's clear that the people living in my town have been very badly affected by the environmental problems. Among all the other regions of the Vologodskaya oblast, our region ranks first in the statistics for cancer diseases. Looking at oncological morbidity at the federal level, here too, our situation is very dire. Respiratory disease ranks highly with respect to causes of morbidity and is prevalent. This gives a clear indication of the excessive amount of harmful substances in the air. Diseases of the digestive tract also rank highly, and this is connected to the quality of the water. (female, 49 years old)

Scientific experts are increasingly concerned by their findings, which are based on their own reflections, hypotheses and investigations, as well as specialist information on pollution. An ecologist respondent expounded on the problems of complexity in water pollution, how reliability of results is affected, and the extent to which much remains unknown.

> If you do an SEI (Sanitary and Epidemiological Inspection) or the Hydrometcentre (Hydro meteorological center water and air monitoring inspection), they can provide up to 40 indices, but amongst them are substances which we are often not even aware of, substances that react with one another causing chemical reactions which then react with each other and cause further chemical reactions. The results from the analyses are themselves very varied depending on the time at which the tests are carried out, be that in the morning, afternoon or evening. Substances tend to react most when they first come into contact with the water and at this point their toxicity also increases.
>
> Secondly, they begin to precipitate; the precipitation either accumulates or is absorbed by various algae, organisms or fish. So what we actually see are the residual concentrations, which remained in the water and not the whole chain of chemically reactive events. Water pollution and the level of pollution are two different things. In order to

> give an indication of the true level of pollution, therefore, we should investigate not only the water, but the riverbed, zooplankton, zoobentos, algae, macrophites and also the fish. Only after taking all of these into account can we begin to speak about the level of pollution. As things stand, our observations are inadequate. (Professor of the Vologda State Pedagogical University)

Scientific research has shown that the negative effects of the pollution from the town of Sokol on the ecology of the Sukhona River extend over long distances. Between 1989 and 1991, more than 60 tributaries of the Sukhona were investigated, including those that cross Sokol (situated at a river outlet) and those that flow through the town of Velikiy Ustyug at the border with the Arkangel'skaya oblast' at the mouth of the river. Concentrations of poisonous substances – typically sulfonates – from the chemical waste and polluted emissions from Sokol were found as far away as the town of Tot'ma, midway along the Sukhona River.

The results are summarized by a professor from the Vologda State Pedagogical University:

> Fish is an upper level, a terminus. All one would need to do to be able to draw conclusions about the state of the water is to monitor the conditions of its existence. One can't estimate whether the maximum permissible concentration levels are better or worse than expected. The fish are sick. They have toxicosis, large liver mass, renal-calculus, and various pathologies including mutation of cranial bones. We have even come across fish without scales, just bare-skinned with articulated fins and backbones. Our fish acquire various diseases and the diseases are themselves an indication that the situation is very dire: that conditions in this natural environment are such that it cannot support natural life, normal functioning. We could yet find ourselves in exactly the same situation as the fish. (female, 40 years old)

Accounts and expert opinions concerning the connection between illness and the pollution are, on the whole, not readily available to the local residents and information is not released in the media. As such, the local community has little or no idea about what is happening.

LOCAL RESIDENTS' PERCEPTIONS OF ENVIRONMENTAL POLLUTION AND ENVIRONMENTAL RISKS

Many of our respondents compare the state of the environment today to that before the environmental reforms. Some noted that the state of the environment has improved (depending on the respondent's age) during the last 10 and 20 years. That is, there was a wide sphere of reference in terms of the answers given and generally people were very optimistic in their understanding of environmental trends. Respondents related improving environmental conditions to the suspension of production in a large number of industrial enterprises. They specifically cited two indices of ecological health. The first was citing the appearance of crayfish in some natural ponds as an indication of the purity of the water. And, second, the prolific growth of mushrooms and berries available for picking in the forests nearby their dwellings.

Some respondents also noted an improving social environment, as indicated by the variety of goods and foodstuffs available in numerous stores, new work and training opportunities, chances to travel abroad, computerization, the advent of ecological education in schools, etc. Suggested is that perception of improved ecology reflects a more general perception of an improved quality of life during the period of economic reform; their experiences reflect the fervor with which residents have embraced new opportunities in their lives. These same respondents usually work as "subbotniks," participating in the spring scavenging and clean-up efforts organized by the town. They are also responsive to requests from the authorities and public organizations for residents to actively preserve the environment in which they live.

These same respondents opposed the closure of the Sokol and Sukhonsk pulp and paper mills and instead supported their rapid modernization (if paid for by the owners). These views suggest a balancing, on one hand, of a concern for a safe environment with a concern to maintain positive advances in social and economic health. They are willing to be active on behalf of these goals.

Other respondents argued that the environmental situation remained as bad as before:

> I don't think anything has changed...something's still not quite right. The air is polluted and there are cases of disease caused by the water...[Things are] neither better, nor worse, but not in the same ways as before. (an employee of the pulp and paper mills: male, 20 years old)

At the same time, perhaps reflecting their physical and social well-being, such residents acknowledge some improvement. It is as if they are trying to optimize their existence as best as they can in the midst of the present environmental and socioeconomic conditions. Thus, they are keen to discuss, in depth, not only the evidence of deterioration, but of improvement as well. As with the previous group, they are opposed to the closure of the

Sokol and Sukhonsk pulp and paper mills and are ready to take part in schemes organized to help improve them on a voluntary basis. While they do not expect any prompt improvements in either the ecological or social situation, they take pains to ensure that these do not deteriorate.

Still a third group of respondents believed that the situation had worsened, citing expert evidence of ecological deterioration:

> The situation is worse everywhere. We're doing the best we can and still everything is going downhill! (female employee of the pulp and paper mills, 49 years old)

These concerns echoed by other respondents are alarmed by what they see as dismal ecological and social conditions in Sokol. Such alarm involves a concern for air quality.

> The air has also become worse. It really is impossible to breathe. (male, 55 years old)

Water quality was also a concern.

> The water never used to contain so much chlorine. There's no denying how much dirtier the water has become. Up till around 1991, we would swim in the river all the time and did not experience any skin problems. Now swimming is categorically forbidden and people are frightened of it. It seems absurd, but we are forced to tell children to stay away from it and yet here we are living on the river-bank. In the summer, you still get that urge to go for a swim, but it's strictly prohibited. (female, 57 years old)

Soil was yet another focus of concern.

> I don't like the land as it tends to be damp and has a putrid smell. There is filth everywhere. (female, 19 years old)

There was an additional concern that conditions were toxic to gardens and houses.

> We really felt the effects of the chlorine emitted from the Sukhonsk Pulp and paper mills when we had a dacha on the other side of the river. You could often come into the garden and see leafless cucumbers, squashes and cabbage, because the leaves had just died. In fact all the leaves are dead. All it takes is for a strong wind to blow in from the north and all the chlorine hits the garden plots. (male, 70 years old)

> When the garden plots have been drenched in water from the treatment plant, what kind of harvest can one expect to reap? Of course, this doesn't just affect the harvest, but the buildings, foundations of the houses and so on. And here's one thing that happens quite often now – I remember at least, two or three occasions when the waste water from the treatment plant flooded the gardens; as a matter of fact all our suburban streets are in a terrible state. But there's no mistaking the effect it has upon farming. Gardeners suffer losses eternally; the sewage has corroded all the house-footings. And that's not to mention the adverse effect is has upon health and the air we have to breathe. There is also a whole list of economic problems stemming from it. (female, 67 years old)

Finally, a pessimism for health was clearly evident.

> I've become particularly prone to bronchial and pulmonary diseases since I've lived in this town...and all the members of my family have a cough. I'm absolutely convinced that this is connected with the town's ecology. I also think this has a negative effect on one's state of mind and mood. One feels constantly depressed from living in such poor, dirty conditions. (female, 57 years old)

> Living in Sokol, has made my mother's ulcer worse. She'd be much better if she lived somewhere else. The conditions really took a toll on her health ...(She later died in connection with her ulcer and acute asthma). On the whole, for all the time we spend living in the village we don't get ill. We only normally get ill in the town. (male, 14 years old, a pupil who lives in the region where the air and water are clean)

> I don't know, perhaps the environmental institutions are just afraid of spreading panic. You know, the people understand that life in Sokol is bad, that there are so many diseases and that this is mainly oncological or allergic ones. It's highly likely that women will soon start to bear mutant children. The ecological situation really is dreadful. It's just that nobody is shouting about it as they don't want to start a public panic, that's all. People need to know how bad the situation is though, how bad it is, is something they must decide for themselves. (male, 36 years old)

It is these pessimistic respondents who particularly take an active stand against ecological degradation and poor social conditions, sometimes utilizing destructive means of protest and desperate acts as part of their political and economic struggle. Viewing the situation as hopeless, these respondents believe that it will be practically impossible to live in Sokol in the next one or two decades. By that time, Sokol will be known as an "ecological disaster zone" and "eco-catastrophe." Many of those perceiving active danger try to visit the countryside more often to get fresh air, drink clean water, and grow produce grown on untainted soils. Particularly people younger than 30 years old would like to leave the town of Sokol altogether. But even they are against the idea of closing the Sokol pulp and paper mills and other industrial enterprises in Sokol – despite the consequences for the natural environment and their health – for fear of the social disaster that would occur in the present if this town-shaping enterprise is lost. They are unwilling to trade present well-being for future well-being.

We also noted that people who had recently moved to Sokol perceived the pollution more acutely than those already living there. However, their concerns dampened over time and apathy set in.

> I walked along the river bank and could see all this water flowing into it from the sewer. It was just awful. It looked like acid. Now we've become accustomed to it. At first it was a shock to see this acid flowing freely down into the river, it was a horrible sight. But now, no one is surprised at all. (male, 48 years old)

Many respondents shared pollution "tales" and "myths" during their interviews that were in effect "ecological horror stories" (strachilki) based upon observations of ecological change. Here are some examples:

> This happened last year. The potato crops had failed and it had been raining, but the rain was not usual rain and left green circles everywhere. It smelt like a blossoming potato plant and the whole vegetable patch looked as if it was covered with clinker (i.e., unburned residues). That's why there was bad harvest of potato. We haven't seen anything as strange as this since then, but this did happen last year. The rain was awful, lots of black clouds gathering and these green stains everywhere. (female, 77 years old)

> Some people say that about 30 years ago a green spot appeared on the surface of the river. Apparently, it was to do with the settling mud. Fishermen were the first who noticed it. They were pulling in the fishing-line and saw that it was green. Later on, other green spots began to appear further downstream. And of course all the air coming from these patches was released into the atmosphere – so we were forced to inhale it. A similar thing happens when it rains. Everywhere yellow puddles appear. And why? Where do they come from? It's the contaminants that settle on the trees. When it pours with rain, these are washed on to the ground leaving these yellow puddles. (male, 55 years old)

> One winter some time back, if you looked at the snow you would see that it was black – yes, black – covered with soot (male 19 years old).

> I've never swum in the river – never had the urge to. Many people, some of them quite close to me, have swum in the river and then ended up in the hospital. That I can do without. I'm not exactly sure what it was, but apparently they'd been poisoned. They didn't tell me exactly what was wrong. They just said that they had a headache or abdominal pains or just felt really ill. It was like a person had a swim one day and had taken ill the next. (female, 14 years old)

Again, it must be noted, that despite the fact that many could offer horror stories of their own, all the respondents supported the continued operation of the Sokol plants irrespective of their understandings about pollution.

Social Organizing

Respondents tended to see those connecting pollution and health as unprepared to take an active stand (or rally around others who did) to improve the ecological situation. In fact, respondents identified no one who could take such a stand or organize others to do so. Respondents were only prepared to become involved if someone else rallied them. This willingness to respond to organized protest was clearly evident in 2003. Back then, the management of the pulp processing works unilaterally and illegally sacked a few hundred workers. In the months that followed, Sokol shook with strikes, demonstrations, and picket lines organized by the trade unions. This shows that in high-risk situations, given the presence of an organizer who they trust, people are prepared to fight for their rights.

REFLECTION: WAYS OF DEALING WITH POLLUTION SUGGESTED BY THE RESIDENTS

As a result of our analyses, we can distinguish four approaches to addressing pollution evident among Sokol residents, varying between those that address the problem and those that address how people respond to the ecological threat. The list is of interest for the approaches identified, as well as for what is not found.

1. *Disregarders*: Residents evidencing disregard ignored information about pollution and continued to adhere to their former way of life. Often middle aged or approaching retirement, they tended to be people unable to envision opportunities to change their lives or the situation. They saw no choice but to continue their involvement in the existing system of relations. They continued to work for the enterprises regardless of occupational health hazards and the consequences for nature. They also continued to catch and eat poisoned fish from the river Sukhona and to grow and consume "poisonous green-stuff" grown in the buffer zone of the pulp and paper mills. Teenagers who did not believe the hazards also ignored the negative information and proceeded to swim in the Sukhona in summer despite all the bans of SEI.

   This disbelief and disregard entailed distrusting officials such as the head of the Regional Fishing Inspectorate, who observed:

   > You can eat fish caught in the Sukhona if you desire to depart this life, but don't give this fish to children. (Sokolskaya Pravda, May 4, 2002)

   Because of bacterial contamination, the Sanitary Epidemiological Service has forbidden bathing and fishing in the rivers in Sokol. Irregardless, local inhabitants continued to bathe in these rivers. The clear contradiction between the experts' findings and residents' behavior suggests a deep distrust of authorities and scientists by residents who preferred to trust their own senses and ignore the risks.

   For example, although water quality conditions remained very poor according to local and regional environmental authorities, a perceived improvement was cited as a basis for risking consumption of local food. Local and regional environmental authorities question the water quality, citing accidents at the Sukhonsk mill that dumped acids into the river, effluents released into the river Pelshma that contain lignosulfonate and lignin and chemical reactions between phenol and chlorine dioxin. Nevertheless, inhabitants have noticed an improvement in the quality of fish consumed from the river. Earlier, according to the residents, the fish had such a strong taste of acid that it was impossible to eat. Now they have learned to tolerate a less overwhelming acid taste.

2. *Risk Reducers through Personal Behavior Change*: This group of respondents became actively aware of environmental deterioration and the importance of preserving environmental conditions favorable to their lives through conversations with neighbors, colleagues, and schoolmates, and through exposure to the mass media, ecologists, and public institutions concerned with the environment. And they also understand that only they can help themselves. They either cannot or do not wish to move away from the area. Therefore, they take it upon themselves to develop ways to reduce the ecological risks they are faced with and the effect that pollution has on them. But the scope of their action is to largely change their own behaviors and lifestyle in order to reduce risk to themselves and their families.

   Risk reducers refuse to use Sokol water for drinking and cooking. Most cannot afford to buy bottled water. Many carry in water drawn from springs on the outskirts of the town. Others use domestic filters, believing that once the water is filtered it improves so much that it does not need to be boiled before drinking.

   Risk reducers seek to reduce the effect of polluted air on their health. Because much of the toxic-smelling air pollution occurs from a mixture of chemicals released at nighttime, these residents close their window shutters at night. However, none of the respondents owned an air ionizer, air conditioner or any other kind of air purification device as, in their estimation, such appliances were too expensive.

   Risk reducers are certain to have a *dacha* (a summer weekend home) to which they can travel to at the weekends or on holiday that is sufficiently far away from the town. Some even stay there all summer, emphasizing the goal of avoiding toxic fumes and preserving their health. Residents suffering from asthma or allergies noted that they felt fine when they stayed at the dacha. Food was grown for consumption at the dacha rather than in urban gardens to avoid eating contaminated vegetables and fruits. Gardeners who associated pollution from the pulp and paper mills with their plants' health now grow sensitive plants under plastic film.

Such practical steps were viewed as helping to reduce the risks from pollution. Risk reducers, however, then ignored the prospect of remaining risks. They did what they could to reduce their exposures to hazards and made peace with the consequences of exposure to conditions over which they have no control.

3. *Outside Activists*: Unlike the personal activism of the risk reducers, another much smaller group of residents evidenced environmental activism. They sought to change the polluting conditions rather than adjust to them. However, these activists tended to have a very partial "backyard" view of the scope of the problem. Furthermore, environmental activists tended not to be permanent residents of Sokol, but rather visited Sokol because their dachas were located there. The lack of drinking water in the suburban part of the town where dachas were located forced a few thousand people to take and use untreated water straight from the river Sukhona downstream from Sokol and the pulp and paper mills. The outside activists focused on the problem that there was no safe water source for this suburban region of Sokol.

   Several of these out-of-town activists pursued the problem through to solution. They achieved acclaim in the press, were the subject of radio programs and were even able to convene a meeting of the Ecological Commission of the Volgodsk regional authority. In the end, they acquired the resources to build two deep wells to serve the suburban section nearby their dachas. Instead of garnering support from Sokol residents, however, this successful activism was discounted. First, the activists were distrusted as "strangers." Officials fearing for their jobs abetted this distrust by discrediting the activists. And activists' success may have solved the immediate problem that affected them, but it did so at a larger cost of drawing attention to the pollution caused by Sokol enterprises upon which most city residents depended for jobs.

4. *Aesthetic Activists*: Yet another group of locals espoused aesthetic ecological causes irrelevant to the big issues. Their strategy was to ignore the major pollution risks by shifting attention to comparatively insignificant ecological problems, as illustrated by the following statement:

> In my opinion, our ecology, the ecology of our town, is far from ranking first, but in some other places the situation is even worse. I just don't like to go along the street and see the broken trees with broken limbs, and piles of garbage here and there. It is rather unpleasant, and you begin to feel uncomfortable. I just arrive at a wish to clean it all. Yes, it is a shame for my native town! However, when we celebrated the 70th anniversary of our town, everything was tidied up, painted, and put to rights beforehand. Finns who visited our town at that time were very pleased to be there.

> But if they only had seen what was here before their visit, I don't know whether they would be so pleased. (female, 15 years old)

People who share this young lady's opinion turn their attention to such issues as the litter in the streets, parks, and on the banks of the river, to the uprooting of trees in alleys and side streets, and to the noise pollution from traffic reaching their flats. These aesthetic problems are their highest priority. While their focus serves to distract from the major ecological risks, these targets are solvable "small" problems. Many "aesthetic activists" are thus able to be successful in making a difference in achieving such goals. They were empowered to successfully participate in the collection of litter, planting trees and plants by homes, and keeping their own yard clean and tidy. This strategy gave people moral satisfaction and the sense that they can make positive changes on behalf of their beloved town and Mother Nature.

CONCLUSION

It should be noticeable from the above list of roles that what is missing in Sokol are people who actively target the principle ecological risk to the local environment and population, namely pollution stemming from the major economic backbone of the region, the local paper and pulp enterprises. Thus, while groups acknowledging environmental issues attended to smaller aesthetic issues, solving local backyard threats, or changing personal behaviors to reduce exposures, it was as if the source of the most significant ecological threat to the region was given a "pass."

We have seen that a concept of ecological pollution did not really develop in the minds of the inhabitants of Sokol before the 1970s. Without the broader ecological awareness advanced during that milieu, people were not really aware of the risks and dangers associated with local industry. This can be explained by the fact that basic ecological knowledge, materials and information was previously absent from the social sphere.

But what is notable is the effect of local adaptation to the polluted conditions caused by long functioning enterprises, particularly the Sokol and Sukhona paper and pulp mills. As these polluters have expanded gradually toward their full capacity, it is striking that they continue to escape local awareness. There are now many generations associated with these companies. And so there has been acceptance of the pollution, of poor air and water quality and of noxious conditions associated with the plants. These source conditions are effectively ignored even by those who are sufficiently aware to try to modify their lifestyle to minimize exposure to the direct pollutants.

We have shown that a new enterprise causing noxious conditions but ecologically benign is not given this same deference. Rather, it can become the focus for activism and action without shaking up the foundation of the community, what is sacred and protected by silent consensus.

The significance of the pulp and paper enterprise to the region and country, supported alike by authorities and townsfolk, has given rise to a sense of complete dependence of the town on the continued functioning of the plant. One could even say that there is an almost paternalistic construct in place, whereby the enterprise is essentially conceived of as the "father of the town."

This merging of the identity of the town and the plant is further reflected in the fact that the town and the plant shared a prestigious award for the plant's work. People are proud of this accomplishment. The plant became a symbol of the town, an important part of the history of the country and a reflection of residents' own lives and entrepreneurial successes. The life of the town has been intricately woven into that of the plant. It is looked upon as more than an employer, but as a source of social good.

The negative construct of pollution that for a time clouded people's minds about the paper and pulp enterprise has been refashioned in a positive light. Discourse about the ecological pollution and risks, stemming from the activity of the Sokol works, was connected with the general theme of ecological problems that swept the Soviet Union as a whole in the 1970s. That time has passed, and with it the attention to these issues as "problems."

We note that a considerably stronger impression of the pollution today is to be found among those traveling through the town and among young specialists from other towns. These were the first people to notice the polluted environment. They were more acutely aware of the ecological problems even though after some time they too got accustomed to it. In the public sphere, the formulation of a notion of ecological degradation and of public discourse on the ecological risks and dangers coincided with the years of perestroika and with the beginning of reform. This was also due to a rise in the rate of production at the time. An active, fervent protest movement was supported by the authorities and organized by new public leaders, who earned themselves political capital on the wave of protests. Some of them even became regional deputies.

Further along the line, economic collapse led to a lull in pollution, which in the minds of the townsfolk was conceived as an ecological upturn.

However, this "upturn" had a negative impact on people's lives. And so lower ecological risk became associated with higher societal risk in the minds of Sokol's inhabitants.

In the 1990s, what Yanitsky called "de-modernization" occurred, along with the "socialization of societal risk." A public concept of "ecological pollution" had been formed in the earlier decade; but it was not matched with a public concept of "ecological risk." The idea of ecological risk continues to be portrayed by specialists via the media, but there is a great deal of indifference and distrust of such specialist information in the general populous. As a result, people are very guarded in their reactions. They do not believe that it is possible to improve the ecological situation. They feel that any improvements, in any case, are not in any way dependent on their efforts. People have once again become accustomed to ecological pollution. Conditions of pollution are seen as being in the same vein as the other "givens" of everyday life. People either ignore ecological risks altogether or they change their everyday practices to accommodate to the reality of these risks by lessening their personal exposure in some ways.

On the whole, contemporary reflections on ecological pollution are centered on personal or individual strategies and distrust toward the authorities and owners of the Sokol plant. Reflections on the ecological and social risks, even considering the negative consequences, have brought the townsfolk to the conclusion that they must be prepared to reconcile themselves with a poor ecology. There is a consensual recognition and agreement that the town is dependent on the paper and pulp plant. People can live with pollution but not with a town in which there is no work.

NOTE

1. Editor's Note: U.S. readers are likely to be unaware of District Heating, employed throughout Russia. A regional heat source is distributed through pipes to interconnected buildings. As a result, there is no need for boilers or furnaces in individual buildings and the energy distribution system that such individual building units demand.

REFERENCES

Beck, U. (1998). *World risk society*. Cambridge: Polity Press.

Beck, U. (2000). *Risk society: On the road to another modernity* (Phillipov, A., Trans. to Russian). Moscow: Progress Tradition.

Beck, U., Giddens, A., Lash, S. (Eds). (1994). *Reflexive modernization policy*. Cambridge: Polity Press.

Edelstein, M. R. (2004). *Contaminated communities: Coping with residential toxic exposure* (2nd ed.). Boulder, Colorado: Westview Press.

Giddens, A. (1999). Sociology: A textbook. *Moscow: URSS Editorial*, 505–508.

Kelley, D., Stunkel, K., & Wescott, R. (1976). *The economic superpowers and the environment: The USA, USSR and Japan*. San Francisco: W.H. Freeman & Co.

Krasnyy Sever. (February 2, 1999).

Kuliasova, A., & Kuliasov, I. (2002a). Reorganization of environmental administration. In: J. Kortelainen & J. Kotilainen (Eds), *Environmental transformations in the Russian forest industry: Key actors and local development* (pp. 72–82). Joensuu: Karelian Institute, University of Joensuu.

Kuliasova, A., & Kuliasov, I. (2002b). Local case study 1: Sokol'skiy pulp and paper mill. In: J. Kortelainen & J. Kotilainen (Eds), *Environmental transformations in the Russian forest industry: Key actors and local development* (pp. 85–97). Joensuu: University of Joensuu, Publications of Karelian Instinute.

Loshchilov, O. (1999). *My Sokol*. Vologda: Polygraphist.

State report on sanitary-and-epidemiological situation in Vologda. (2000).

State report on sanitary-and-epidemiological situation in Vologda. (2001).

Sokol'skaya Pravda. (July 18, 1999). Accident in the river.

Sokol'skaya Pravda. (October 23, 2001). Consistently bad quality water.

Sokol'skaya Pravda. (March 10, 2002). Vologda region centre of hydrometeorology, information.

Sokol'skaya Pravda. (May 4, 2002). Eat fish if you don't want to live.

Sokol'skaya Pravda. (July 10, 2002). The life of Sokol and the region in the opinion of the regional authorities.

Vologodskaya Nedelya. (March 21, 2002). Official reference.

Yanitsky, O. (1998). *Russia: Risks and dangers faced by a transitional society*. Moscow: Institute of Sociology RAN.

Yanitsky, O. (1999). Ecological sociology as risk-reflex. *Sociological Investigations*, *6*, 50–60.

Ziegler, C. (1987). *Environmental policy in the USSR*. London: Frances Pinter. http://www.russiatrek.com/r_vologda.shtml

THE CASE OF THE PINEWOOD LANDFILL: THE POLITICS OF RISK, RATIONALITY, AND THE DISPOSAL OF HAZARDOUS WASTE

Nicholas D. Martyniak, William K. Hallman and Abraham H. Wandersman

INTRODUCTION

Local landfills in communities across the US are the battlegrounds in the conflict between our desire to consume goods at an extraordinary rate and our inability to deal with waste that is a by-product of this consumption. Despite efforts to reduce the amount of wastes generated through source reduction, in 2003, US residences, businesses, and institutions produced more than 236 million tons of municipal solid waste (trash and garbage), approximately 4.5 pounds of waste per person per day (EPA, 2003a). Also in 2003, 16,694 generators of regulated hazardous waste accounted for more than 30 million tons of hazardous wastes, more than half a pound of hazardous wastes per person per day (EPA, 2003b).

In 2003, there were just over 1,700 municipal landfills in operation across the United States with an additional 10,000 municipal landfills were no

Cultures of Contamination: Legacies of Pollution in Russia and the U.S.
Research in Social Problems and Public Policy, Volume 14, 73–91

ISSN: 0196-1152/doi:10.1016/S0196-1152(06)14004-1

longer in operation but were still maintained as closed facilities (EPA, 2003a). In fact, the US Environmental Protection Agency (EPA) suggests that the total number of closed or inoperable landfills probably well exceeds 10,000 (EPA, 2003a). There is no accurate information on closed landfills because waste, for the greater part of US history, has been handled at the local level. In contrast, the vast majority of the waste produced during the latter half of the twentieth century to present is rarely disposed of in the same community, let alone in the same state, from which the waste was produced (Goldstein & Madtes, 2001).

The preponderance of the waste in the United States is still deposited in the ground. In terms of municipal solid waste, 30 percent is recycled, recovered, or composted, 14 percent is burned in combustion facilities, and 56 percent is still deposited in landfills across the United States (EPA, 2003a). Most hazardous waste,[1] especially liquid hazardous waste, is still placed in injection wells below ground in facilities that are supposed to meet federal hazardous waste disposal standards as dictated by both Resource Conservation and Recovery Act (RCRA) and the Safe Drinking Water Act (EPA, 2003a).

There is currently a conflict between society's increasing demand for goods and services and the decreasing number of places where waste produced by the market can be disposed. Inevitably, waste has to go somewhere; but fewer states, and ultimately fewer communities, are willing to accept the risks of the by-products of a consumer economy. Because most waste continues to end up in landfills, it finds its way quite literally into someone's backyard. The number of states accepting hazardous waste where dumping is the primary method of disposal is less than 20 (see Fig. 1).

In the United States, the federal government has typically been the intermediary in this conflict, creating legislation and regulations concerning how waste should be safely disposed to minimize the risk to surrounding communities. In the 1970s and 1980s, with the rise of the environmental movement in the US and the legacy of environmental disasters such as Love Canal continuing to make headlines, the federal government responded to public pressure with sweeping environmental legislation designed to better control the disposal of the nation's waste, and the landfills that most Americans knew began to disappear. Waste, like many other consumer goods, began to be treated and disposed of through a process of specialization. Landfills became specialized in terms of which types of waste they were most effective or efficient at treating and/or disposing. Stricter regulations specifying *how* different kinds of wastes were to be treated and those governing the construction, maintenance, and monitoring of landfills,

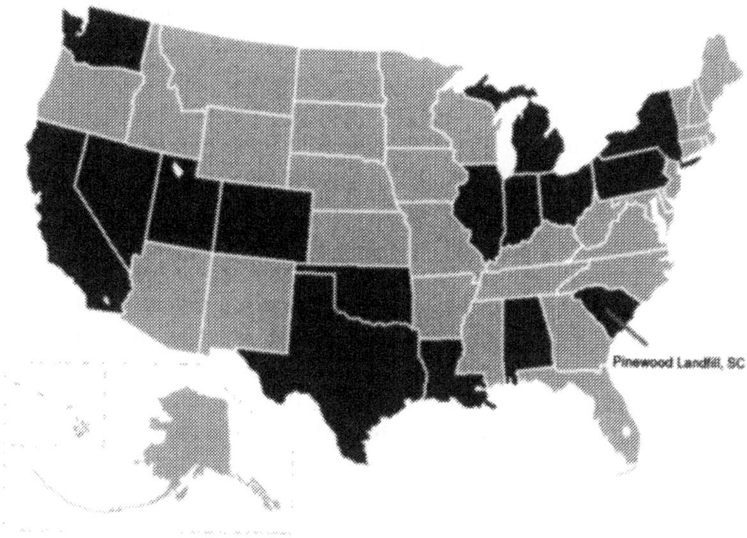

Fig. 1. Current US Hazardous Waste Landfills: States in Black are those which Currently have One or More Commercial Hazardous Waste Landfills. *Source:* Adapted from Environmental, Health, and Safety Professionals' website http://www.ehso.com/cssepa/tsdflandfills.php

especially those dealing with hazardous wastes, had positive environmental benefits.

Enhanced regulation, the filling of existing facilities, increasing costs and liabilities, community opposition to new facilities or expansion, and consolidation within the waste industry that favored large facilities reached by rail or long truck haul caused a major drop in the number of waste facilities. As a result, according to the EPA, the number of landfills in the United States that accept municipal solid waste continues to steadily decline; from nearly 8,000 in 1988 to 1,767 in 2002. To compensate, new landfills are much larger than they had been in the past. Although this increased scale may lead to some efficiencies, their larger sizes make them even more undesirable to communities, and therefore much more difficult to site.

Although the federal government creates the standards, the actual siting process of landfills that accept hazardous waste, as well as the associated permits for disposal, is under the discretion of the individual states. The history of federal involvement in the disposal of all types of wastes has been relatively brief; since the late 1970s, both the federal and state governments have attempted to regulate waste facilities to protect the public health and well-being of the communities that surround them (Guntermann, 1995). However, a fundamental tension exists between the need to dispose of the nation's waste at a low cost and the resources required to ensure adequate protection of the health of the surrounding communities. Some research suggests (Groothuis & Miller, 1997; Kasperson, Golding, & Tuler, 1992; Kunreuther & Patrick, 1991; Pastor, Sadd, & Hipp, 2001) that because it lacks the will or ability to do both, a state's priority is the disposal of waste. In part, this may be because the impacts of failing to deal with wastes are visible, tangible, and immediate, creating mounting pressures on politicians and regulators to do something about them. In contrast, the problems faced by communities surrounding waste facilities can be both literally and figuratively invisible, because they are often delayed, chronic in nature, and frequently blamed on the poverty, lifestyle, or lack of access to healthcare experienced by those who live there.

Southern states accept a disproportionate percentage of waste sent from municipalities not found within their own borders (Boer, Pastor, Sadd, & Snyder, 1997; Bullard, 1990; Davidson & Anderton, 2000).

Some researchers have pointed to a phenomenon known as the "race-to-the-bottom" when trying to explain the migration of waste across state borders (Bullard, 1990; Downey, 1998; Kasperson et al., 1992). They specifically argue that the states with the lowest environmental standards receive the highest percentage of national waste.

In theory, US environmental legislation provides municipalities a basis for fighting to close existing landfills if they could prove that these facilities were negatively impacting the public health of local communities. But this process is a difficult one and the burden of proof is extremely high. The responsibility to show adverse effects on public health typically lies with the affected communities rather than with the landfills accepting the waste or the state responsible for monitoring landfill operations. Moreover, the standard of proof for causality is set at such a high level that even if a near perfect correlation exists between local health problems and proximal levels of pollution or contamination, many communities still cannot prove causality (Williams, 1998). Indeed, some critics of this community approach (Canter, 1991) contend that it is often easier to show negative environmental consequences to an endangered or threatened species related to a landfill than to show negative public health consequences for communities.

Many communities contend that despite federal and state right-to-know legislation, they are often kept in the dark, either intentionally or unintentionally, not just as to the creation of new facilities but also regarding the day-to-day activities of facilities that are already in operation. In addition, even when a community discovers that a landfill is to be sited in their backyard or when they discover that there may be potential health problems associated with existing facilities, a tension exists between the state's desire to put the waste somewhere with a "safe" level of assessed risk and the community's desire to have the waste go somewhere else (Kunreuther & Easterling, 1990). This divergence between the state and local communities concerns the legitimacy of creating new facilities and the maintenance of existing ones – the conflict is over the assumed irrationality of perceived risk and the assumed rationality of risk assessment. This battleground is the predominant theme in the following case history of a hazardous waste landfill in South Carolina.

Efforts to limit import or export of waste have been blocked by US Supreme court decisions protecting the Interstate Commerce Clause of the US Constitution, including direct bans of out-of-state waste (*Philadelphia* v *New Jersey*, 1977) and charging higher prices for out-of-state wastes (*Chemical Waste Management, Inc.* v *Hunt*, 1992). The only remaining strategy left to states is to limit the number and size of permitted landfills. Although effective within their own states, these policies essentially perpetuate the race-to-the-bottom of waste in the US for those states and regions without strict environmental standards.

Compliance with federal and state regulations has made facility siting a matter for expert judgment. However, expert risk assessments that are typically associated with such siting decisions are often different from the assessments of the risks made by the residents of the potentially impacted communities (Edelstein, 2004). Experts for the government presently assess risk by determining the probability of a negative outcome resulting from a particular event (e.g., the probability of the contamination of a community's water supply by a hazardous waste landfill) compounded by the severity of that event (e.g., how many people would die or be injured as a result of a contamination) (Hallman & Wandersman, 1989; Wandersman & Hallman, 1993). Although scientific risk assessors typically focus on the physical systems involved with landfill operations and the probabilities that these systems will operate correctly under stress, communities consider a whole host of other factors in their assessments of risks related to waste disposal. These include the potential for future health and environmental impacts (Hallman & Wandersman, 1995), the possibility that their community has been specifically targeted for the state's waste other than for purely economic reasons, the potential for political corruption, the high probability that their property values will fall, and other factors (Wandersman & Hallman, 1993). Because communities are not limited by the restricted factors used by experts, communities' perceptions of risk are often higher than the risk assessed by experts – thus facilitating a disconnect between the community and the state. This chapter explores the rationality of this disconnect by considering a community whose high level of perceived risk regarding a landfill sited in their backyard was considered to be irrational by both the state and the private sector until the risk became a toxic reality.

What if communities' risk perceptions are right from the start? Factors like political corruption in the siting process, the disproportionate distribution of environmental hazards like landfills, the threat of decreasing property values, and most importantly, the possibility of negative health problems may all legitimately point in the direction of a higher assessment of risk than that which is formulated by experts. The likely disconnect between the community and the state then becomes a policy question of not how to narrow the gap between risk assessments and perceived risk in favor of experts, but how to include community risk perceptions in the siting process. The following history of the Pinewood landfill is a case-in-point of how the present policy of trying to convince "irrational" communities to trust expert risk assessments can have potentially devastating consequences.

A CASE HISTORY: THE PINEWOOD LANDFILL, SOUTH CAROLINA

South Carolina's Pinewood landfill operated for nearly 30 years, receiving hazardous and toxic material from forty states. Pinewood landfill was only one of four facilities on the East Coast permitted to accept hazardous wastes and one of two located in the Southeast. More than 70 percent of the hazardous material transferred and deposited to the Pinewood landfill came from states outside of South Carolina (*The State* 05/10/89). This importation of wastes from the outside is at the core of what locals view as the injustice of the Pinewood story, making their local area into a dumping ground for outsiders and residents into "the recipients of all of the radioactive, hazardous, and now contagious wastes…from all over the East Coast" (Editorial, 1988).

It should be noted that Pinewood landfill went by several different names (GSX, SCA, Lake Marion, and Laidlaw), as its ownership changed over the course of its nearly 30-year history. The Pinewood label derives from the mailing address of the facility and was predominantly used by locals. The

280-acre facility is located approximately forty miles southeast of Columbia, South Carolina, in southwestern Sumter County, a rural community in South Carolina with population characteristics consistent with other small southern agricultural counties in "the Deep South." The total population of Sumter County was only a little over 95,000, with Rimini, the town closest to the landfill, having a population of about 1,100. Per capita income in 1986 averaged only a little over $6,000 per person. At that time, more than 40 percent of the population in the local census tract was Black. Back in 1976, with little economic activity beyond that of agriculture and some tourism, this sleepy little community had no idea what was to be its future when a "kitty litter' mine was opened (ATSDR, 2000).

From Kitty Litter to Toxic Wastes

A geologist named Dr. Paul Bennet had found opaline claystone near Pinewood, South Carolina. Given its ability to absorb up to twice its own weight in liquid, it was an ideal material for cat litter. Excavation of the claystone was relatively inexpensive, as most of the material could be found close to the surface. However, the claystone was saturated with water because the site was located just a few feet above the Tuscaloosa Aquifer nearby Lake Marion. The US oil crises of the late 1970s drove up the price of fuel needed to dry the claystone to the right consistency for shipping, prompting Bennett to look for waste oils from local manufacturing plants to use as fuel.

Although Bennett and the manufacturing companies both benefited from the arrangement, to avoid paying to transport the industrial waste fuel, Bennett agreed to take other industrial wastes. These were solidified by mixing them with lower grade left over "kitty litter" and placed into the pits created from excavating claystone. Within a year, the facility had shifted to accepting industrial wastes full-time. In the fall of 1977, the facility was issued an industrial waste permit by the South Carolina Department of Health and Environmental Control (DHEC) on the presumption that the claystone was resistant to leakage. Subsequently, the state permit was transferred to a new owner, Services Corporation of America, Inc. (SCA). The transfer of the permit grandfathered the new owner from the just passed RCRA.[2] SCA began operations even before the permit was officially transferred and without the knowledge of the local community and local and state officials. No records were retained of the quantity and types of wastes actually dumped in the facility during this period. This chain of events meant that no public hearings were held prior to industrial waste disposal at the site and issuance of the permit (Chapelle, 2000). This illusive and suspicious process was at the foundation of many community concerns about the landfill and likely contributed to the perception of risk (see Slovic, 1987).

Regulation and Public Comment

When the landfill subsequently sought to expand its disposal activities, it required a new state permit for the disposal of industrial waste under RCRA. Before the permit application was even submitted in 1980, the landfill began upgrading the existing disposal pits by installing clay and synthetic liners, a leachate collection system, and groundwater monitoring wells. Wastes were transferred from old, unlined pits to the new ones. SCA clearly intended to meet the new regulations and avoid delay having to remediate preexisting conditions. These activities generated additional community suspicion. They were initiated without community knowledge or input and no records were kept of how much existing waste was excavated, if there were any spills during this process or what types of hazardous materials were moved around during this overhaul (ATSDR, 2000).

In 1979, hearings on the expansion application finally afforded an opportunity for the Sumter County community to be heard at public hearings, where it was argued that the landfill was poorly sited and should not have been permitted for hazardous waste in the first place. Without regard to these comments, the permit was granted in September 1980 (ATSDR, 2000). The message sent by the issuance of this permit was that residents' fears were viewed as invalid and irrational in contrast to the claims made by experts for the EPA and landfill.

Confirmed Fears

There followed a series of events that seemed to confirm community concerns. These included the following:

- Testimony of excess particulate matter (PM) at the landfill exceeding Occupational and Safety and Health Administration standards.
- EPA verification of further violations made a matter of public record in April 1982.
- A bulk waste spill occurred on a major stretch of highway leading into the landfill, one of several toxic spills during transportation of hazardous materials to the site (ATSDR, 2000).

Such "accidents" did not affect landfill compliance with EPA standards and regulations, nor did they change risk assessments by EPA experts, which failed to take into account the probabilities of human error or dishonesty. Communities, in contrast, weigh these possibilities in anticipating greater probabilities for potential accidents than are expected by the experts.

However, more problems confronted the landfill. A new expansion application filed in 1983 was deemed deficient by EPA. Then in November 1984, the owners of the landfill reported detection in on-site monitoring wells of chromium exceeding the National Interim Primary Drinking Water Standard. Subsequently, a statistically significant change in groundwater quality was reported.

The impact of these revelations was mooted, however, when the landfill was purchased by the GSX Corporation in October 1984. Although EPA had recently deemed the groundwater monitoring system inadequate, it transferred the operations permit to a new owner (ATSDR, 2000). These perceived failures further eroded community trust, increasing public perceptions of risk (see Slovic, 1987).

When Pinewood landfill again sought expansion to meet increasing demands for hazardous waste disposal, it sought to mollify EPA's groundwater concerns by undertaking monthly "voluntary" monitoring of the sedimentation ponds. It also expanded the landfill's infrastructure in anticipation of the expansion permit, creating additional disposal ponds with double-lined cells and leachate collection and detection systems (ATSDR, 2000).

From 1978, when the landfill began operations, to 1984, it was estimated that over one billion pounds of hazardous wastes had been buried at the site. The community was definitely aware of the amount of waste coming into the landfill just by the sheer number of trucks and railcars going in and out of the Pinewood facility (ATSDR, 2000). With little direct communication by the owners of the landfill or the state concerning the activities at the facility, the community was left to imagine how much waste and what types of waste were actually being dumped. As a result, the community suspected that the worst types of hazardous waste were being dumped in their backyard.

Confirmation of these fears came in 1984, with release of a list of wastes accepted by the facility that year. It indicated acceptance of 8,204 tons of waste contaminated with lead, 1,257 tons containing mercury, 1,653 tons containing creosote, 3,170 tons contaminated with chlordane and tons of solidified wastes containing such halogenated solvents as tetrachloroethylene, trichloroethylene, methylene chloride, and chlorinated florocarbons; tons of corrosives such as phenol; hydrocarbon solvents such as toluene, benzene, and xylene; heavy metals such as cadmium and arsenic; and pesticide components such as aldicarb. As of December 1984, the landfill was receiving more than 100,000 tons of chemical garbage each year from more than two dozen states. Eighty percent of the wastes came from out-of-state sources (Associated Press, 1984).

Health Concerns Arise

In 1984, the Sumter County community became more divided over this issue than any other in its history. Up to this point, much of the controversy over Pinewood landfill dealt with the safety of its design and operations. Now residents spoke out at hearings about possible links between the landfill and community health problems. One resident described miscarriages and premature births that had occurred over the first five years of the landfill's operation. The South Carolina Chapter of the American Academy of Family Physicians voiced their opposition to the dump by passing a resolution opposing hazardous waste dumps in general and Pinewood in particular. Landfill employees defended the company insisting that the risk to the community was low. The manager of the water-monitoring lab said he would be willing to drink the water from a test well at any time (Associated Press, 1984).

In January 1984, a Sumter County task force looking into the Pinewood landfill recommended tighter restrictions on the amount of waste to be buried there and more intensive health and environmental monitoring. State Senator Phil Leventis called for an independent study, challenging the experts by asserting "it's not a question of *if* landfills will leak, but when." He noted that many Pinewood residents were using bottled water in response to the risk and many more could not afford this luxury. At the same time, the federal EPA announced that it was fining the landfill $25,000 for failing to install and maintain an adequate groundwater monitoring system. Nevertheless, landfill officials heatedly defended themselves. Facilities' Vice President Joseph Boren charged that people had the facts twisted and their fears were misplaced. He blamed the residents' emotionalism on publicity over Love Canal (Pope, 1985a).

Nevertheless, the case against Pinewood landfill continued to build. In May 1985, *The State* newspaper reported that, according to state records, waste at the landfill ranged from slightly dangerous pesticides to lethal toxins and insecticides. The report further indicated that toxicologists know little about what happens when different hazardous substances are combined. Around the same time, it was revealed that the landfill's clients were among some of the worst polluters in the US, including Union Carbide, known as the polluter of Love Canal, as well as Monsanto, Diamond Shamrock, American Cyanamid, IBM, Exxon, and Sperry (Pope, 1985b). In September 1985, environmentalists considering the ecological future of South Carolina publicly discussed their concerns over bioaccumulation of contaminants in the state's food chains. At the same time, DHEC estimated that Pinewood cleanup costs could reach $832 million at the current stage of the landfill and more than $2 billion were it is to reach full capacity. State Senator Leventis charged that the agency was being unreasonably lenient with Pinewood and that the unusual relationship between regulators and the facility operators could be corrupt. He alleged a revolving door and exchanges of money between the DHEC and Pinewood landfill. State Representative Alex Harvin added his concerns that Pinewood landfill threatened both water quality and the development of Lake Marion (Pope, 1985c). And then, in June 1986, DHEC announced the discovery of contaminated groundwater in an aquifer beneath the landfill, although it claimed that Pinewood might not be the source (Pope, 1986b).

A Local Organization

The media attention attending these problems fostered the organization of a grass roots organization called Citizens for a Safe Environment (CASE). CASE undertook a health survey of 300 residents living along Lake Marion in an effort to spur an official investigation, finding that a quarter of the families were suffering from open sores, hair loss, or nausea (Associated Press, 1986). The effort succeeded. A subsequent state study faulted both the landfill and the DHEC, lamenting the lack of an accurate inventory of buried chemicals and inadequate environmental monitoring and chemical analysis (Pope, 1986a).

However, the Pinewood landfill appeared to be exonerated by several reports released in 1986. A DHEC report concluded that contaminated groundwater was the result of landfill operations in the late 1970s and was unassociated with the activities of the present owner (Pope, 1986b). EPA reports indicated no significant acute toxicity of surface water samples taken near the landfill and found no contaminants in a major drain leading from the landfill. And a study of Rimini residents by the Medical University of South Carolina concluded that several health problems found to be present in the community were not related to the landfill (ATSDR, 2000).

Concurrently, however, there was a rash of new indications of trouble. The DHEC issued a consent order in late June 1986 for Pinewood to conduct corrective action for groundwater contamination. Around this time, pollution was found in Lake Marion near Pinewood and in four residential wells and an irrigation well. EPA found volatile organic compounds (VOCs) in on-site groundwater at the facility in November. Meanwhile, experts hired by the community criticized the landfill site location, environmental monitoring, and lack of medical exams. And a massive fish kill occurred in Lake Marion around the same time that blood analyses taken a year earlier were found to have high VOC levels. Although DHEC officials attributed the fish kill to "natural" causes and labeled the blood results as "inconclusive" (ATSDR, 2000), these occurrences carried weight in the public's assessment of risk.

These events brought allies from the University of South Carolina and public officials. But the effort to rally the legislature behind the community failed when a proposal to double costs for wastes dumped by out-of-state sources did not pass (Pope, 1988a). Lacking any alternative means of being heard, CASE adopted civil disobedience tactics to draw more public attention. Protests and demonstrations were held outside the facility and in the state capital. An attempt to block the road to the landfill in March 1988 resulted in the arrest of several members for trespassing and injury to personal property. One of the arrested protesters summed up her concerns this way (Pope, 1988b).

> We've tried everything else. We've been to public hearings; we've been in Columbia time and time again. We're tired of doing that, we're tired of writing letters. We need to get our point across that this [dumping] has got to stop. We have had it, we have taken enough. This thing [the landfill] has been down here ten years, we want it closed and cleaned up.

When the State Health and Environmental Control Board subsequently recommended a hiked levy on out-of-state waste disposal that would increase costs fivefold, state officials rejected the measure because of the large amount of revenue that was generated for the community by the disposal of waste at the landfill. They feared that opposition to the landfill in South Carolina might result in the loss of business (i.e., waste) to Alabama, where a competing facility existed. And others argued that the environment was already contaminated at Pinewood, so there was no reason not to locate the facility there. Pinewood landfill representatives argued that the fee change was a violation of the commerce clause and thus illegal (Pope, 1988c).

In this situation, anger developed against North Carolina, a major exporter of hazardous wastes to South Carolina. In 1989, South Carolina governor signed an executive order mandating an end to out-of-state shipments of toxic waste from states like North Carolina that ban disposal within their own boundaries. Challenged as a "blacklisting" of North Carolina, the order was quickly ruled illegal in a federal court challenge. One attorney for the waste industry went as far to make an analogy to the effect that "the state's use of the blacklist is [like] putting a Berlin Wall between North and South Carolina…It would remind me of the Soviet Union, with all the republics fighting one another" (Pope, 1990).

Meanwhile, Pinewood applied for an extension of their RCRA permit for continued dumping of hazardous waste. For the first time, a public hearing was held on the RCRA permit, with the outcome that the community overwhelmingly opposed the permit (ATSDR, 2000).

At that same time, citizens of Sumter County petitioned the Agency for Toxic Substances and Disease Registry (ATSDR) to conduct a public health assessment. A meeting scheduled with ATSDR to discuss a public health petition for a medical examination of the community's health drew a large crowd. ATSDR representatives visited Rimini in Sumter County to tour the Pinewood facility, observing the heavy truck volume transporting waste to the landfill, strong chemical odors and potential leaks through cracks and holes in the rail cars at the waste transfer station (ATSDR, 2000).

At this time, scandal shook the South Carolina statehouse. House Speaker John Rogers was indicted for criminal racketeering and extortion in January 1991 and accused of taking money from Pinewood on three separate occasions in 1988, 1989, and 1990. The last payment was made under yet another new owner, Laidlaw Environmental Services.

Residents continued to complain about their health. They expressed concerns about the lack of landfill monitoring; the possible connection between discovery of VOCs in samples of residents' blood and releases of VOCs by the landfill into Lake Marion and into the air by the former burning of waste oil; continuing health problems that included open sores, skin rashes, hair loss, and nausea; the possible relationship between fish consumption from Lake Marion and infant illness and the possible effects that could be associated with long-term, low-level exposure of contaminants from Pinewood. Their efforts paid off but were not rewarded. Residents finally succeeded in winning an ATSDR study to examine health effects from Pinewood. However, the Petitioned Public Health Assessment completed in 1993 concluded that, given the paucity of data, no illness found in the Rimini community could conclusively be connected to the activities by the

Pinewood landfill (ATSDR, 2000). The report offered little comfort to the community and failed to alleviate or directly address community concerns. In fact, it was viewed as helping the Pinewood landfill.

Despite the ATSDR report, however, the landfill had found itself under substantial legal pressure from a slew of lawsuits, including a challenge by the community that reached the South Carolina Supreme Court. Put on the defensive, Laidlaw Environmental Services, owner of Pinewood since 1992, agreed to establish a $100 million dollar trust fund to cover any future damage from the landfill. Although Pinewood officials touted their corporate responsibility (Pope, 1992), community experts estimated landfill cleanup to actually cost close to $1 billion.

The state's cooperation in the creation of the trust fund rekindled a debate in the local community. Some Sumter County residents were now ready to accept that the landfill would never be closed by protest. Given that the waste still had to go somewhere, and that the owners of the landfill had agreed to cover any potential costs due to contamination, the waste was going to find its way into their backyard. Others rejected Laidlaw's argument that the landfill was completely safe and that any contamination, if it were to occur, would take years to manifest itself. These advocates rejected such "low-risk" claims, arguing that the landfill should never have been sited in Rimini in the first place because of its location in close proximity to Lake Marion (Pope, 1993) and only a few feet above the Tuscaloosa aquifer, a major source of water for the Southeast US. Proximity to the community's water source was the major source of the community's fear.

At this time, the environmental justice movement had just started to gain momentum with the ground-breaking publication of Bullard's (1990) book, *Dumping in Dixie*, which gave a voice for a movement that was seeking legitimacy. Community and environmental activists in South Carolina made efforts to create environmental justice legislation at the state level. Several bills were introduced in the S.C. General Assembly in 1991 and 1992, which required DHEC to consider race and economic status when issuing new permits for polluting industries. One bill went so far as to have mandated DHEC to perform a demographic study when issuing any new permits for pollution-generating industries. Neither of the bills ever came to fruition, but they reflected the growing concern by communities of color that race and income are at least predictive of the present siting decisions of industrial polluters made by state authorities. The changed focus moved beyond the purely economic explanations of where landfills are sited into the political economic realm of public and private officials who are making conscious decisions as to where the nation's waste is to be located (Cushman, 1993).

While the community continued to protest the landfill over the next several years, the facility continued to accept over 135,000 tons of hazardous waste per year. In the late 1990s, Safety-Kleen, Inc. purchased the landfill and operations continued into the twenty-first century. The trust fund that was to be set up under Laidlaw only accumulated one year of contributions, amounting to a little over $14 million of the more than $100 million that was initially promised. In 1995 the DHEC decided not to require the trust fund. Safety-Kleen opted instead to take out a $120 million insurance policy to cover any major leaks that might occur under its ownership. Unlike the trust fund, however, claims after a contamination event might be delayed and even denied by an insurance company attending to its own economic incentives. The state initially supported the use of an insurance policy, but in July 2000, after significant leaks of VOCs were detected, the DHEC required the owners of the landfill to again pay into the trust fund.

Safety-Kleen initially fought the DHEC decision in courts, but then filed for Chapter 11 bankruptcy, seeking protection for an amount exceeding $1.6 billion. In January 2000, the South Carolina Court of Appeals ruled that the landfill "is full and no longer can take shipments of toxic waste…The company must put $133 million in the bank by 2004 to pay for a major environmental cleanup, if necessary" (Fretwell, 2000). Although the courts have basically settled the issue concerning the closure of the landfill, Pinewood, under Safety-Kleen, is currently fighting its way through bankruptcy court facing $1.6 billion in potential payouts, including cleanup costs close to $1 billion and additional related lawsuits estimated in the hundreds of millions. Executives of Safety-Kleen also face accounting fraud charges in federal court as well as dozens of civil and class-action lawsuits.

In the end, it appears that the community had been right all along. Local residents perceived the risks associated with the siting of a landfill near a major aquifer as a dangerous and threatening prospect, arguing that a leak and the subsequent contamination of the local water supply was not a question of *if*, but *when*. Despite assurances to the contrary, the landfill ultimately did leak. The fact that the community accepted the expert assessments, which predicted that there was a very low probability that the landfill would fail, caused the state to be ill-prepared to deal with the necessary cleanup and remediation of the site.

CONCLUSION

The state, the county, and the local municipality are the authorities responsible for siting decisions that determine where hazardous waste landfills are to be located, and therefore, how they are distributed across the US. Most communities, faced with the prospect of having a hazardous waste landfill in their backyard, would probably opt to not have such a facility in its midst. One of the primary reasons is that communities often associate high levels of risk with landfills—especially hazardous waste landfills (Wandersman & Hallman, 1993). These high levels of perceived risk are tied to specific concerns about harmful health effects, a decrease in property values, and other consequences, which are more generally tied to a fear of contamination of the community's environment (e.g., air, soil, water, etc.).

These community concerns are often branded as irrational by many within the state and the private sector because experts deem the risks to be much lower, they view landfills as a source of profit and jobs, and they favor the concentration of wastes in a few communities rather than their dispersal to many communities (Wandersman & Hallman, 1993). Expert assessment of risk is based on the assumed probabilities of failures of physical systems. But what if it takes years or even decades for contamination to occur or to be detected when it does occur?

Community's risk perceptions of hazardous waste landfills are often much higher than the risks produced in the expert assessments conducted on behalf of the private sector or the state. The goal of policymakers is often to narrow the gap between the level of risk determined by the expert risk assessments and the level of risks perceived by local communities. Usually the strategy is to try to align perceived risk with the level of risks determined by the formal risk assessment (Wandersman & Hallman, 1993). However, this policy goal encapsulates several critical assumptions: first, public risk perceptions are irrational as they are based on "nonscientific" variables; second, risk assessments are accurate in their formulations of probability and severity; and third, risk analysis, in general, should be outcome-based. It should be noted that this last assumption is, in fact, a normative one.

One of the major reasons that community risk perceptions are often so different from expert risk assessments is that communities take "process" into consideration when evaluating risks. They consider whether or not environmental hazards like landfills are evenly distributed across the state or the nation; they consider how much involvement or voice the community has in the siting process; and they consider the potential for "dirty" politics that can go into the siting process. Communities often cite concerns of political corruption as a major variable in the determination of where landfills are to be located. These considerations, in addition to other factors, create community risk perceptions that can starkly contrast with the risk assessments formulated by experts. The state and industry argue that these "external" procedural variables produce inaccurate formulations of risk, which facilitate irrational behavior (i.e., protests, demonstrations, etc.).

But what if some of the factors that communities take into consideration in their assessments of risk are, in fact, correct? What does it take for community risk perceptions to be legitimized? At least from the viewpoint of the state and industry professionals, there often needs to be some form of proven contamination for these perceptions of risk to be legitimized. In the case of the Pinewood landfill, it took almost 25 years after the landfill first started its operations for the community's concerns and perceptions of risk to be legitimized by the eventual leaking of the landfill. What was for decades considered to be irrational behavior in the form of protests, demonstrations, and other types of community activism linked to irrational risk perceptions, is now considered a heroic display of community dedication. Why this change? Why are a community's predictions of the likely failure of political, regulatory, and fiscal systems not privileged in the same way that expert's assessments of physical systems are given credence? Why can't they also be incorporated into formal risk assessments? Perhaps this too is a case of not if, but when.[3]

NOTES

1. Hazardous wastes are defined by the US Environmental Protection Agency (EPA) in regulatory terms as those which embody one or more of four characteristics: ignitability, corrosivity, reactivity, and toxicity (EPA, 2003).

2. RCRA placed strict standards on new landfills requiring rigorous "cradle-to-grave" standards. Any waste accepted by landfills had to be traced from the waste's inception at the pipe, to the transport and to the final disposal in the ground.

3. This has been such an interesting chapter to write, and there is so much more that could have been written about the Sumter County community and its battle for environmental equity and a safe environment. Particular acknowledgment needs to be given to Francis Chapelle for his work on the early history of the landfill. Charles Pope, a writer with *The State* newspaper, whose dedication to the community and the State of South Carolina is most appreciated. His efforts to keep this issue in the mind of the public through his intimate and personal writing over the last twenty years most definitely contributed to the eventual closing of the landfill. Finally, special thanks to the people for whom the Pinewood Landfill was a personal concern, with particular gratitude to CASE whose tireless, grassroots fight proves that communities that bear the brunt of the nation's waste have a voice that needs to be heard.

REFERENCES

Agency for Toxic Substances and Disease Registry (ATSDR). (2000). *Petitioned Public Health Assessment of GSX Landfill*. Pinewood, Sumter County, South Carolina.

Associated Press. (1984, December 20). Doctors' group opposes hazardous waste dumps. *The State*, p. 9C.

Associated Press. (1986, August 15). GSX neighbors complain of ills. *The State*, p. 2C.

Boer, J. T., Pastor, M., Jr., Sadd, J. L., & Snyder, L. D. (1997). Is there environmental racism? The demographics of hazardous waste in Los Angeles county. *Social Science Quarterly*, *78*(4), 793–810.

Bullard, R. D. (1990). *Dumping in Dixie: Race, class, and environmental quality*. Boulder: Westview Press.

Canter, L. W. (1991). Environmental impact assessments for hazardous-waste landfills. *Journal of Urban Planning and Development*, *117*(2), 59–76.

Chapelle, F. H. (2000). *The hidden sea: Ground water, springs, and wells*. Tucson: Geoscience Press.

Chemical Waste Management, Inc. v Hunt. (1992). *United States Supreme Court case*. 504 U.S. 334.

Cushman, J. H. (1993, November 22). Pollution-racism to be investigated. *The State*, p. 1A.

Davidson, P., & Anderton, D. (2000). Demographics of dumping: A national environmental equity survey and the distribution of hazardous material handlers. *Demography*, *37*(4), 461–466.

Downey, L. (1998). Environmental injustice: Is race or income a better predictor? *Social Science Quarterly*, *79*(4), 766–778.

Edelstein, M. (2004). *Contaminated communities: Coping with residential toxic exposure* (2nd ed.). Westview Press: Boulder, Co.

Editorial. (1988, May 23). S.C. Now nation's burying ground. *The State*, p. 6A.

EPA. (2003a). Municipal solid waste generation, recycling, and disposal in the United States: Facts and figures. Retrieved February 26, 2006, from municipal solid waste in the United States. Website: http://www.epa.gov/msw/pubs/msw05rpt.pdf of Garbage in America.

EPA. (2003b). National biennial RCRA hazardous waste report.

Fretwell, S. (2000, June 16). Safety Kleen loses major battle. *The State*, p. A1.

Goldstein, N., & Madtes, C. (2001). The state of garbage in America. *Journal of Composting & Organics Recycling, December*, p. 42.

Groothuis, P. A., & Miller, G. (1997). The role of social distrust in risk-benefit analysis: A study of the siting of a hazardous waste disposal facility. *Journal of Risk and Uncertainty*, *15*(3), 241–257.

Guntermann, K. L. (1995). Sanitary landfills, stigma and industrial land values. *The Journal of Real Estate Research*, *10*(5), 531–542.

Hallman, W. K., & Wandersman, A. H. (1989). Perception of risk and toxic hazards. In: D. L. Peck (Ed.), *Psychosocial effects of hazardous toxic waste disposal on communities* (pp. 31–56). Springfield, Illinois: Charles C. Thomas.

Hallman, W. K., & Wandersman, A. H. (1995). Present risk, future risk, or no risk? Measuring and predicting perceptions of health risk associated with a hazardous waste facility. *Risk: Health, Safety & Environment*, *6*, 261–280.

Kasperson, R. E., Golding, D., & Tuler, S. (1992). Social distrust as a factor in siting hazardous facilities and communicating risks. *Journal of Social Issues*, *48*(4), 161–187.

Kunreuther, H., & Easterling, D. (1990). Are risk-benefit tradeoffs possible in siting hazardous facilities? *The American Economic Review*, *80*(2), 252–256.

Kunreuther, H., & Patrick, R. (1991). Managing the risks of hazardous waste. *Environment*, *33*(3), 13–36.

Pastor, M., Sadd, J., & Hipp, J. (2001). Which came first? Toxic facilities, minority move-in, and environmental justice. *Journal of Urban Affairs*, *23*(1), 1–21.

Philadelphia v New Jersey. (1977). *United States Supreme Court case*. 437 U.S. 617.

Pope, C. (1985a, January 30). Sumter County Task Force calls for tighter restrictions on dump. *The State*, p. 3C.

Pope, C. (1985b, May 26). What's in Pinewood landfill? *The State*, pp. 1A, 12A.

Pope, C. (1985c, November 22). Leventis calls for ethics investigation. *The State*, pp. 1A, 15A

Pope, C. (1986a, July 8). GSX pollution surprises EPA. *The State*, p. 2C.

Pope, C. (1986b, December 16). Study says contamination not linked to GSX landfill. *The State*, pp. 1C, 10C.

Pope, C. (1988a, March 1). Don't dump on S.C., citizens say. *The State*, p. 3C.

Pope, C. (1988b, March 15). Five arrested at dump protest. *The State*, p. 1C.

Pope, C. (1988c, September 22). GSX gears up to fight "unjustifiable" tax increase. *The State*, p. 1A.

Pope, C. (1990, December 22). Judge rules against ban on N.C. Waste Campbell blockade. *The State*, p. 1A.

Pope, C. (1992, October 6). GSX to establish trust fund. *The State*, p. 1B.

Pope, C. (1993, May 16). The waste flows downhill. *The State*, p. 1D.

Slovic, P. (1987). Perceptions of risk. *Science*, *236*(4799), 280–285.

Wandersman, A. H., & Hallman, W. K. (1993). Are people acting irrationally? Understanding public concerns about environmental threats. *American Psychologist*, *48*(6), 681–686.

Williams, A. N. (1998). Cancer clusters: What role for epidemiology? *Minnesota Medicine*, *81*(5), 14–28.

FEAR OF CANCER IN A RURAL APPALACHIAN COMMUNITY FOLLOWING NOTIFICATION OF AN ENVIRONMENTAL HAZARD

Janice L. Hastrup, Sherilyn N. Thomas and Michael R. Edelstein

INTRODUCTION

In this chapter we explore some of the intriguing questions raised by contaminated communities. Is there a connection between exposure to environmental hazards and psychological distress? If yes, how best can it be measured? What kinds of psychological problems are aggravated by this kind of life stress? How do we know that victims are truly experiencing increased problems such as anxiety, depression and fears about their health?

Several studies have shown increased rates of psychological problems in the wake of exposure to environmental hazards on a wide range of standardized psychological measures (Bachrach & Zautra, 1986; Davidson, Baum, Fleming, & Gisriel, 1986; Greve, Bianchini, Doane, Love, & Stickle, 2005; Unger, Wandersman, & Hallman, 1992). These studies reveal increases in measures such as concern for health (Gibbs, 1986), depression (Gibbs, 1986), psychological distress (Baum, Fleming, Israel, & O'Keeffe, 1992; Davidson et al., 1986; Markowitz & Gutterman, 1986; Smith, North,

Cultures of Contamination: Legacies of Pollution in Russia and the U.S.
Research in Social Problems and Public Policy, Volume 14, 93–115

ISSN: 0196-1152/doi:10.1016/S0196-1152(06)14005-3

& Price, 1988), neuropsychological impairment (Gatchel & Newberry, 1991) and sociological measures such as alienation (Kroll-Smith & Couch, 1993). The psychological effects of technological disasters such as the Three Mile Island nuclear accident in Harrisburg, Pennsylvania were shown to have uncommon persistence over time (Baum et al., 1992).

Various studies on psychological problems and environmental hazard exposure suggest that the adverse impacts are greater from human-caused and technological disasters than from natural disasters such as earthquakes (Davidson et al., 1986; Kroll-Smith & Couch, 1993). One study compared exposure to contamination, a flood and a third community not experiencing any disaster (Baum et al., 1992). Greater symptoms of psychological distress were found among individuals living near a toxic hazard than among those exposed to a flood or residing in a comparison community. Psychological distress was measured using two surveys, a checklist of emotional health symptoms, the SCL-90-R, and an inventory of depressed feelings, the Beck Depression Inventory. In addition to such paper indicators of stress, biochemical alterations observed during a stressful task suggested more severe and enduring psychophysiological differences between the toxic hazard-exposed community and both the flood and the comparison communities. All participants were paid for participation and were also told in advance that the results were not part of and were unlikely to affect pending litigation.

Litigation

Some cases of community contamination in the United States result in litigation, where a "toxic tort" is filed on behalf of victims against parties thought to be responsible. Litigation presents an interesting challenge and opportunity for understanding the psychological consequences of toxic exposure. On one hand, it may present an organized opportunity for study resulting in expert testimony in court with unparalleled access to those affected. On the other hand, it places the plaintiffs, the people bringing the lawsuit, in a unique and stressful situation. They are expected to fully disclose private information about themselves to lawyers, judges, jurors and experts. Their self-interest is served if they can make clear how they have been harmed. But they must expect that they have to defend any claim that they make against hostile questioning.

Given the special context of litigation, it is notable that findings of adverse psychological impacts have been documented both in data collected for lawsuits (Edelstein, 2004; Gibbs, 1986) as well as those not involving litigation (Davidson et al., 1986). Some investigations of populations previously involved in litigation (e.g., Green et al., 1990) provide evidence of enduring changes in psychological functioning long after financial settlements are completed.

Measuring Psychological Impacts

There are multiple ways to obtain psychological measurements. Face-to-face clinical evaluations have provided results similar to those obtained from the standardized measures employed here. Of greater interest in the current project is the difference between quantitative and qualitative approaches.

Edelstein (2004) contrasts quantitative and qualitative research approaches as searching for "statistical versus logical validity." These varied approaches reflect deep epistemological as well as methodological differences. The power of quantitative psychological research is its ability to determine whether differences on some indices between varied conditions have significance or occur merely by chance. The search for statistical validity dictates the research approach, forcing it to conform to conditions of scientific comparison or control, often simplified and contrived even if in the field. The rationale of the method, therefore, reflects the need to meet the conditions for valid research even at the cost of ecological validity. Findings of significance under those conditions are useful for answering a specific question: are there significant differences determined by comparing results to norms or benchmarks for standardized measures or to the results found with control or comparison groups? This is a potentially important basis for concluding that effects of contamination have been validly substantiated.

Conditions of logical validity, on the other hand, occur when the researcher gains a sufficient understanding of the context, within which the contamination victims have been affected, to understand informant reports of a wide range of impacts. The validity of these observations is established, not by the statistical significance of comparisons, but by the logic – or the frame of understanding – created by the whole context. In Edelstein's (2004) approach, a careful review of all relevant documents is used to establish a chronology or "eco-historical context" that allows for the baseline conditions of life in the affected community to be compared with the life conditions after contamination has become known and accepted. Stress due to forced accommodations and adjustments to the changed conditions (i.e., stressors) is examined along an entire continuum of "social process" encompassing not only the individual, but families, neighborhoods and communities, relevant institutions and the society as a whole. Each of these levels of process frames the experience of higher and lower levels and all interact in complex ways. The chronology serves as the basis for a detailed interview format to be developed that guides an intensive discussion of the baseline (pre-contamination) life (behaviors, cognitions and emotions) of victims contrasted with life after discovery and acceptance of the contamination event. Ideally, whole families are interviewed in their homes, where the researcher not only questions and records answers but witnesses the interaction of narrative and place. Other group interviews with unrelated victims from the affected community are used to discuss emergent questions from the study. The response of individuals and groups in these interview settings can be judged valid in light of its "fit" with the emergent understanding or "logic" of the contamination context. Thus, in contrast to measuring differences between groups, as does quantitative research, the qualitative approach asks a very different question: what changes have occurred in the context that have impacted the affected group and with what consequences?

The match of the two approaches is ideal, allowing for a global measurement of change and impact, stressor and stress, while also allowing the strength of the cause/effect relationship to be measured for a sample of the impacts. The present report describes the use of both qualitative and quantitative methodologies to evaluate the impacts of an environmental hazard on a community with a distinctive culture and eco-history.

While findings can be expected to vary with contextual factors, culture, population characteristics and the chronology of events in a given community, a basic pattern of findings has been identified in nearly all contaminated communities studied regardless of the research approach utilized. Accordingly, there is a general convergence of findings between quantitative studies and qualitative studies of contamination (Edelstein, 2004; Erikson, 1994; Kroll-Smith & Couch, 1993). Therefore, beyond a global test of the overall pattern of causes and effects found with a contamination incident, there is ample basis in the literature to predict that a known hazardous waste site would cause significant increases in the incidence of specific indices of psychopathology in proximate residential populations.

Fear of Cancer

A primary concern for residents of contaminated communities is the fear of increased risk of cancer for themselves and their families (Edelstein, 2004;

Erikson, 1994; Vyner, 1988). Cancer is a dreaded disease with life-changing and often fatal consequences; feelings of helplessness and the long latency periods associated with the development of cancers contribute greatly to its being almost universally feared. While well documented in qualitative studies, there is a tailored quantitative approach available to examine the fear of cancer.

One may question individuals directly about their fear of cancer; for example, members of a community may be asked to estimate their likelihood of developing cancer at some point in their lifetimes. An indirect measure has also been developed, the Fear of Cancer Index (Berman & Wandersman, 1992). The index is derived from two ratings, the first identifying a respondent's knowledge of cancer symptoms and the second reporting on symptoms actually experienced. The 25 items of the Fear of Cancer Index combine the American Cancer Society's historical list of eight cancer warning signs (American Cancer Society, 2006) with an additional 17 items from the Symptom CheckList-90-R (SCL-90-R; Derogatis, 1983). The Fear of Cancer Index can be used to assess beliefs about cancer and also its potential occurrence. It not only allows for measurement of a given population but also allows for a comparison of groups exposed to environmental hazards and those without such exposures.

The use of both a qualitative and quantitative approach in the same community allowed for a rare examination of the convergence or divergence of the findings between the two independent studies. If the findings were found to converge, then greater confidence would be gained for each. In contrast, major contradictions in findings would require explanation, potentially suggesting a basis for invalidity in one or the other, or both of the approaches.

CONTAMINATION IN AN APPALACHIAN COMMUNITY

The present chapter describes psychological problems as well as direct and indirect measures of fear of cancer in an Appalachian community facing an environmental hazard. The quantitative study employed a second comparable Appalachian community that did not face a known environmental hazard. The hazard community is set in a mountainous area of the southern Appalachians in a rural hollow about a mile from a main highway. Beginning in the mid-1960s, a local farmer hauled trash and liquid wastes from a nearby industrial plant, including such hazardous materials as the solvents

TCE (trichloroethylene), TCA (111 trichloroethane), Freon (trichlorotrifluoroethane) and Varasol, as well as various coolants, cutting oils, plating wastes, etchants, paint wastes, solder fluxes, ammonium hydroxide, methylene chloride, alcohol, PCBs (polychlorinated biphenyls), acetone, toluene and the heavy metal chromium. Apparently unaware of the hazards, the farmer poured most of the wastes into unlined pits on his family farm, sometimes igniting the liquids. In addition, barrels of wastes were strewn about and eventually leaked; other waste oils were spread onto the dirt roads of the area as a dust suppressant. Although neighbors in the hollow daily witnessed the disposal of these materials and breathed acrid smoke from the burn pits, they were also unaware of the potential hazards. They trusted their neighbor and firmly believed in his right to do what he pleased on his private property.

It was not until 1982 that the dump was discovered accidentally during an environmental inspection of another site. By the late 1980s, after several years of cleanup and testing, government officials had gradually come to view the situation as a serious health threat for nearby residents. Government documents refer to the various contaminants as probable human carcinogens, compounds capable of causing liver and pulmonary cancer and lymphomas, as well as potential causes of mutagenic and somatogenic effects. Among the potential somatic effects that officials cited as a basis for concern were liver and kidney damage, skin and eye problems and such other general symptoms as anemia, nausea, fatigue, headaches and drowsiness. Subsequent testing showed the presence of both dense nonaqueous phase liquids (DNAPLs) and volatile organics in the ground and surface water, both on and off the disposal site. These contaminants also appeared in nearby residential wells. While state officials and industrial representatives played down the danger, some residents began to associate the pollution with health problems in their families. Word of the problem spread through the community, even as site workers began to wear "moon suit" protective clothing that seemed to confirm the rumors.

By the summer of 1988, residents had taken matters into their own hands, initiating a series of meetings to which state Superfund officials and a local environmental leader were invited. These experts confirmed fears, stating that both spring and well water were unsafe to drink, and that even skin contact with the water was to be avoided. Officials also told residents to avoid consuming produce as well as fish and game from the area, even though generations of residents had subsisted largely on these food sources. Organized citizens pressed successfully for an alternative water source; a lawsuit was brought against the industry that had produced the wastes.

Although remediation under Superfund led to the erection of a plant to pump and treat groundwater, residents doubted the cleanup and considered the treatment plant itself to be a source of danger. Their distrust was fed by the plant's intrusiveness on the landscape, frequent alarms set off at the facility and the fact that the industry that produced the wastes also built and operated the treatment plant. Residents also reasoned that the plant would take a very long time to clean the water, if it ever could.

Two studies were undertaken to provide evidence for plaintiffs in a lawsuit who were residents of this community. The first and second authors evaluated traditional measures of psychopathology and both direct and indirect measures of fear of cancer among the residents of the community and a comparison community. The third author used a qualitative approach, a standard formal interview, which is an accepted form of inquiry in social science, and may be particularly well suited for the evaluation of the impact of environmental hazards (Edelstein, 2004; Kroll-Smith & Couch, 1993). Such interviews provide an assessment of such factors as stigma, community disruption, family integration, lifestyle changes and the overall meaning of events to respondents which are not usually measured with purely quantitative approaches to psychological distress and psychopathology.

QUALITATIVE METHODS USED TO MEASURE FEAR OF CANCER AND OTHER IMPACTS

Using the clan as his basis of sampling, Edelstein conducted intensive group interviews with a substantial sample of extended families involved in the litigation. Interviews with nonrelated groups of adults and teenagers were also used. The interviews were conducted in the community, mostly in residences affected by the contamination. The study was used to document significant impacts for residents in terms of their lives in the community, community conflict, environmental stigma, family conflict and personal functioning.

Such interviews are designed to generate data for further assessment, as well as to evaluate specific hypotheses. In environmental hazard situations, it is impossible or extremely difficult to obtain random assignment, standardization, pre–post comparisons and other methodological advantages of laboratory research. Qualitative field researchers rely heavily on statements made by individuals. Respondent bias in such situations is dealt with by systematically comparing responses, evaluating different family members' and different families' perspectives and the concreteness of details which participants provide. Family interviews allow participation of some members (e.g., children) who are unaccustomed to public speaking or who would be unusually stressed by it (e.g., testifying in court). Edelstein (2004) notes that the logical validity attained through qualitative research may be even more valuable than the statistical validity attained through quantitative research for actually understanding what has occurred in the contaminated community.

Method

Interviews and Sample

The qualitative field study involved a combination of intensive family interviews and interviews with groups of unrelated individuals.

1. *Intensive Clan Interviews.* Intensive interviews were held with a spatially diverse sample of six extended families (i.e., clans) living in family compounds or along family-owned geographic hollows. The interviews, ranging upwards of four and one half hours, were held in the clan compound, usually at the home of the elders, with between 11 and 18 family members. The six families were chosen through an elaborate sampling plan. Clan locations were placed on a map showing concentric circles at half-mile intervals outward from the dumpsite. Demographics of the extended families were then reviewed with the goal of creating a diverse sample, biased in favor of the largest (excluding families with fewer than five adults still living in the community) and closest clans (within one mile of the site). Among the other factors considered were the presence and age of children and length of residence. Where families were spatially clustered, one family at a given location was chosen randomly.
2. *Group Interviews.* To represent those excluded from the intensive interviews, a group interview was held with 15 individuals from 10 smaller families. The group was roughly equal in gender and represented families living closest and farthest from the site of contamination. To get the perspective of teenagers, a second group interview was held with eight unrelated residents aged between 14 and 18, representing all families having children in that age group.

Analysis

Transcribed clan interviews were analyzed both longitudinally as a complete narrative discussion of a family's impacts and as cross-sections according to key indicators of psycho-social impact. For each of the six clans studied, a within-clan analysis was made that took into account the history, family structure and impacts for each extended family of a particular place. The approach allowed a between-clan comparison that examined commonalities and differences in impact according to such factors as location (closer or further from the contamination site and which hollow the clan occupied), size of the clan and length of time in the community.

Findings

Despite some important differences among the families, an overarching and consistent pattern of psycho-social impact was evident.

Eco-historical context

Prior to the point of alarm, life in the hollow was closely tied to the land for all families. Many residents had witnessed the dumping or burning of oil at the dumpsite, but they thought it to be harmless. Besides, they firmly believed that it was none of their business what happened on other people's private property. At the time, before Love Canal, like all Americans, they had no concept of toxic contamination. Later, they erected a fence of denial around their memories of the dumping so that they could safely believe that they were immune from consequences. Natural changes were blamed for an evident decline in the number and health of local game and fish, and state officials were believed when they said that it was safe to drink water from polluted wells. Nevertheless, a contamination belief system slowly developed for hollow residents as details of pollution spread by word of mouth, the media and at public meetings. A major turning point was a blunt assessment of risks by a state Superfund official at a key meeting in the late 1980s, subsequently confirmed when workers appeared at the dumpsite wearing "moon suits" and when the polluting corporation paid to hook people up to new water sources.

Perceived exposure

All residents interviewed, regardless of their geographic location in the community, believed that they had been exposed to contamination from the dump. Some had direct impacts from visiting the dumpsite, touching oiled roads and breathing burning wastes. All had indirect contact from consumption of local game and fish, local milk, produce and water, breathing the air and touching the soil.

Changed behavior

Knowledge of contamination caused dramatic behavioral changes. Water was carried for drinking and cooking; children's outdoor play was restricted; fishing, hunting, gardening and other subsistence activities were stopped or limited. There were meetings to attend. Residents faced stressful and frustrating adjustments to their country lifestyle and values.

Lifescape changes

Lifescape changes also occurred for hollow residents, entailing shifts in fundamental life assumptions about health, control, home and environment and trust.

An evident shift in resident perceptions of health occurred given the likelihood of their exposure to toxic chemicals, reinforced by symptoms and illness identified in the local area. Past and current health problems were viewed as having been caused by exposure to the chemicals. Expectations and fears for future health problems were expressed by each of the families interviewed, including fears of cancer and genetic problems. Given the poverty, lack of insurance and historic distrust, doctors were generally not consulted; some residents discussed the use of folk remedies to ward off the feared illnesses.

People's sense of control over their lives was also affected. Appalachians are known for their quiescence and, given their Baptist faith, for their acceptance of fate (Gaventa, 1980). However, hollow residents were decidedly unaccepting of what had occurred. Nevertheless, they were generally helpless to affect a situation that had entrapped them and their kin and families. They had lost the freedoms of their prior lives in the hollow, the individuality and free will afforded by the independence of their subsistence livelihood. Lacking education and qualifications to compete in the outer world, they were doubly disabled because their livelihood and homes were both affected. If relocation entails breaking up the family, their community and social supports would also be diminished. And, in a fatalistic vein, they perceived themselves to have been exposed to hazards. Residues of that toxic exposure will remain no matter what decisions that they now made about their future. They thus have a certainty that the chemical exposures will at some point cause health problems and that the land is dead. At the same time, the details of any effects and what to do about them are highly uncertain. The residents are left with a decisional dilemma. Not trusting the cleanup, and not wanting to move, they fear staying. There is no desirable course of action.

The positive meanings of home and environment have now been inverted. For many of the hollow residents, home and environment are one concept. Accustomed to the outdoors, the land was their home. The hollows are home. The knobs are home. The river is home. Thus, unlike the conventional victim of residential toxic exposure, many of the residents of the hollow are less focused on the house as home. House is not the sphere of living. Home is an amorphous concept associated with a range of attached places. The clan compound is home. Home also meant ancestral home, siblings' homes, kin's homes, as well as their private home. Thus, there is a merger of inversion of home with the inversion of the land. What was once safe, valued, tended and loved is now feared, reviled and neglected. The inversion of home and land also reflects the environmental stigma through which the land is devalued by virtue of its association with the contamination. It is now tainted. It has no economic value. Thus, on top of all their other problems, the devaluing of the home and land serves to trap the victims in place by removing the major equity that they hold.

Trust was the final casualty of contamination. The contamination incident in the hollow led to blaming. The corporation whose waste was dumped was blamed both for causing the problem and then not helping. Government was blamed for failing to prevent the problem and then covering it up and for not being helpful to residents. Interestingly, there was no anger at the kin of the local farmer who had accepted the wastes on his property and then burned or buried them; it was recognized that they were now in the same predicament as everyone else. Significantly, the distrust of the responsible polluter and government contributed to distrust of the cleanup effort. In fact, a facility built to air strip solvents from the groundwater was viewed not as a solution, but as a new source of contamination. With the potential for restoring the land viewed as hopeless, the potential for restoring life on the land was similarly hopeless.

Similar results are common for the U.S. contaminated communities. Several specific findings underscore the unique character of this toxic incident.

Place identity

Place identity plays a uniquely important role for residents of the affected community. For the people of this remote hollow, self-definition and family definition follow place definition. The direct bonds to nature developed from the personal experience growing up in this environment are here magnified by life within extended families rooted historically to place.

Historical ties

History was another factor. Newcomers and oldtimers were impacted by the contamination event in generally comparable ways. However, oldtimers have a temporal as well as spatial dimension to their place relationships based upon historical ties to the place where their kin are buried in the now contaminated land. The resting place of their forebears must be honored and protected. Their very definition of self and family is place-dependent. Their story is not one of mobility and migration, movement and discovery, but of staying put, being immobile, being rooted in place and avoiding change. If newcomers are defined by the place they chose to live, oldtimers are defined by the place they are already at.

Kinship ties

Kinship ties were also central to residential impacts. Most residents lived in family compounds, clusters of nuclear homes on commonly owned land. Elders were central to knitting together the family structure. In families where the elders made it clear that they would refuse to leave the hollow, children indicated that they could not leave their elderly parents. The challenge of how to keep the family together during and after a move were seemingly insurmountable, giving one more excuse to stay put.

Subsistence

As a rule, the more that place defines interconnected spheres of life – home, land, food, work, family – the greater the vulnerability of the person, the family, the kin group and the community to some threat to the land (Edelstein, 2004). Residents of this community, dependent upon the land for at least partial subsistence, were highly vulnerable to harm.

Community

Community in the hollow consists of the interconnected kin networks of the key extended families. These extended-family groups were a ready-made network system already in place for efficiently communicating, disseminating information and eliciting support. Although normally serving to protect privacy, these networks provided the hollow with a basic norm of neighborly concern and interdependence, enabling a remarkable ability to approach the contamination event cohesively, even under the leadership of the newer community members. Little dissension was evident in the community.

Location

Location was another factor. Some clan compounds directly overlook the treatment plant. Others lie on low ground along the creek flowing from the site. Others are on high ground not prone to flooding. Still others lie along roads "oiled" with contaminated oils. And some residents live on the back side of the hollow outside the areas where assistance with a new water source was provided. The most proximate residents were the most likely to have experienced direct impacts from the dumping at the site, smelling and viewing the burning "oil," having identified pollutants in their water or garden, and later on, overlooking the prominent treatment plant. Given these differences, it was notable that the basic psycho-social impacts were common to the entire community. There were several key reasons for this commonality. First, no one was free of exposure. Several generations of children throughout the area played directly in the landfill site and used the river for recreation. All hunted and fished the common area. Second, entry to the hollow was past the dumpsite and treatment plant, making it visible to all. Third, the creek flooded much of the hollow, uniting areas near and far. Finally, the very shape of the mountains creates the feeling of being inside a bowl, giving an incredible sense of coming together inside the hollows. All were touched in some direct way by the threat.

QUANTITATIVE METHODS USED TO MEASURE FEAR OF CANCER AND OTHER IMPACTS

Subsequent to the qualitative study, the quantitative study was undertaken.

Participants in the Study

A sample was drawn from the Hazard Community consisting of 174 individuals (88 men and 86 women). A second sample was drawn from a Comparison Community that included 143 individuals (71 men and 72 women). Because the educational levels and cultural experiences of the Hazard Community differ from the samples used to construct the US norms for both the MMPI-2 and the SCL-90-R, this Comparison Community was matched in educational background and cultural experiences to the Hazard Community in order to account for preexisting population differences (Anastasi, 1988).

The Comparison Community was also a socio-economically and demographically similar rural community. It was located approximately 50 miles away from the contaminated community in a region with no recent experience with environmental hazard notification. This unaffected community was particularly important because rural Appalachians may differ from national normative samples with respect to kinds of responses to stressful environments and interpretation of questionnaire items (cf. Keefe, 1988; Keefe, Hastrup, & Thomas, 2005). Socioeconomic status is also related to psychological distress (e.g., Cockerham, 1990; Hudson, 2005); the selection of a similar community and collection of information on educational background made it possible to adjust statistically for small differences between the two communities. The Comparison participants received a donation toward the repair of a local community building in exchange for their participation.

Measures

Participants were asked to provide direct estimates of perceived likelihood of developing cancer and four additional health problems for comparison purposes, both for themselves and for the average person. It was expected that the estimates of fear of cancer and a number of subscales on two traditional measures of psychological difficulties (the SCL-90-R, Derogatis, 1983; and the MMPI-2, Greene, 1991) would be elevated in the Hazard Community but not in a nearby Comparison Community, described below. In order to assess psychological difficulties in this setting, a battery of measures was included: two traditional measures of psychological distress (the MMPI-2 and the SCL-90-R); the discussed Fear of Cancer Index (Berman & Wandersman, 1992); estimates of likelihood of developing cancer and four other health problems (deafness, high blood pressure, diabetes and effects of a car accident) for oneself and for the average person; and a question about how often thoughts or fears of having cancer occurred to the individual in the past week.

The Minnesota Multiphasic Personality Inventory (MMPI-2) and the SCL-90-R were selected to evaluate psychological problems. The former has the advantage of providing validity scales which can be used to screen out participants whose responses were likely to be the result of inadequate reading skills, poor comprehension of instructions or deliberate faking (Greene, 1991). The subscale scores of the MMPI-2 and the SCL-90-R are expressed as T-scores, which have an average or mean of 50 and a standard deviation of 10. A T-score of 65 is considered a significant elevation, but some scores below that level may still indicate considerable distress. For example, a T-score of 60 places one in the upper 16% of the US population on that subscale. For purposes of collecting data for the Fear of Cancer Index, eight additional items consisting of the American Cancer Society's warning signs were included in the same format as the other 90 items of the SCL-90-R. The procedure for completing the SCL-90-R requires rating symptoms on a 5-point scale (ranging from 0 = not at all to 4 = extremely).

The eight cancer warning signs from the American Cancer Society were integrated with 17 items from the SCL-90-R as recommended by Berman and Wandersman (1992) to form a Knowledge of Cancer Warning Signs Inventory. Participants were asked to rate each item according to whether it was actually a sign of cancer (0 = not a warning sign to 3 = surely a warning sign). A Fear of Cancer Index was constructed by multiplying the appropriate items from the SCL-90-R with the corresponding items on the Knowledge of Cancer Warning Signs Inventory. With a total of 25 symptoms surveyed, the resulting score could theoretically range from 0 to 300, but actual scores are usually at the low end of this range. In order to determine the source of the elevation in the Fear of Cancer Index, the Hazard and Comparison groups were compared on their total score and also beliefs about cancer warning signs and frequency of occurrence of such signs separately.

Since the community with the environmental hazard had been told that the chemicals to which they may have been exposed could increase their risk of developing cancer, participants were also asked in a direct fashion whether they expected to develop cancer during their lifetimes. In addition, they were asked about expectations of several other negative health outcomes (car accident, deafness, diabetes and hypertension) not commonly associated with exposure to environmental hazards. Answers were expressed as percentages, with a range from 0 to 100. Participants were also asked about the frequency of thoughts or fears of having cancer during the past week as never (coded 0), sometimes (1), frequently (2) or constantly (3).

The two communities were administered the questionnaires in separate group sessions. A small number of individuals in each group were unable to complete the MMPI-2 due to poor health or vision. Those with insufficient reading skills were administered the tape-recorded version of the MMPI-2.

Often in studies of intact communities, it is useful to know whether they differ in average socioeconomic status. Many ways of measuring this involve asking people to report their occupations and highest educational level achieved. Unfortunately, both communities had high levels of unemployment, so only education could be measured. The mean educational level of the Hazard Community was 10.4 years; for the Comparison Community, it was 12.0 years. This difference was statistically significant [$F(1,286) = 31.73$, $p<0.01$]; in other words, it was unlikely to have happened just by chance. Therefore, a statistical adjustment was made.

Results of the Quantitative Study

Differences in results between the Hazard and Non-Hazard Groups will be described in the following order: direct measurement of fear of cancer and other health-related fears, indirect measurement of fear of cancer, using the Fear of Cancer Index, frequency of thoughts about cancer during the past week, the MMPI-2, the SCL-90-R and observations about gender differences.

Fear of cancer: Direct measurement

Table 1 describes differences and similarities between the two groups on estimates of likelihood of oneself or the average person having various health problems.

The Comparison Group's estimate (45.4%) for developing cancer themselves is similar to the actual risk in the U.S. population; the Hazard Group's estimate (66.6%) is significantly and substantially higher, $F(1,269) = 38.23$, $p<0.01$. Several other expectancies were significantly elevated in the Hazard Group relative to the Comparison Group (average

Table 1. Mean Estimates of Probability of Developing Certain Health Problems in One's Lifetime for Self and for the Average Person.

	Hazard Group (%)	Comparison Group (%)
Self		
Car accident	48	40
Deafness	29	24
Cancer	67	45*
Diabetes	42	31*
High blood pressure	63	55*
Average person		
Car accident	52	43*
Deafness	28	25
Cancer	59	48*
Diabetes	36	42*
High blood pressure	60	56

*$p<0.05$ for Hazard Group relative to Comparison Group.

person developing cancer, self developing diabetes, hypertension or having a car accident), but by less than 11%. That is to say, the Hazard Group had higher expectancies of themselves and others having health problems generally, but the differences were particularly evident for cancer risk, where the Hazard Group's estimates were 22% higher than in the Comparison Group.

Fear of cancer: Indirect measurement

Using the Fear of Cancer Inventory, the Hazard Group had an average score of 37.16 vs. 13.90 for the Comparison Group ($p<0.01$). Importantly, both groups had equal accuracy in their knowledge of cancer warning signs. The Hazard Group was similar to the Comparison Group in accuracy of reporting signs of cancer designated by the American Cancer Society and only slightly higher in reporting nonsigns as signs of cancer. The two groups differed greatly, however, in the frequency of reported occurrence of the signs, which in turn created the difference in total Fear of Cancer Index scores.

The direct and indirect methods of assessing fear of cancer were significantly correlated in both groups ($r = +0.42$ in the Hazard Group; $r = +0.28$ in the Comparison Group). A zero correlation means no clear association; correlations range from -1.00 to $+1.00$; scores at the extremes of this range means one can predict one score from the other very well. Both of these correlations are statistically significant; that is, they are likely to be found again. But they are in the moderate range. This moderate level of correlation likely reflects the reduced reliability of using just a single-item measure for direct assessment and the more limited range of scores for the Comparison Group.

Frequency of cancer thoughts

The Hazard Group averaged 1.68 on the 0 to 3 scale of frequency of thoughts or fears of having cancer in the past week, compared with an average of 0.74 for the Comparison Group. This difference is quite large and is also statistically significant, $F(1,289) = 90.45$, $p<0.001$.

MMPI-2

The mean validity scales of the MMPI-2 were all in the normal range. The MMPI-2 F-scale is designed to detect individuals who do not understand instructions adequately or who deliberately present themselves in a negative light; participants scoring above a T-score of 90 on the F-scale were excluded from the analysis. This eliminated 13% of the Hazard Group and 8% of participants from the Comparison Group. When all 10 MMPI-2 clinical scales were considered together in a procedure called MANOVA,

Table 2. Comparison of MMPI-2 Clinical Scale Results from Hazard ($N = 151$) and Comparison ($N = 139$) Groups According to Mean T-Scores (Scales 1, 4, 7, 8 and 9 are K-corrected).

MMPI-2 Scale Group	1*	2*	3*	4*	5	6*	7*	8*	9	0
Hazard	69.4	64.0	63.2	51.7	48.4	53.5	60.1	56.5	49.8	57.6
Comparison	53.5	53.4	49.2	48.0	49.6	48.0	52.0	49.0	47.2	55.7

*$p < 0.05$.

a highly significant difference between the groups was observed, $F(10,278) = 7.69$, $p < 0.01$.

As shown in Table 2, the differences between the means of the two groups are on the order of a standard deviation (10 points) for many of the scales. These data suggest that approximately half of the Hazard Group might be diagnosed as showing at least moderate levels of depression (scale 2), anxiety (scales 3 and 7) and concern with health (scale 1). Smaller but significant differences also occurred on scales tapping social alienation and suspiciousness (scales 4, 6, and 8).

SCL-90-R

Results from the SCL-90-R analyses are presented in Table 3. The raw scores are presented for purposes of comparison with the data of Baum et al. (1992), along with the average T-scores ($X = 50$, S.D. $= 10$) on which the actual statistical analyses were conducted.

Differing slightly from Baum et al.'s (1992) results, the total number of bothersome symptoms (positive symptom total) was not significantly elevated, although the means differed in the expected direction. The intensity of reported symptoms, measured by the Global Severity Index was, however, significantly elevated in the Hazard Group, $F(1,269) = 27.68$, $p < 0.01$. A multivariate analysis of variance on the T-scores for all nine subscales of the SCL-90-R was highly significant, $F(9,252) = 6.22$, $p < 0.01$, suggesting an overall pattern of differences between the two groups. Analyses of each of the scales separately showed significant differences on eight of the nine scales, tapping varied aspects of anxiety, depression, anger and suspiciousness.

Gender differences

Gender differences on the various measures are summarized in Table 4. A number of subscales of the MMPI and SCL-90-R differed significantly between men and women, as did the frequency of cancer thoughts and

Table 3. Means for Raw and T-Scores, Adjusted for Education, on the SCL-90-R.

SCL-90-R Scores	Hazard Group Raw Score	Hazard Group T-Score	Comparison Group Raw Score	Comparison Group T-Score
Positive symptom total	29.75	52.9	24.17	50.4
Global severity index	0.87	62.4	0.51	54.4*
Somatization	1.28	64.6	0.62	54.6*
Obsessive-compulsive	0.99	59.7	0.63	54.7*
Interpersonal sensitivity	0.72	57.2	0.49	53.8*
Depression	0.94	59.9	0.60	54.1*
Anxiety	0.90	61.2	0.46	53.4*
Hostility	0.64	55.8	0.46	52.3*
Phobic anxiety	0.40	56.2	0.26	53.1*
Paranoid ideation	0.66	55.6	0.48	52.6*
Psychoticism	0.44	58.4	0.31	55.5

*$p<0.05$ for T-score comparisons between Hazard and Comparison Groups on this scale.

Table 4. Significant Gender and Gender by Group Interactions.

Measure	Hazard Group		Comparison Group		Significance G = Gender, I = Gender by Group interaction
	Women	Men	Women	Men	
MMPI Scale 1	72.1	66.2	54.6	52.4	G, $p<0.05$
MMPI Scale 2	67.3	60.2	54.2	52.7	G, $p<0.05$
MMPI Scale 3	65.6	60.4	51.0	47.6	G, $p<0.05$
SCL-90-R depression	58.0	62.3	53.1	55.1	G, $p<0.05$
SCL-90-R Phobic anxiety	54.8	57.9	50.9	55.5	G, $p<0.01$
Cancer thoughts	1.83	1.49	0.98	0.52	G, $p<0.01$
Lifetime estimated %:					
Car accident – self	43.0	54.1	44.5	35.2	G, $p<0.01$ I, $p<0.05$
Cancer – self	73.2	59.7	51.2	39.3	G, $p<0.01$
Cancer others	64.8	51.7	54.2	42.5	G, $p<0.01$
Diabetes others	43.6	39.6	41.5	31.1	G, $p<0.05$
High blood pressure – self	68.6	56.3	57.2	52.1	G, $p<0.05$
High blood pressure – others	66.7	52.9	58.7	52.8	G, $p<0.01$

several of the estimates of lifetime estimated prevalence of various health events.[1] Most salient to our discussion is the evident greater degree of health pessimism and cancer pessimism among women. While such findings are not a critical aspect of the current study, they suggest a potential for confounding future studies if gender is not balanced in the groups being compared.

DISCUSSION

The present study shows substantial convergence of quantitative results with an independent qualitative study in the same contamination setting. The qualitative study examined fear of cancer within the overall context of psycho-social impacts to the contaminated community. Key impacts of toxic exposure found elsewhere were found to take on a distinctive signature in this community, driven by the centrality of the land, the family and subsistence elements of the prior "normal" way of life in the community. Of particular note in illustrating the contextual power of qualitative research, a particular focus of the analysis was the significance of the severance of historic clan ties to place forced by the contamination. In examining the threat to the importance of place and community for families with multigenerational bonds to location and surround, it was concluded that place played a major role beyond the inversion of home routinely found in contaminated communities.

With reference to fear of cancer, the qualitative study explicitly documented the occurrence of this fear and others during face-to-face interviews with a substantial sample of community members. But, beyond an individual's concern for their life prospects or a mother's fear for her children, the study allows for an understanding of the larger meaning of this fear. As a function of environmental stigma, fear of cancer was now a property of the landscape. Many residents had come to believe that there was no future in a setting with pervasive groundwater contamination. Yet to move away was to abandon the total, internal community represented by an elder-led clan, family and place ties, the remains of one's ancestors, potential for a family compound that kept the clan intact and the forests and streams that nurtured and connected people to their own nature. Given the close connection of identity and place, the contamination placed residents into a dilemma offering no good choice – either to leave the physical and social place that defined them or to remain on tainted land. They would lose, no matter what their decision.[2]

On their part, the quantitative results suggest that the residents of the Hazard Community have a substantial and pervasive fear of cancer, evidenced through an array of experimental measures, allowing us to place cancer fear into a psychological context. This pervasive intrusion into daily life of the fear of developing some form of cancer may help explain the variety of psychopathological symptoms of stress found on the various subscales of both the MMPI-2 and the SCL-90-R. For example, while about 18% of the Comparison Group (about what is found in most community samples) would qualify as at least somewhat depressed (MMPI-2 Scale 2

T-score of 65 or more); the proportion in the Hazard Group, however, was a striking 55%. This difference occurred despite adjustment for differences in education, which served in this study as a surrogate for socioeconomic status. The increased expectation of personally developing cancer and the increased rate of depression and anxiety are in accord with the scientific literature on the psychological effects of environmental hazard situations, including situations in which no lawsuit was involved.

This quantitative study also showed the usefulness of an indirect measure of fear of cancer. Although significantly correlated with direct measurement, the Fear of Cancer Index appears to provide some additional information on the sources of that fear. While more research is needed to validate this scale, the present investigation showed a substantial difference in the distribution of Fear of Cancer Index scores between the two groups. The Hazard Group residents also believe they have a much greater risk of developing cancer, but only a limited ($\leq$ 11%) increase in their risk for other adverse health events.

The quantitative study is unusual for its rural Appalachian setting, provision of local norms, the variety of differences in MMPI-2 and SCL-90-R scale elevations for psychological problems observed and the percentage (55%) of individuals in the Hazard Community who showed clinically significant elevations in depression. The limited differences in expectations concerning noncancer-related negative events suggest that the problems are specific to the environmental hazard, rather than some overall pessimism.

As expected (Keefe, 1988), the Comparison Group, from a neighboring rural Appalachian community not affected by reports of an environmental hazard, differed somewhat from the national norms for the MMPI-2, but considerably less than the Hazard Group. The comparison of the Hazard Group with a local normative sample thus permits an inference that the differences observed were not the result of cultural variations (e.g., an elevated scale score for anxiety reflecting fear of subways and elevators due to limited exposure).

These further results may also be compared with the intensive family and group interviews obtained from a representative sample of the community of the Hazard Group. This qualitative work clarifies the content and substance of the severe community stresses measured in the study described in this paper. While both studies might stand independently as two approaches to documenting community stress, together they served to strengthen and substantiate their findings. Triangulation of quantitative and qualitative data has been used in other settings to test for convergence with similar success (Edelstein, 2004; Gibbs, 1986; Kroll-Smith & Couch, 1993; also Couch & Mercuri, Chapter 6, this volume). The qualitative interviews provide an additional layer of meaning to the quantitative measures, which may be especially helpful in describing to nonprofessionals the severity of the distress (e.g., over half of the Hazard Group suffering from at least moderate levels of depression).

CONCLUSION

This report demonstrates further confirmation of a causal link between community contamination and psychopathology in affected residents. This causal relationship can be demonstrated to be statistically significant (i.e., likely to be found again) using traditional psychological measures such as the MMPI-2 and the SCL-90-R. These findings suggest that the continued testing and use of the Fear of Cancer Index is also justified as part of a set of standardized measures of community stress following notification of exposure to an environmental hazard. It also suggests that the best use of such quantitative approaches is in conjunction with qualitative studies that identify the meaning of events for those affected and the specific nature of context studied. When these two perspectives are provided, information about the extent of impacts can be combined with insight into why those impacts occurred.

NOTES

1. One, theoretically unimportant, interaction of group and gender (car accidents for oneself) was observed.

2. See Edelstein (2004). Edelstein has a book in progress tentatively called Hazardous Hollows that presents these findings in detail. Pending publication, contact this author for information.

REFERENCES

American Cancer Society. ACS History. Retrieved February 18, 2006 from http://www.cancer.org/docroot/AA/content/AA_1_4_ACS_History.asp

Anastasi, A. (1988). *Psychological testing*. New York: Macmillan Publishing Co.

Bachrach, K. M., & Zautra, A. J. (1986). Assessing the impact of hazardous waste facilities: Psychology, politics, and environmental impact statements. In: A. H. Lebovits, A. Baum & J. E. Singer (Eds), *Advances in environmental psychology, Vol. 6: Exposure to hazardous substances – psychological parameters*. Hillsdale, NJ: L. Erlbaum Associates.

Baum, A., Fleming, I., Israel, A., & O'Keeffe, M. K. (1992). Symptoms of chronic stress following a natural disaster and discovery of a human-made hazard. *Environment and Behavior*, *24*(3), 347–365.

Berman, S. H., & Wandersman, A. (1992). Measuring Fear of Cancer: The Fear of Cancer Index. *Psychology and Health, 7*, 187–200.

Cockerham, W. C. (1990). A test of the relationship between race, socioeconomic status, and psychological distress. *Social Science in Medicine, 31*, 1321–1326.

Davidson, L. M., Baum, A., Fleming, I., & Gisriel, M. M. (1986). Toxic exposure and chronic stress at Three Mile Island. In: A. H. Lebovits, A. Baum & J. E. Singer (Eds), *Advances in environmental psychology, Vol. 6: Exposure to hazardous substances – psychological parameters*. Hillsdale, NJ: L. Erlbaum Associates.

Derogatis, L. R. (1983). *SCL-90-R version: Manual I*. Baltimore: Johns Hopkins University.

Edelstein, M. R. (2004). *Contaminated communities: Coping with residential toxic exposure*. London: Westview Press.

Erikson, K. (1994). *A new species of trouble*. New York: W. W. Norton and Co.

Gatchel, R. J., & Newberry, B. (1991). Psychophysiological effects of toxic chemical contamination exposure: A community field study. *Journal of Applied Social Psychology, 21*, 1961–1976.

Gaventa, J. (1980). *Power and powerlessness: Quiescence and rebellion in an Appalachian Valley*. Urbana: University of Illinois Press.

Gibbs, M. S. (1986). Psychopathological consequences of exposure to toxins in the water supply. In: A. H. Lebovits, A. Baum & J. E. Singer (Eds), *Advances in environmental psychology, Vol. 6: Exposure to hazardous substances – psychological parameters*. Hillsdale, NJ: Lawrence Erlbaum Associates.

Green, B. L., Lindy, J. D., Grace, M. C., Gleser, G. C., Leonard, A. C., Korol, M., & Winget, C. (1990). Buffalo Creek survivors in the second decade: Stability of stress symptoms. *American Journal of Orthopsychiatry, 60*, 43–54.

Greene, R. L. (1991). *The MMPI-2/MMPI: An interpretive manual*. Boston: Allyn & Bacon.

Greve, K. W., Bianchini, K. J., Doane, B. M., Love, J. M., & Stickle, T. R. (2005). Psychological evaluation of the emotional effects of a community toxic exposure. *Journal of Occupational and Environmental Medicine, 47*, 51–59.

Hudson, C. G. (2005). Socioeconomic status and mental illness: Tests of the social causation and selection hypotheses. *American Journal of Orthopsychiatry, 75*, 3–18.

Keefe, S. E. (Ed.) (1988). *Appalachian mental health*. Kentucky: University Press of Kentucky.

Keefe, S. E., Hastrup, J. L., & Thomas, S. F. (2005). Psychological testing in Appalachia. In: S. E. Keefe (Ed.), *Appalachian cultural competency*. Knoxville: University of Tennessee Press.

Kroll-Smith, J. S., & Couch, S. R. (1993). Technological hazards: Social responses as traumatic stressors. In: J. P. Wilson & B. Raphael (Eds), *International handbook of traumatic stress syndromes*. New York: Plenum Press.

Markowitz, J. S., & Gutterman, E. M. (1986). Predictors of psychological distress in the community following two toxic chemical incidents. In: A. H. Lebovits, A. Baum & J. E. Singer (Eds), *Advances in environmental psychology, Vol. 6: Exposure to hazardous substances*. Hillsdale, NJ: Lawrence Erlbaum Associates.

Smith, E. M., North, C. S., & Price, P. C. (1988). Response to technological accidents. In: M. Lystad (Ed.), *Mental health response to mass emergencies*. NY: Brunner/Mazel.

Unger, D. G., Wandersman, A., & Hallman, W. (1992). Living near a hazardous waste facility: Coping with individual and family distress. *American Journal of Orthopsychiatry, 62*, 55–70.

Vyner, H. M. (1988). *Invisible trauma: The psychosocial effects of invisible environmental contaminants*. Lexington, MA: D. C. Heath & Co.

TOXIC WATER AND THE ANTHILL EFFECT: THE DEVELOPMENT OF A SUBCULTURE OF DISTRESS IN A ONCE CONTAMINATED COMMUNITY

Stephen R. Couch and Anne E. Mercuri

INTRODUCTION

In 1990, testing revealed the existence of benzene in the municipal water supply of a community named Three Lakes, a residential subdivision of Houston, Texas. The water was quickly changed to a clean supply, but residents were not notified that there had been a problem until five months later. This provoked much anger within the community, along with concerns over present and future health problems. A grassroots group formed in response to this problem, but lasted only one year. The failure of this social movement organization left community residents to fend for themselves. In the words of one resident, the community reacted "like someone stepping on an anthill – everyone running in different directions."

This paper examines qualitative and quantitative data collected from subdivision residents nearly a decade after their water problem. The data reveal significant psychosocial disruption among members of the

Cultures of Contamination: Legacies of Pollution in Russia and the U.S.
Research in Social Problems and Public Policy, Volume 14, 117–137

ISSN: 0196-1152/doi:10.1016/S0196-1152(06)14006-5

community, including heightened levels of alienation, distrust and demoralization. While paralleling the findings of numerous studies of contaminated communities, these data are striking because they were collected six to eight years after the environmental problem had taken place and because community residents did not know of the water problem until months after it had been resolved.

We begin by orienting this chapter in the literature and introducing the concept, "subculture of distress," to help understand how communities become debilitated by the social process that accompanies many disasters. We then describe what happened at Three Lakes. Next, we describe the methods by which we gathered our data and then examine several components of the subculture of distress and one psychosocial outcome, providing qualitative and quantitative evidence for their existence in Three Lakes. Finally we end with a discussion of some implications of this case.

CHRONIC TECHNOLOGICAL DISASTERS AND CORROSIVE COMMUNITIES

Over the past several decades, numerous sociologists have studied the relationship between physical hazards and disasters and local communities. A lively debate has emerged about whether the type of disaster (e.g., "natural" or "technological") affects the social response (Couch & Kroll-Smith, 1985; Freudenburg, 1997; Quarantelli, 1985).

Regardless of one's position in this debate, scholars have amassed strong empirical evidence that there is a social process that accompanies many (perhaps most) long-term technological disasters that creates debilitating social conflict within the local community. Building on Kai Erikson's seminal study of the Buffalo Creek flood and his later work that notes community trauma in cases of long-term technological disasters (a "new species of trouble") (Erikson, 1976, 1994), Freudenburg coined the term "corrosive communities" to describe what happens in cases of chronic technological disasters (Freudenburg, 1997). Within the community, social conflict develops between people and groups who come to have very different views of what is happening and very different approaches on how to deal with the situation. Chronic technological disasters are usually ambiguous across a number of dimensions, particularly concerning how dangerous the situation is for members of the community, and consequently what should be done about the problem. Lines of conflict also emerge over who is to be blamed and who is responsible for mitigating the situation. These issues are framed by a typical governmental and corporate response that is slow, bureaucratic and often evasive and divisive in itself. An exacerbating factor has been seen as the chronic nature of the disaster agent itself, perhaps resulting in the development of health effects years later from ingestion of toxins today, tomorrow or next week. Therefore, some in the community see an immediate need to mitigate the environmental hazard or move away from it; others note the ambiguity of the evidence and see the problem as much less serious.

The social process accompanying chronic technological disasters has been described as a "secondary victimization" of the community and its population. Already subjected to the direct impacts of an environmental hazard, people become victims of a breakdown in the social community, often affecting families, friendships and community networks and adding considerable stress to an already difficult situation (Couch & Kroll-Smith, 1985; Edelstein, 2004; Kroll-Smith & Couch, 1990).

The Three Lakes community differs from most other disaster communities that have been studied. By all official and expert accounts, the hazard agent was mitigated shortly after discovery and months before the community found out it ever had a water problem. Therefore, since there is no credible evidence of a continuing problem, a major line of conflict is missing. Nevertheless, significant community conflict over hazard-related issues, and social and negative psychosocial impacts of that conflict, were evident years after the hazard was removed. In the absence of an ongoing disaster agent, how were the conflict and psychosocial impacts maintained for so long?

We suggest that part of the answer can be found in the development of a "subculture of distress." A subculture of distress is a set of beliefs, created by and circulated among members of a collectivity and resulting in and reinforcing affliction, suffering, anxiousness and dread. Because they are social, the beliefs in a subculture of distress are more difficult to change once they become established than are individual perceptions. Unlike many cases of natural disaster in which a disaster subculture arises which helps residents prepare for, cope with and recover from the calamity, the subculture of distress exacerbates and prolongs the suffering of people who are trying to cope with an environmental hazard.

In some ways, this study parallels Baum and colleagues' studies in which psychological impacts of the 1979 Three Mile Island nuclear disaster near Middletown, Pennsylvania were found to persist years after the accident took place (see, e.g., Baum & Fleming, 1993; Baum, Gatchel, & Schaeffer, 1983). In this chapter, we focus on the psychological and social effects (psychosocial) and social conflict, adding a social dimension to Baum et al.'s conclusions.

THE STORY OF THREE LAKES[1]

Developers began construction in the community of Three Lakes in the mid-1980s by filling in the lakes after which the community was named. The subdivision was planned to accommodate about 1,000 homes; by 1990 however, only 240 were completed. The mainly white middle and working class community drew its water supply from an underground aquifer. The water was provided and maintained by the local municipal utility district, known as "MUD." Residents of Three Lakes elected property owners to the MUD board of directors. Some board members resided in Three Lakes, while others owned homes in which they did not reside, but rented to other families. This organization carried with it high expectations of trust, since the board members would presumably share a vested interest in maintaining a safe and plentiful supply of water. The board members who resided in Three Lakes became the target of much anger by the other residents when the contamination became known.

Residents generally agreed Three Lakes was a friendly community and a good place to raise children. Politically, property owners were represented by the Community Improvement Association (CIA), which collected dues, maintained communal property such as the swimming pool and made sure homeowners followed property regulations. The developer of Three Lakes, who carried three votes to every homeowner's vote, originally controlled the CIA. The homeowners later organized and changed the vote ratio to one-to-one, giving the homeowners active in the CIA a sense of accomplishment and solidarity.

An oil refinery was originally located near the rear of the Three Lakes' property with disposal wells throughout the entire area. Many residents immediately noticed problems with the water. It appeared cloudy, smelled bad, tasted foul and was often white from gas bubbles. Initially, nobody was overly concerned. Some attributed it to "city water," and one noted she was not seriously worried, as "you have to trust people who provide things." Several homeowners did complain to the MUD board. It was not until seven years later, however, that the MUD board meeting minutes reflected any of the residents' concerns regarding the quality of water at Three Lakes.

In 1990, benzene, a known human carcinogen, was added to the list of chemical contaminants tested for in public water supplies. The test results in

Three Lakes were shocking. Benzene at 11 times the amount allowed by the Environmental Protection Agency was found in the Three Lakes water supply. Additional testing confirmed the original findings and two weeks later the supply was shut down. Residents were switched to a temporary alternate water source from a neighboring district. Since the water was now safe, the MUD board made the momentous decision not to inform residents that they had been using contaminated water. This would later cause much anger and social conflict among the residents of Three Lakes.

Almost immediately, rumors began circulating through the community. A local reporter discovered the contamination and ensuing cover-up. Immediately before he reported his findings, the MUD board sent a notice to residents informing them of the prior contamination and later switch to a clean water source. This took place five months after the initial discovery of benzene in the water supply.

In the midst of this upheaval, a few members of the CIA wanted the organization to take action. This was met with resistance by some and led to conflict among the members. A grass-roots organization, WELL, formed in response and at its height was comprised of about 60 residents. Members wrote to Congress and other officials, investigated MUD records and the Board's actions, prompted soil and water sampling and received media coverage. While WELL initially generated considerable support, it could not withstand the eventual conflict, both social and personal, among its members and folded about a year after its formation.

Eventually, the situation in Three Lakes stabilized. The various lawsuits, which were brought in response to the contamination, were consolidated under one firm, which believed it had determined the party responsible for the contamination. The original MUD board members were replaced by new ones. A permanent clean water supply was found and property values in the development stabilized somewhat. Nearly a decade after the initial discovery of the contamination, new homes were being built in Three Lakes, although the community's name was being changed. The irony of the name change and the attempt to disassociate the community from its toxic history was not lost on many of the residents.

METHODS

The senior author gathered the data for this chapter between October 1996 and June 1998. He was employed by a law firm representing plaintiffs who were suing Exxon Corporation for allegedly causing the contamination of

Three Lakes' water. The senior author reviewed documentary material on Three Lakes, including census data, newspaper articles, selected depositions and other case documents. He also conducted several site visits to the community and interviewed 28 current or former residents of Three Lakes who were participants (litigants) in the lawsuit against Exxon. Almost all of the interviews took place in respondents' homes and lasted an average of an hour and a half, with the residents being interviewed individually, as couples or in a group of three. The second author conducted a thematic content analysis of these interviews, with some of the results being presented below.

In addition, the senior author conducted a telephone survey of a sample of plaintiffs. The simple random sample of 120 was drawn from a list of the 450 case plaintiffs. Of this sample, nine were eliminated because they were deceased, or represented the estate of others, or were already interviewed during an earlier stage of the research. Of the 111 plaintiffs in the sample, 82 usable interviews were collected, representing a response rate of 74%. The interviews included questions about respondents' views of Three Lakes as a place to live, their opinions about the safety of Three Lakes' water, their exposure to water contamination, their opinions on the relationship between the contamination and their health and the health of their family, any group response in which they were involved, and demographic information. More generally, they were also asked what, if any, effects they believe the contamination had on themselves, their family and their community.

Finally, respondents were asked questions that allowed the construction of several scales measuring levels of distrust, environmental worry, alienation, and demoralization. The "trust scale" is a three-item scale widely used to measure the level of trust or distrust a respondent holds toward other people (National Opinion Research Center, 1994).[2] The "environmental worry" scale (short form) is composed of eight Likert-scale items used to measure a respondent's anxiety over various aspects of environmental problems (Bowler & Schwartzer, 1991).[3] The anomie scale, perhaps the most widely used scale measuring alienation, contains five agree–disagree items that tap a respondent's estrangement from social life and institutions (Srole, 1956).[4] Finally, the demoralization scale is comprised of nine (mainly agree–disagree) items perhaps best described as measuring pessimism and despair (Schuessler, 1982).[5]

Since the sample for this paper was drawn from a list of litigants rather than a complete list of people who lived at Three Lakes at the time of the contamination, we need to ask the extent to which our sample is representative of the population of potentially affected residents. Perhaps litigants exaggerate the extent to which they are affected because they believe they have something to gain by doing so. Or perhaps those most affected are the people who join the suit. While we have no way to directly measure these concerns, studies in the literature that have examined this question have found few differences between litigants and non-litigants (Bowler, Mergler, Huel, & Cone, 1994; Erikson, 1993; Green et al., 1990; Murphy & Keating, 1995). In addition, few differences were found between demographics of our respondents and demographics of all those living in the census tract that includes Three Lakes.[6] The frequency distributions of the survey questions were normally distributed, rather than being skewed in the direction of a biased response. The qualitative interviews and documentary data provide different forms of data and provide the authors with additional information by which to judge the validity and honesty of survey responses. The fact that so many of the findings parallel studies carried out in other contaminated communities bolsters the evidence that the data in this study are valid.

A SUBCULTURE OF DISTRESS

A subculture of distress is a set of beliefs that is created by and circulated among residents of a community, which results in and reinforces affliction, suffering, anxiousness, and dread. Such a subculture of distress emerged from the contaminated community of Three Lakes. Like similarly affected communities, Three Lakes suffered a breakdown in the ordered routine of its social life that once made it a cohesive family-oriented community. Despite individual awareness of poor water quality and complaints regarding the water, prior to the public knowledge of benzene contamination, neighbors talked to each other, children played together and the residents held communal activities such as neighborhood parties and yard-of-the-month contests. When the contamination moved from a possibility to confirmed public knowledge, a subculture of distress began to form in the community of Three Lakes.

An important first element in the formation of a subculture of distress is the uncertainty people experience over the extent of the contamination and what the present and future health effects might be. Several residents expressed "fear of the unknown," and virtually all the respondents articulated uncertainty. The insidious nature of chemical contamination leaves victims feeling unsure over how extensive the problem is or will grow to be. Victims often feel there is no identifiable end to their dilemma and often there is not. Litigation may take years and long-term health effects are still unknown for many chemical contamination victims.

Most respondents in the telephone survey believed that they were exposed to contamination from Three Lakes' water and each and every one believed that their health may be affected in the future or were unsure of future effects. Mothers expressed worry over future health effects for their children and even their daughters' future pregnancies. One young man reported he was not sure if he would ever father children due to the uncertainty. Some mothers reported feeling guilty for having bathed their children in the water and having allowed them to swim in the community pool and drink the tap water, even though they were not aware of the danger at the time. One woman reported being "horrified" and is still afraid to drink any water, even bottled water. She noted that water is "the very life-giving essence... then it turns on you and you are poisoned by it. It's like having a sadistic mother."

In-depth interviews constantly referred to human and animal diseases and deaths that residents feared were linked to the water contamination, but the connection was uncertain. Maladies reported by residents included various kinds of cancers, rashes, lupus, high blood pressure, menstrual problems, respiratory problems, miscarriages, and attention deficit disorder. A mother who lost her son to leukemia lamented, "You thank God for giving you healthy children, but then.... If it's an accident, that's one thing. But this is unnecessary.... How can you do that to people?" Her family doctor told her that benzene is known to cause leukemia.

The uncertainty is an expected precursor to a sense of powerlessness. Residents of contaminated communities often feel powerless to change their circumstances. Often there are valid reasons for their feelings of powerlessness. Many who would like to move cannot afford to do so and often there is a belief (often valid) in loss of property value. Three Lakes was no exception. Several of the respondents from the in-depth interviews expressed concern over falling property values and their inability to sell their homes. "Ruined lives," "lost hopes and dreams," "being beaten," and "everything we worked for is gone," are all statements made by residents who had worked hard to achieve "the American Dream" of owning their own home. Interviews underlined the sense of loss – for one family who had hoped to build a new home, thanks to the contamination, "all hopes of selling (their current home) went out the window." Another couple believed they lost out on refinancing their home because of the benzene problem. They had put a lot of hard work and money into their home and it was hard to accept the fact that the house "wasn't worth a damn."

Most of the homeowners stopped maintaining their homes and yards. A "why bother?" attitude was expressed by several homeowners. A local shopping center once patronized by the community was abandoned and vandalized; the yard-of-the-month contest was no more and residents felt powerless to stop the deterioration of their once vital neighborhood. Houses that were being sold were selling "real cheap" to the "trailer park crowd," and one street became known as "skid row," a reference to urban areas known to be frequented by vagrants. Some residents simply moved away leaving the mortgage companies "stuck with the homes." Some of those who chose to stay did so for moral reasons. "Conscience has kept us here. We'll not sell it to anyone without telling them [about the contamination]."

Environmental stigma is another element in the culture of distress found in contaminated communities. Stigma involves changing perceptions of someone or something from a positive to a negative based on that someone or something being marked by an observable undesirable or deviant characteristic (Goffman, 1963). Stigma in environmentally contaminated communities is referred to as "environmental stigma" (Edelstein, 1991, 2004) and involves the discrediting of a place such as Three Lakes. Residents reported feeling stigmatized by friends and family members from outside the community. Many nonresidents refused to allow their children to play there; others insisted on bringing their own drinking water even after contamination was no longer present. One woman reported people at church saying, "You're in the subdivision where the benzene was."

Perceived environmental stigma was found in the majority of the interview transcripts of Three Lakes. Three Lakes became nicknamed "Benzene Acres." One resident reported, "Even now, I cringe when a new person brings my daughter home. They will think, oh, it's *that* neighborhood."

Social isolation occurs in contaminated communities affected by stigma, adding to the culture of distress. In Three Lakes, some of those interviewed noted they had withdrawn from their community, were less social or felt isolated. Neighborhood events like pool parties would draw fewer and fewer people, until eventually they stopped altogether. Neighboring "went down hill" – "neighboring didn't work when people were hollering at one another." Some respondents recounted growing hostility among community members in public encounters on Three Lakes' streets. It was noted that even next-door neighbors who were once close friends would no longer even say "hello" to each other, and children who once played together were no longer able to do so. One noted that things got "real ugly."

The disillusionment experienced by those who trusted others to provide safe water to their homes may lead to the alienation of people from their community and social institutions. Distrust becomes pervasive in the lives of those who live through environmental contamination. More than half of the interview participants reported losing trust in local, state and federal officials and becoming disillusioned in the process. One disenchanted woman stated, "People don't care anymore what happens to other people. It's a me-me world.... The government's just as bad. The whole society, the whole world has changed." Several mourned the loss of their faith and trust in others, noting they were not so distrustful before. One man noted, "No amount of money can restore trust."

Distrust became so rampant that, even though all expert opinion and repeated water testing indicated that there has been no water quality problem since 1990, several of the survey respondents believed that Three Lakes' water was still unsafe, and half were unsure. One woman noted, "The same people kept telling you the water was fine who told you that before."

Distrust extended from the institutions that provided the water or were expected to deal with the problem, to friends, neighbors and family members. People thought others knew more than they did and were keeping secrets for their own gain. Nobody knew whom to trust. One respondent summed up how the contamination has changed his view: "The older I get, the less I believe people." Regarding officials and water authority boards, one woman bemoaned, "You're at their mercy."

Anger and conflict are inescapable and elemental facts of life for many victims of contamination. A majority of the in-depth interview respondents expressed anger towards whomever they deemed responsible. Anger towards those responsible for the contamination and ineffectual government officials and agencies was high. Residents were angry over issues such as property loss, health effects and the loss of their community. When people found out about the MUD Board's "cover-up," they were "infuriated." "How can you do that to people?" one asked. "You trust people to build houses that are safe. It makes me very angry." Others were angry with fellow community members for their views and actions. Some parents reported their children were angry over the changes in their lives since the disclosure, such as parents arguing or being emotionally preoccupied and routines being disrupted. One woman noted that her husband, who had never "raised his voice at the children," had now been yelling at them. Neighbors were angry at each other over the uncertainty and what to do about it. Often husbands and wives argued incessantly and were angry at each other for months at a time. One couple reported not speaking to each other for a week. Lost friendships and divorces were two reported results of the conflicts concerning the contamination and what to do about it. One woman summed up the anger she experienced rather succinctly; "I feel like a bombshell waiting to explode."

The subculture of distress that formed in Three Lakes triggered serious consequences for the residents. Their health was undoubtedly impacted. A local medical group conducted a survey and estimated 35–40% of the participants had diagnosed medical problems. According to one resident, a federal agency continues to phone yearly to "ask if we're still alive." The social consequences of the contaminated community of Three Lakes were severe and included lost relationships, estranged families, severed friendships, and even divorce. The economic effects were equally harsh and included financial problems, lost savings, repossessions, lost equity in homes, and ultimately for some, bankruptcy. Uncertainty, powerlessness, environmental stigma, social isolation, disillusionment, distrust, and anger are elements of a subculture in which individual reactions are raised to the level of shared beliefs, reinforced through communications networks and given validity by a continuing series of events that become part of the collective history of the community. All of these elements found in a subculture of distress are more than individual perceptions and emotions. The result is the creation and maintenance of a subculture of distress that contributes to the disempowerment of the community as a collective and to the long-lasting nature of the psychosocial effects of experiencing life in a once contaminated community.

QUANTITATIVE EVIDENCE

The qualitative data we have provides rich evidence for the development of a subculture of distress among present and former Three Lakes' residents. This is supplemented by quantitative data drawn from the telephone survey.

As discussed above, we used three scales to quantitatively measure aspects of the subculture of distress. The first was Bowler's Environmental Worry Scale. Our expectation is that Three Lakes' residents will worry much more about environmental matters than people who have not lived with contamination and about the same as others who have been exposed. We compared the mean score at Three Lakes for this scale with groups from Bowler and Schwartzer's study of chemically exposed workers in New Mexico and Texas (see Table 1) (Bowler & Schwartzer, 1991). The mean score of 26.26 at Three Lakes was significantly higher not only compared with Bowler and Schwartzer's nonexposed worker comparison group (20.71) and a comparison group of students (19.67), but also compared with Bowler's exposed workers (24.24).

Table 1. Environmental Worry Scale.

	Three Lakes	Exposed Comparison	Non-exposed Comparison	Students
Mean	26.26	24.24	20.71	19.67
SD	4.32	3.87	3.55	3.22
N	82	291	259	250
		$T = -3.88$	$T = -11.33$	$T = -14.33$
		$p < 0.0005$	$p < 0.0005$	$p < 0.0005$

Table 2. Distrust.

	Three Lakes	Sunburst	1994 GSS
Mean	1.78	1.64	1.56
SD	1.17	1.27	1.15
N	81	97	1716
		$T = -0.78$	$T = -1.91$
		NS	$p = <0.05$

The second quantitative summary measure was a measure of distrust (see Table 2). We expect that people who live through the social processes surrounding environmental contamination have heightened distrust compared with others. Here, the Three Lakes mean score was compared with a chemically exposed sample of residents of a midwestern U.S. town (we'll call it Sunburst) (Couch, Kroll-Smith, & Wilson, 1997) and respondents to the 1994 General Social Survey (GSS) (National, 1994). The Three Lakes mean of 1.78 was significantly higher than the GSS mean (1.56) and higher, although not significantly so, than the exposed residents of Sunburst (1.64).

We also included Srole's anomie scale as a measure of alienation (Srole, 1956). As with distrust and environmental worry, we expect that people living through a contamination experience will be more alienated from social institutions than others. We compared the Three Lakes mean score with those of Sunburst and a non-contaminated comparison community used in the Sunburst study. The Three Lakes mean of 2.91 was significantly higher than that of Sunburst's comparison community (1.7) and nearly identical to the mean found in Sunburst (2.9) (see Table 3).

Finally, we used Schuessler's demoralization scale as a personal outcome measure (Schuessler, 1982). We expect that people who are worried about the physical environment, and are distrustful of others and are alienated, are

Table 3. Anomie.

	Three Lakes	Sunburst	Sunburst Comparison
Mean	2.91	2.9	1.7
SD	1.51	1.6	1.6
N	78	97	308
		$T = -0.04$	$T = -5.76$
		NS	$p = <0.0005$

Table 4. Demoralization.

	Three Lakes	National Comparison
Mean	2.79	2.4
SD	2.64	1.99
	80	1522
		$T = -1.70$
		$p = <0.05$

more likely to be personally demoralized than others. The Three Lakes mean of 2.79 was significantly higher than the mean of 2.4 found in a large national comparison group used by Schuessler (see Table 4).

INTERPRETATION

How can we explain these findings? Why, many years after the contamination had ended, do we still find highly elevated levels of various elements of psychosocial distress? We suggested above that a subculture of distress developed that included elements that exacerbated rather than ameliorated residents' distress. How might we conceptualize what happened in Three Lakes to help us understand how such a subculture could be developed and maintained over a period of years?

We suggest that a subculture of distress is sustained primarily through three interrelated components. The first is that information streams necessary to clearly assess the situation and assuage fears and uncertainty are often blocked or intentionally diverted by those in positions of authority. Residents were angry and demanded information regarding the present and future health effects of benzene contamination. The initial attempts by the community to gather information were met with resistance. The only thing people were certain of was that they had ingested benzene through their water. They were unsure of how much was ingested or what the possible health effects would be. Information regarding the source of the contamination and who was responsible was also lacking. Residents began to wonder if a number of inexplicable illnesses in the community might be due to the contamination. One resident asked, "How could they have played with our lives for six months?" Others reported being "furious." An informational meeting was held, but did nothing to assuage the residents' fears. The MUD board members were defensive, the "experts" on hand gave contradictory information and community members were angry.

Adding to the increasing social conflict and blocked information was the rapid influx of lawyers into Three Lakes when the contamination became public knowledge. Various legal firms agreed to represent certain residents and instructed their clients not to share discoveries or new information with their neighbors. This pitted neighbor against neighbor and effectively blocked information streams, adding to the already escalating levels of social conflict in Three Lakes.

The second component is related to trust expectations. The residents had high trust expectations of those on their local municipal utility district and their Community Improvement Association. After all, some of the members and officials of both also resided in Three Lakes. A breakdown of trust, especially where there are high trust expectations, contributes to the producing and sustaining a subculture of distress. The residents expected that their elected officials were maintaining a safe water supply for the community and were both devastated and angry when that trust was violated.

The third factor integral to the development and also the sustaining of a subculture of distress is the differing and conflicting frames adopted by the various participants in a contaminated community. From a sociological perspective, the social construction of this case by its participants is significant. Three Lakes' residents came to view the situation in very different manners. These socially constructed frames facilitated conflict, not collaboration. The residents active in WELL had adopted an "environmental justice" frame (Capek, 1993) and advocated for accurate information, democratic participation and compensation from those responsible. This led to conflict between those residents who framed their experience in terms of property values and the fear of losing the equity and value in their homes. Also in conflict were residents who adopted a "health frame" and were concerned primarily with the present and possible future health impacts on themselves and their children.

Uncertainty created by a lack or blocking of information, the violation of trust expectations, and competing frames helped to create and sustain a subculture of distress in Three Lakes. This subculture helped to tear down the social fabric of the community and exacerbated the pain and distress suffered by Three Lakes' residents. It represents a sociocultural legacy likely to last as long as, and arguably, for most, be at least as pronounced as, physical effects of the contamination itself.

DISCUSSION

Contrary to the old adage, time does not necessarily heal. Three Lakes provides a case in point. Six to eight years after the water contamination problem had been solved, residents and ex-residents were still dealing with psychosocial pain and suffering caused by the contamination incident. While a high level of uncertainty surrounds the physical effects of the contamination, evidence of the social and psychological effects is significant. When the contamination became public knowledge, a social process began that tore apart the social fabric of the Three Lakes community. A subculture of distress developed which exacerbated the social and psychological problems of community residents. If left to run its course, this subculture of distress has debilitating effects even long after the physical contamination problem has been resolved. In the absence of interventions or extenuating events, uncertainty, distrust, powerlessness, stigma, disillusionment, anger, alienation, and competing frames take their toll on the community and its members.

Students of natural disasters have studied disaster subcultures for some time. They have found that in areas prone to certain types of natural disasters, a disaster subculture develops which helps residents to prepare for and respond to disasters. For example, in natural disasters, a therapeutic community often forms by which neighbor helps neighbor to respond to the catastrophe (Fritz, 1961). At Three Lakes, nothing of the sort happened. Instead, there was a high level of social conflict within the community. This difference was not lost on community residents. More than one resident recalled that a tornado had hit the subdivision a few years ago. At that time, neighbors came through for one another, doing all they could to help those in need. For the water problem, respondents reported that the reaction was just the opposite, with everybody looking out only for themselves. Instead of the community being solidified, it was "like someone stepping on an anthill."

This study adds to the growing body of evidence showing a distinct and debilitating pattern of social response to technological hazards and disasters (Couch & Kroll-Smith, 1985; Erikson, 1994; Finsterbusch, 1989; Freudenburg, 1997; Kroll-Smith & Couch, 1990, 1993). We suggest that part of the reason for the salience of this pattern is the development of a subculture of distress that reinforces beliefs that are detrimental to a positive collective response. At Three Lakes, a grassroots group (WELL) and a homeowners association (CIA) did respond to the problem. However, WELL lasted only one year, and the CIA's response was ineffectual. Part of the reason for this failure has to do with micro-sociological factors such as social networks and information flows (Couch, 1999). But part, we suggest, has to do with a subculture that reinforces uncertainty, distrust, alienation, and conflicting and individualized responses to the problem. It is extremely challenging for social groups to respond effectively in the face of these kinds of cultural beliefs.

This study has practical implications for those wanting to help communities respond to technological hazards and disasters. Response will have to come without the benefit of a "naturally developing" positive disaster subculture. Instead, a positive response may be possible only when for some reason a subculture of distress does not develop (Aronoff & Gunter, 1992) or when active efforts are made to combat or overcome the subculture of distress (Cole & Stevens, 1996; Kindler, 1996; Picou, 2000). We will learn much from studying positive examples of how intervention strategies have worked in practice and in attempting to devise new ways for communities to avoid the debilitating effects of a subculture of distress.

Finally, it is interesting to note that the literature on recovery from trauma also contains stage models (Herman, 1997; pp. 155–156). While differing in detail, Herman argues that most models can be condensed into three stages:

1. safety – where the safety of the victim is reestablished;
2. remembrance and mourning – wherein meaning is attached to the trauma and the lost parts of one's life and self are mourned; and
3. reconnection – where the victim reconnects to social support systems and social networks (1997, pp. 155–213).

If people living in contaminated communities suffer in similar ways to victims of more traditional traumas, their recovery should take a similar shape as well. However, the stages of recovery as presented by Herman presume that the trauma itself is over and that the person can get on with the work of recovery. In Three Lakes, the trauma is ongoing and, in some ways, never-ending. Due to an extreme lack of trust in water authorities, a high percentage of residents are unsure of the safety of their water. Even those who believe the water is safe live daily with the fear that former contamination will result in serious illness sometime in the future. This fear never ends. Safety can never be reestablished.

People in Three Lakes did formulate meanings of what happened and mourned the loss of community. However, they did this in very different ways, seeing what happened through the several conflicting frames discussed above. Rather than have recollection function therapeutically to help in recovery work, the construction of several sets of meanings resulted in social conflict and breakdown and secondary psychological trauma.

The third stage of recovery is reconnection, through which the victim reconnects with the social world. Once again, we find reconnection an especially difficult task in the aftermath of toxic contamination. Reconnection presumes a relatively stable social world with which a victim can reconnect. This was not the case in Three Lakes. Even though buildings remain standing, the community's social cohesion was torn apart, rendering the usual formal and informal support systems inadequate to deal with people's trauma.

More is implied by these observations than the difficulty individuals and communities have in recovering from chronic technological disasters. We think that the ongoing, dynamic nature of this process makes "recovery" itself a suspect concept. "Transformation," a change in form, nature or character of an individual and a community, is a term that seems better suited to describing and studying the reality of chronic technological disasters.

Survivors of such disasters cannot recover what was lost but must reformulate (transform) their selves and their lives given their relationship with contamination and experiences related to it. The effects of contamination and its related experiences will never end. The perpetrating problem cannot be put behind the victim, but must be a part of the victim's life forever. Reconnection not only must take place to a transformed social setting but must also be accomplished by a self forever altered by a dynamic ongoing relationship with the contamination and its experiences.

Transformation better captures the dynamic, evolving nature of survivors' adjustments to changing social and environmental structures and conditions. It allows for the fact that some people view their sufferings living through a chronic technological disaster as ending in a positive change in themselves, while others see the opposite. And it better captures the experience as expressed by chronic technological disaster survivors themselves, who so often state that their lives have been forever changed, or that thanks to the disaster, they will never be the same.

Arguably, transformation is a term that better describes a chronic technological disaster community's experience as well. Community structures and culture are forever altered, its history forever changed and shaped, by the experience of a chronic technological disaster. The community does not return to what it was before, but becomes something significantly different and new – if it survives at all. Recovery is a relatively static concept, implying a return to a state once possessed or inhabited. Transformation is dynamic, relational, emphasizing change, open to a dialectical interpretation of individual and social life, a type of interpretation arguably better able to capture not just the nature of life during extreme events, but everyday life as well (see e.g., Harvey, 1996).

Finally, we need to work through the practical implications of this conceptual approach. Models of individual and group recovery from trauma are built around a return to stability and normality after the source of the trauma has been removed. Even in chronic trauma such as abuse, the abuse must stop before recovery can proceed: "...no therapeutic work can possibly succeed if safety has not been adequately secured" (Herman, 1997, p. 159). With chronic technological disasters, recovery/transformation must often take place in the midst of ongoing danger, or at the very least, amidst the perception of it. Even when removed from the source of harm, contaminated individuals must live their lives with the fear that the contamination will cause health problems sometime in the future. And, as discussed above, all of this takes place within a social environment that, far from providing support, often is racked with destructive social conflict and the breakdown of the social community itself.

Assisting communities and individuals to deal with this type of problem requires imaginative, innovative models adjusted to meet these unique circumstances. These models must eschew the temptation to view contamination cases as primarily technical problems and must give ample credence to the social and psychological issues discussed above. While responding to technical issues, a focus on community building, conflict management and social as well as technical education, may help individuals and their communities hold together in the face of the centrifugal forces set in motion by the contamination. Involving the local community in decision-making processes and providing them with social as well as technical assistance promises to give the community some measure of power over its ultimate fate.

NOTES

1. Data for the following section were drawn from in-depth interviews, newspaper articles and court documents.

2. Respondents choose one forced-choice response for each of three questions. The response indicating higher distrust is scored "1," while the higher trust response is scored "0." Scores range from 0 to 3. A sample question is:

> "Generally speaking, would you say that most people can be trusted or that you can't be too careful in dealing with people?"

3. For each item, the scoring ranges from 1 to 4, with the higher score indicating higher environmental worry. The scale is a summary of scores for the eight items, with scale scores ranging from 8 to 32. Sample questions include:

> "The image of being poisoned often pops into my mind."

> "Toxic waste is a smaller problem than the media make it out to be."

4. Each item is scored "1" for the high anomie answer, and "0" for the low anomie response. The scale scores range from 0 to 5. Sample questions are:

> "It's hardly fair to bring children into the world with the way things look for the future."

> "There's little use writing to public officials because often they aren't really interested in the problems of the average man."

5. The scale score is the sum of responses to nine items. Higher scores indicate higher demoralization. Sample questions are:

> "Compared to others, my life is not too good."

> "I can't do much for other people."

6. A comparison of demographic data between the Three Lakes sample and census tract figures showed that the Three Lakes sample's demographic characteristics closely matched census tract demographics. The one major difference is that females constituted 61% of the Three Lakes sample while comprising only 49% of the census tract population. However, in the Three Lakes sample, females and males showed no significant differences on any scale scores and on questions of perceived exposure or health impact.

REFERENCES

Aronoff, M., & Gunter, V. (1992). It's hard to keep a good town down: Local recovery efforts in the aftermath of toxic contamination. *Industrial Crisis Quarterly*, *6*, 83–98.

Baum, A., & Fleming, R. (1993). Implications of psychological research on stress and technological accidents. *American Psychologist*, *48*, 665–672.

Baum, A., Gatchel, R. J., & Schaeffer, M. A. (1983). Emotional, behavioral, and psychological effects of chronic stress at Three Mile Island. *Journal of Consulting and Clinical Psychology*, *51*, 565–572.

Bowler, R., Mergler, D., Huel, G., & Cone, J. (1994). Psychological, psychosocial and psychophysiological sequelae in a community affected by a railroad chemical disaster. *Journal of Traumatic Stress*, *7*, 601–624.

Bowler, R., & Schwartzer, E. (1991). Environmental anxiety: Assessing emotional distress and concerns after toxic exposure. *Anxiety Research*, *4*, 167–180.

Capek, S. M. (1993). The environmental justice frame: A conceptual discussion and an application. *Social Problems*, *40*, 5–24.

Cole, H. S., & Stevens, M. A. (1996). *Learning from success: Health agency effort to improve community involvement in communities affected by hazardous waste sites.* U.S. Department of Health and Human Services, Public Health Service, Agency for Toxic Substances and Disease Registry.

Couch, S. R. (1999). Toxic water, toxic knowledge and social breakdown. Paper presented at the annual meeting of the American Sociological Association, Chicago.

Couch, S. R., & Kroll-Smith, J. S. (1985). The chronic technical disaster: Toward a social scientific perspective. *Social Science Quarterly*, *66*, 564–575.

Couch, S. R., Kroll-Smith, J. S., & Wilson, J. P. (1997). Toxic contamination and alienation: Community disorder and the individual. *Research in Community Sociology*, *7*, 95–115.

Edelstein, M. R. (1991). Ecological threats and spoiled identities: Radon gas and environmental stigma. In: S. R. Couch & J. S. Kroll-Smith (Eds), *Communities at risk: Collective responses to technological hazards* (pp. 205–226). New York: Peter Lang.

Edelstein, M. R. (2004). *Contaminated communities: Coping with residential toxic exposure* (2nd ed.). Boulder: Westview.

Erikson, K. (1976). *Everything in its path: Destruction of community in the Buffalo Creek Flood.* New York: Simon and Schuster.

Erikson, K. (1993). The view from East Swallow. *Research in Social Problems and Public Policy*, *5*, 23–46.

Erikson, K. (1994). *A new species of trouble: Explorations in disaster, trauma, and community.* New York: W.W. Norton.

Finsterbusch, K. (1989). Community responses to exposures to hazardous wastes. In: D. L. Peck (Ed.), *Psychosocial effects of hazardous toxic waste disposal on communities* (pp. 57–80). Springfield, IL: Charles C. Thomas.

Freudenburg, W. R. (1997). Contamination, corrosion and the social order: an overview. *Current Sociology*, *45*, 19–40.

Fritz, C. E. (1961). Disaster. In: R. Merton & R. Nisbet (Eds), *Contemporary social problems* (pp. 651–654). New York: Harcourt, Brace and World.

Goffman, E. (1963). *Stigma: Notes on the management of spoiled identities.* New York: Simon & Schuster.

Green, B., Grace, M., Lindy, J., Gleser, G., Leonard, A., & Kramer, T. (1990). Buffalo Creek survivors in the 2nd decade: Comparison with unexposed and nonlitigant groups. *Journal of Applied Social Psychology, 20*, 1033–1050.

Harvey, D. (1996). *Justice, nature and the geography of difference*. Cambridge, MA: Blackwell.

Herman, J. (1997). *Trauma and recovery: The aftermath of violence – from domestic abuse to political terror*. New York: Basic.

Kindler, J.D. (1996). *Evaluating a community discussion in a technological disaster: Using Antheil's comprehensive evaluation model as a conceptual guide*. PhD thesis, Temple University, Philadelphia, PA.

Kroll-Smith, J. S., & Couch, S. R. (1990). *The real disaster is above ground: A mine fire and social conflict*. Lexington: University Press of Kentucky.

Kroll-Smith, J. S., & Couch, S. R. (1993). Technological hazards: Social responses as traumatic stressors. In: J. P. Wilson & B. Raphael (Eds), *The international handbook of traumatic stress syndromes* (pp. 79–92). New York: Plenum.

Murphy, S., & Keating, J. (1995). Psychological assessment of postdisaster class action and personal injury litigants: A Case Study. *Journal of Traumatic Stress, 8*, 473–482.

National Opinion Research Center. (1994). *General social survey*. Chicago: University of Chicago.

Picou, J. S. (2000). The "talking circle" as sociological practice: Cultural transformation of chronic disaster impacts. *Sociological Practice, 2*, 77–97.

Quarantelli, E. (1985). What is disaster: The need for clarification in definition and conceptualization in research. In: B. J. Sowder (Ed.), *Disasters and mental health* (pp. 41–73). Washington, DC: US Department of Health and Human Services, National Institute of Mental Health.

Schuessler, K. (1982). *Measuring social life feelings*. San Francisco: Jossey-Bass.

Srole, L. (1956). Social integration and certain corollaries: An exploratory study. *American Sociological Review, 6*, 709–716.

PART 2:
CLOSED CITIES, SECRET ACCIDENTS AND SOCIAL ADAPTATIONS

SECTION 2: INTRODUCTION

In a seminal essay, Mumford (1963) argued that society is forced to adapt to the demands of the technological systems it chooses. He distinguished between what he called "Authoritarian" and "Democratic" technics. Societies choosing the former path select technological and energy systems that are high risk and therefore demand authoritarian, centralized, secretive and closed social controls. In contrast, the second low-risk path allows for decentralized, open, participatory and democratic control. The two directions, and thus the resulting social forms, could not be more contradictory.[1] Democratic technics, not Authoritarian, are necessary for a sustainable future.

This section is a study of authoritarian technics and their consequences. It focuses on the closed world created by the Cold War and the introduction of nuclear (and biological and chemical) weapons and support technology, as well as their civilian parallels. The USSR and the U.S. embarked on essentially parallel paths to pursue nuclear reactions first for military and then for civilian purposes. Here we examine the social consequences of this authoritarian and hard path.

In the first three chapters in the section, our focus is upon the closed world of Minatom, the Russian nuclear agency. We focus on one particular Closed City, Ozersk, and the consequences of the world's second worst nuclear accident, Kyshtym 57, the impacts on downwind and downstream residents of the region and the resulting social movement. The next two chapters examine the consequences of the world's worst nuclear accident to date, the Chernobyl accident. The final two chapters examine the issues of closed cities that become open, but where the toxic legacies and social norms linger, one in Russia and one in the U.S.

Chapter 7, Nadezda Kutepova and Olga Tsepilova's *Closed City, Open Disaster* takes the reader on a rare trip behind the fence to give us an insider's introduction to the ZATO, the Closed Administrative Territorial Establishment. These Closed Cities, omitted from maps during the Soviet period, housed scientists and workers employed by Russia's secret facilities. The focus here is on the Mayak military nuclear complex in the Chelyabinsk region of the Ural Mountains and one of its ZATO, the city of Ozersk, where Kutepova has lived her entire life. After an historic review of Mayak and the ecological crises it has caused, the authors focus upon the many civil rights violations for ZATO residents that stem both from exposures to environmental hazards and to the secretive and closed nature of the community. Although privileged in some respects, ZATO residents are denied even the basic level of freedom found throughout Russia and, for reasons that include the norms, social pressures and history of totalitarian control, ZATO populations are uncommonly submissive to these rights violations. As if to demonstrate the closed nature of Ozersk, the authors themselves faced threats from state security in continuing their ZATO research. Their fascinating discussion of the effort of ZATO to find their advantage in the new global economy as high technology centers reminds one of comparable efforts at U.S. nuclear facilities (see Edelstein this volume, Chapter 12).

Chapter 8 takes a different twist on the Mayak story. In *The Most Contaminated Place on Earth: Community Response to Long-Term Radiological Disaster in Russia's Southern Urals*, Natalia Mironova, Maria Tysiachniouk and Jonathan Reisman offer a thorough rendition of Mayak's history of environmental contamination. Of greatest fame is an event long regarded as the world's worst nuclear release, the Kyshtym 57 accident (named after a local village and the year of 1957), which contaminated a wide swath of the region. Releases from Mayak to the Techa River have been another major source of contamination. The consequences of these events has been to place residents of downstream and downwind villages at peril, to bring about weak protective efforts, and, over time, to force a massive resettlement of people. However, not all the at-risk population was removed. The chapter provides important insights into the mobilization of the remaining residents, most notably in the contaminated village of Muslumovo. An analysis of interesting adaptations by residents to the situation is offered, including the idea of "Toxic Tourism," the practice by some residents to earn money based upon international interest in their victimization.

Mironova and her colleagues go on to place these historic issues into the context of the new social movements and new democracy that arose during Perestroika and after Chernobyl. Mironova is in a unique position to describe these subsequent events with the authority of a key player in shaping them. The chapter describes the rise and fall of green politics through the brief democratic period after Perestroika. We are given a detailed view of the anti-nuclear movement that emerged during this period and continues under Mironova's leadership. This is no local grassroots effort focused on one event, but rather a national movement with ties to an international movement. Even the powerful movement that evolved, however, was unable to turn the tide on the most recent nuclear battle in Russia, over the planned importation of spent nuclear fuel for reprocessing, a move that may be also intended to kick start a new phase of the Russian nuclear industry and revitalize Mayak into the future.

Chapter 9 on the *Psycho-Social Consequences Due to Radioactive Contamination in the Techa River Region of Russia* offers a third perspective on the same places and events. With Maria Tysiachniouk, I examine those living at the epicenter of the Mayak facility's major nuclear disasters. Residents of the small village of Muslumovo remain there some 50 years later, but at what cost? The chapter provides an opportunity to apply psycho-social event analysis in a Russian context. It further examines Environmental Justice, a central construct in how U.S. contamination issues are now viewed, in this context.

Secrecy and information control were key issues in the aftermath of the Chernobyl disaster, disclosed not by the Russians, but by the detection of radioactivity as a cloud of radioactivity lofted over Sweden. In Chapter 10, *Environmental Risk Perception after Chernobyl*, Irina A. Zykova, provides a rare look at the April 26, 1986 Chernobyl accident through the eyes of a primary scientist who spent much of her subsequent career researching the disaster. Zykova emphasizes the impacts of psychological stress due to the accident, accenting the impacts of unseen contaminants. Delays in admitting that the accident had occurred and mapping its extent meant that residents exposed to radioactive iodine were denied the easy protection of iodine pills that would have blocked the body's uptake of the contaminant. Zykova reports on her extended participant observation of one of the affected regions. She describes how many affected populations were held in limbo for a decade while decisions were made about relocation. Zykova explains that efforts to make these people give up local foods failed because they could neither find nor afford alternatives, and besides they initially did not believe that there was a major problem. A half million Russians continue to live in contaminated areas; many continue to use local foods despite long lasting threats from radioactive Cesium. She goes on to compare perceptions of risk by lay people and experts and media coverage of radiation issues in Russia that shape understanding and response to otherwise invisible threats. Zykova takes the position that anxiety is a much greater contribution to health problems than was exposure to radiation. Addressing this anxiety would require much improved government communication. However, rather than learning lessons about communication from Chernobyl, she observes Russia continues to avoid open discourse about radioactive threat.

Our final focus in this section is upon Closed Cities that become open, raising the question of whether closed cultures become open, as well. In Chapter 11, *Habitual Risk Taking: Life in Dzerzhinsk—"the Capital of Soviet Chemistry,"* Alla Bolotova asks, "Why do Russians acquiesce to risk? Can they admit their mistakes after a disaster? Or must they look the other way?" She searched for an answer by plumbing the evolution of risk and risk-acceptance within the eco-history of Dzerzhinsk, a former Closed City. Once the prototype for the socialist "Radiant City," Dzerzhinsk is now described as the "dirtiest city in Russia." Guided by theorist Ulrich Beck in her explorations of this culture of secrecy and acceptance, Bolotova presents an insightful critique of the Soviet mode of industrialization. She explains people's acquiescence by examining a complex mix of possible contributing causes: the deep cultural roots of secrecy in a Closed City, peoples' desensitization to risk because of the well established culture of ignoring worker and environmental hazards, the cultural barriers to discussing problems and lack of public spaces where people can communicate, the failure of an informative media to develop, the lack of opportunities for direct political action, the pressure for quotas to be met and the development of secret investigative and repressive organizational systems. As a prototype of the Soviet image of "the Radiant City" envisioned under socialism, to admit Dzerzhinsk's problems is to tarnish its accomplishments, stigmatize its prospects and thus dim the radiance. But, Bolotova asks, perhaps one has to address these problems to create a radiant city.

In the final chapter of this section, Chapter 12, *Hanford, A Culture of Secrecy, Denial and Impact*, this writer reprises work done in the 1980s as part of the Washington State effort to understand the consequences should the Hanford Nuclear reservation be selected as the U.S. high-level nuclear waste repository. Hanford had long become open – but only to a degree. Information about the facility and the hazards it had caused were still classified as secret. The qualitative study contrasts the perspective of downwinders, those living in the impact zone of the complex, and residents of the nuclear community of Richland, glimpsing how their views of the world diverged. Research was done primarily before the U.S. government began to declassify files documenting contaminant releases from Hanford. So at the time of the study, perceptions of possible release events were not substantiated and, therefore, easily contested. This chapter provides a glimpse of a U.S. nuclear culture that in major ways parallels the Soviet culture of Ozersk and Mayak. The consequences are of a similar type as well, although not as dire.

In fact, the elements of uncontrolled environmental hazards, poor historical management of hazards, poor public disclosure of risks and creation of a separate and secret system for protecting toxic facilities are common elements in the U.S. and Russia. One cannot go so far as to claim that the situations are identical; they are not. However, they reflect common cultures of contamination. And they support the view that authoritarian technique shapes society in an authoritarian form.

NOTE

1. Lovins later characterized hard (nuclear) versus soft (solar) energy paths along similar lines (1977).

REFERENCES

Lovins, A. (1977). *Soft energy paths: Toward a durable peace*. New York, NY: Friends of the Earth.

Mumford, L. (1963). Authoritarian and democratic technics. *Technology and Culture*, 5(1), 1–8.

Michael R. Edelstein
Editor

CLOSED CITY, OPEN DISASTER

Nadezhda Kutepova and Olga Tsepilova

> Dear Daughter,
>
> By no means should you tell anybody where you are from. Otherwise you will come back home and will not find your father or mother. They will be taken by the Black Raven.[1] (Warning by the first author's mother).

INTRODUCTION

Russia represents one of the world's most dangerous ecological risk zones. Yet, the risk is not evenly distributed within the country. Certain areas of Russia show a disproportionately heavy concentration of pollution and present an even higher ecological risk than the country as a whole. Making a major contribution to the list of areas at greatest risk are the "ZATO." These unique "Closed Administrative Territorial Establishments" are Russian settlements or cities containing large nuclear enterprises; they are held in a state of extreme secrecy and security. Despite the end of the cold war, ZATO persist and their production activities continue to cause terrible damage to the environment and to the health of their residents. Additionally, the sites of closed former ZATO represent lasting ecological and health threats.

In this article, we summarize the establishment and history of the ZATO. We then provide an in-depth analysis of the ecological, socio-economic and human rights issues that surround one specific ZATO, the city of Ozersk, Chelyabinsk oblast. Ozersk has a population of 82,000 people. Typical of ZATO, Ozersk is organized around a series of nuclear production facilities.

Cultures of Contamination: Legacies of Pollution in Russia and the U.S.
Research in Social Problems and Public Policy, Volume 14, 147–164

ISSN: 0196-1152/doi:10.1016/S0196-1152(06)14007-7

A SHORT HISTORY OF THE ZATO

After the Second World War, the Soviet Union sought to develop its military-industrial complex in order to compete with the scientific, technical and military potential of the West. In the late 1940s, the Soviet government selected a series of sites in Russia for the creation of unique establishments for national defense intended for the industrial development of nuclear weapons. The industrial facilities included scientific complexes, engineering and construction bureaus, experimental plants and testing sites.

Initially these territories were referred to as "Mail Boxes" and later as "Closed Cities." The majority of the Mail Boxes appeared in the 1940s and the 1960s. The designation "ZATO" (i.e., Closed Administrative Territorial Establishment) came into use much later, at the end of the 1990s and well after the dissolution of the Soviet Union.

The locations of ZATO were selected for key geographic characteristics. They are situated deep within the territory of the country, out of reach of enemy warplanes. Existing civilian settlements populated by indigenous ethnic groups provided construction labor. The sites contained rivers and lakes, as large amounts of surface water are required to cool nuclear reactors and for other nuclear technology. These same water sources historically supported the agrarian livelihoods of the local population.

A third group of ZATO residents were drawn from the citizens evacuated during WWII from places where there were military battles. Practically all who arrived brought their families with them.

There were major restrictions for being accepted into the ZATO. Migration to the ZATO from outside was heavily regulated by the administration of the territory's nuclear enterprise. Admission depended on the needs of the enterprise, as well as certain family circumstances, such as marriage. The "trustworthiness" of potential workers and staff was assessed through an intense screening process conducted by the NKVD[2] (later renamed the KGB, and now called the FSB) (Evseev, 2003). Those who passed the screening were proven loyal and recognized as "reliable."

At the same time, departure from the ZATO was even more difficult. Migrating from the ZATO back into the country was prohibited from the very beginning. Later on, this action was allowed only in special cases, for instance, a young adult attending a university and not returning. In addition, a citizen might be forced to leave the ZATO for engaging in an "immoral lifestyle" or for committing a crime.

Currently there are 42 "Closed Cities" within the territory of Russia with an approximate total population of 1.5 million people. Approximately 75%

of these are administered by the Ministry of Defense and 25% by the Ministry of Nuclear Energy (Minatom). Some 64% of Russia's "closed population" lives in just 10 cities under Minatom. These 10 cities receive an extreme level of protection. They are surrounded by barbed-wire fences and defended by military personnel. They have stringent systems to limit access, including an application process and special permission for entry.

Each ZATO contains one small city and one or several towns or villages. In the majority of cases, the city was founded around one big nuclear enterprise, where most of the working citizens are employed. The immediate grounds of this enterprise are defended by an additional circle of defense and security. Smaller industrial enterprises, not defense related, are also likely to be found on the territory, including those involved in forestry.

Historically "Closed Cities" were created as places for those building nuclear weapons and their families to live. It was initially impossible to predict the number and type of problems to be caused by this model of enclosed communities. For example, it was not anticipated that little outposts would grow into big cities. Initially, the nuclear authorities of the area appointed by Moscow supervised the small towns (Gubariev, 2001; Lappo & Polyan, 1998). Later, given the increasing number of facilities and their growing communities, these management roles were separated. However, all the way through to the 1990s, during Perestroika, the leaders of both the city and the enterprise were appointed in Moscow. It was not until the beginning of democratic change in the later 1990s that the rules were modified and the position of the leader of the city became an elected one.

OZERSK: A TYPICAL NUCLEAR ZATO

In Chelyabinsk oblast there are three ZATO of the Federal Agency of Atomic Energy. They are Ozersk (formerly Chelyabinsk 65 and Chelyabinsk 40), Snezhinsk (formerly Chelyabinsk 70) and Trekhgorny (formerly Zlatoust 36). It is striking to observe these three cities. Despite their distance from each other, they look alike enough to be triplets, reflecting their social construction, the mentality of the population and their social problems. Our point is that Ozersk is a typical ZATO. How was it created?

On December 1, 1945, USSR Decree No. 3007-697 called for the construction of a factory for the production of plutonium. A site was subsequently chosen in Chelyabinsk province in an area of the Ural Mountains offering large amounts of surface water for reactor cooling. Under a top secret 1947 order by Chairman Stalin, facility number 859 was built with a special security regime; territory up to a radius of 25 km from the plant was protected with a passport system.[3]

Mayak is the Russian word for lighthouse. This name was selected for the new nuclear facility. From start up, "PO Mayak" manufactured nuclear armaments requiring uranium, plutonium and tritium. In the early 1950s, Mayak began producing weapons-grade plutonium. By 1976, the plant had further expanded into nuclear fuel reprocessing.[4] Eventually it would expand to become one of the biggest radiochemical production facilities in the world (Evseev, 2003, pp. 77–78).

Today, under the terms of the Nuclear Test Ban Treaty, all of Mayak's uranium–graphite reactors are shut down, although not decommissioned and dismantled (Larin, 2001). With only 40% of its aging equipment functioning, nuclear reprocessing operations are no longer profitable. The only economically viable industry remaining is nuclear waste disposal. Each year, some 150–200 metric tons of spent nuclear fuel arrive from Ukraine and Bulgaria (About Mayak, 2003a). Mayak is currently taking steps to expand its capacity to accept foreign spent nuclear fuel (About Mayak, 2003b).[5]

ECOLOGICAL PROBLEMS OF OZERSK

Table 1 lists current and closed operations at Mayak. Several conclusions can be stated upfront about the ecological threat represented by the several separate enterprises which make up Mayak today. First, one must conclude that the facility collectively represents the highest category of ecological threat and resulting insecurity. Secondly, this conclusion is reached despite the fact that current levels of pollution have decreased significantly compared to the Soviet period for the simple reason that the facilities are not working at full capacity. And third, contemporary concern must not only consider the potential for new disasters, but also incorporate the ecological legacy of past practices, and particularly the consequences of three radiological catastrophes which took place at Mayak. We will examine these historical crises before returning to current and planned operations.

Discharge of Radioactive Waste to Waterways (1949–1956)

The startup of the first reactor at Mayak, and its initial work to extract plutonium for Russia's first nuclear bomb, were set against the rush of the armament race (Larin, 2001). Deadlines were continuously shortened and

Table 1. The Mayak Facility (Kyshtimsky Okrug, p. 5).

Ozersk (Chelyabinsk-65)	"MAYAK" Production Plant
Types of Activity	• Former production of weapon plutonium • Production of weapon components • Recycling of spent fuel • Storage of fissionable materials • Testing production of MOX fuel elements • Tritium production
Nuclear Materials and Equipment	• Five disabled reactors for plutonium production, which were completely stopped between 1987 and 1991. • Two reactors to produce tritium based on highly enriched uranium fuel, which were also used to produce isotopes. • RT-1 facility to reprocess spent plutonium, including temporary and long-term storages for plutonium dioxide and highly enriched plutonium. • Plant no.1, which is a facility to oxidize and purify highly enriched uranium. • The "Granat" experimental facility to produce MOX fuel. • The "Komplex-300" facility to produce MOX fuel (under construction). • An industrial plant for the acidification and cleaning of highly enriched uranium. • Storage facility for temporary and long-term storage of fissionable materials (dioxide plutonium and highly enriched uranium) intended for reprocessing, extracted from dismantled nuclear weapons (under construction) • Stockpile of more than 1,000 kg of highly enriched uranium, and more than 1,000 kg of produced and enriched plutonium. Non-military recycling activities at the RT-1 facility have resulted in about 30 tons of plutonium dioxide with another ton of plutonium dioxide extracted annually from the spent fuel. • The contaminated waste ponds, including the TKV holding ponds, Lake Karachai, and another pond referred to as "Old Wetland."

corners cut (About Mayak, 2003c). The ecological threats of the nuclear cycle were ignored despite ample evidence of the dangers that radioactivity presents for living organisms. Under pressure from the Communist Party for immediate results, the Soviet nuclear physicists failed to create appropriate safety measures to protect the staff of the combine and the local population.

In December 1948, the first batch of radiated uranium went through the plutonium-239 extraction process. The extraction of plutonium employs large amounts of dangerous chemicals and compounds such as acids, salts and bases. In addition it needs large amounts of water. Water filled with radioactive and toxic substances was discharged into the environment. Continuous discharge of liquid radioactive waste into the Techa River thus began (Larin, 2001).

An especially large discharge to the Techa River involving more than 200,000 curie of radioactivity occurred in September 1951 (Evseev, 2003). Radioactive exposure resulted for the entire watershed of the Techa River (including Lake Kiziltash and the rivers Tobolo, Iset, Irtish and Ob), extending approximately 1000 km downstream and affecting a resident human population of approximately 124,000 people. Because radioactive dose decreased with distance from the Mayak site, the greatest impact to the ecosystem and human population occurred in the Techa River region of Chelyabinsk and Kurganskaya oblasts. Some 24,000 people received a harmful dose of radiation according to the medical standards in effect.

In reality, however, this one large spill was merely the apex of the crisis. The inability of Mayak to contain its radioactive waste was a chronic problem stretching from March 1950 through 1956. Ninety-five percent of this mostly liquid radioactive discharge entered the river system by November 1951. The discharge totaled about 76 million cubic meters of waste water with a total radioactivity of 2.75 million curie. The average daily discharge during this period was 4.3 curie. About one-quarter of this discharge consisted of long-living radio-nuclides, including Strontium-90 with a half life of 28 years and Cesium-136 with a half life of 30 years (Larin, 2001).[6]

High levels of radioactive pollution resulted in protective action by government. From 1952 until 1958, limits were placed upon swimming and use of soils within a 452 square-kilometer territory, including 38.5 square kilometers on the industrial grounds of PO Mayak. A "Sanitarian Protection Zone" was established within 195.7 square kilometers and a "Zone of Observation" for the remaining 217.9 square kilometers. While the relatively high levels of radioactive pollution were recognized as limits for human habitation, it is interesting to note that wildlife subsequently flourished in these areas given the absence of people.

Currently, the greatest threat comes from radioactive waste holding ponds called the TKV ("Techa Cascade of Water") situated within the Mayak sanitarian zone in Chelyabinsk oblast. It is feared that a major flood in the TKV would destroy the dam impounding the waste and separating it from the combined Techa-Iset-Tobol-Irtish-Ob watershed (Kyshtymsky Okrug, 2003). TKV is not in compliance with Russian laws governing ecological security. Radioactive wastes are not safely isolated from the environment and disposal of the wastes there violates rules for safe waste management.[7]

Of particular continuing concern in terms of human exposure have been 9,500 residents of four settlements along the Techa River within Chelyabinsk oblast. Since the mid-1950s, the use of Techa River water for drinking and agrarian activity has been prohibited. For several years radio-nuclides have been stable in the river at three of these settlements and residents are not considered to be actively in danger. However, the village of Muslyumova, situated in the middle current of the Techa River and closest to the TKV holding ponds, reveals 2.5 times the acceptable levels of Strontium-90 (5 bk/l) in tests of river water.

The 1957 Radioactive Discharge and the East Ural Radioactive Footprint

On September 29, 1957, 250 cubic meters of high-level radioactive waste exploded in a storage area at Mayak. The accident occurred when two system failures coincided. Highly radioactive liquid wastes generated at Mayak were collected and stored in special reservoirs made from concrete and steel, called "bottles." Because the contents of each bottle contained a self-heating solution of nitrates and acetate salts, they had to be cooled to remain stable. A simultaneous failure of the cooling and ventilation system and of the monitoring system occurred. As the liquids heated up, water evaporated from the solution, leaving the highly explosive sediment uncooled.

The explosion was estimated to be equivalent to 70–100 metric tons of TNT and to have released 20 million curie of radioactivity into the environment (Evseev, 2003).

However, secret official documents at the time acknowledged only a 2 million curie release. This recognized plume of two million curie was blown to a height of approximately 1,000 meters and carried by southeastern winds at a speed of more than 10 m/sec. The radioactive cloud traveled 100 km in the first 4 hours; by 10 or 11 hours fallout from the explosion had enveloped an area of some 23,000 square kilometers to the northeast of Mayak, subsequently known as the "Eastern-Ural Radioactive Footprint." This footprint covered a band 30–50 km wide and stretching some 350 km almost to the large Siberian city of Tumen.

The footprint covered parts of the oblasts of Chelyabinsk, Tumenski and Sverdlovski containing some 27,000 people living in 217 settlements. The region was primarily agricultural, consisting of a mix of wheat cultivation and animal husbandry (Evseev, 2003). Within this acknowledged region of exposure, a 2 curie/square kilometer limit was set as the maximum permissible level for human habitation.

However, 90% of the radioactive pollution (18 million curie) remained in the immediate vicinity of Mayak. There, the greatest threat from the disaster occurred for the residents of adjacent villages, staff of the combine Mayak, military construction personnel, prisoners and soldiers defending the plant. Fortunately, families of Mayak workers, comprising the 100,000 residents of the city of Ozersk, were spared this level of intense exposure.

Given the secrecy and the attempt to hide the scale of the disaster, the consequences of this accident at Mayak in 1957 have never been adequately assessed and continue to pose a real threat to the health and life of the people in the east Ural region of Russia.

Tragedy at Lake Karachai

From the onset, most of the liquid waste from Mayak's plutonium production has been released to a natural lake, Lake Karachai. Over time Lake Karachai accumulated 120 million curie of radioactivity. In the spring of 1967, a severe drought dried out the lake, exposing 45 hectares of its polluted banks to wind erosion that carried off some 600 curie of radioactivity (Evseev, 2003, p. 79). Currently, even as workers are burying Lake Karachai, however, the plant continues to discharge liquid radioactive waste into it.

Lake Karachai has come to be considered as one of the most significant sources of radioactive pollution in the southern Urals. Pollution continues there both as the wind carries aerosols up into the air and then deposits them in the upper level of soil and as polluted water seeps into the groundwater. The maximum zone of pollution at Lake Karachai extends outwards from the lake for a radius of 300–1,000 meters beyond the lake's edge (Glavolenko et al., 2003).

NEW THREATS

The city of Ozersk and adjacent settlements are located in a region of high ecological risk due to these past and continuing incidents. In addition, ecologically dangerous enterprises and pollution of the territory continues to bring new ecological risks.

The expansive industrial grounds and developed infrastructure of Mayak continue to bustle with activity, evidencing continuing plans for nuclear and research activities. The primary focus of current activities is the production of radioactive isotope products. These products are very dangerous and their production process very dirty. Mayak is a significant player on the world isotope market, producing about one-fifth of isotope products worldwide (About Mayak, 2003b).

Adding to the overall risk is the storage of waste materials from nuclear fuel. Russia decided to build a nuclear waste complex at Mayak as part of its effort to comply with the Strategic Arms Initiative to reduce and limit strategic nuclear weapons and end the use of some warhead systems all together (SNV-2). Anticipating the potential to one day reprocess weapons-grade plutonium into a type of fuel called MOX, Russia also wanted to stockpile plutonium (Wolfston, Dougherty, & Astred-Chuen, 2002). At the 41st General Session of the International Atomic Energy Agency in October 1997, Russia announced its plan to collect up to 500 metric tons of highly enriched uranium and up to 50 metric tons of plutonium from its nuclear program for this purpose. The storage of weapons-grade plutonium is to be done at Mayak in a facility funded by the United States under international control with the participation of the International Atomic Energy Agency.

While these developments underscore the urgency to liquidate nuclear weapons, they also suggest the attractiveness of Mayak as a project site already located in a territory overrun by ecologically hazardous enterprises. This site suitability is further aided by the already active import of foreign spent nuclear fuel from Soviet Pressurized Water Reactors (VVER-440) and the planned reconstruction of the RT-1 reprocessing plant at Mayak, which will add the ability to additionally reprocess spent nuclear fuel from western-style reactors (VVER-1000) (About Mayak, 2003b).

Finally, adding to this buzz of nuclear enterprise are plans to build the South Ural Nuclear Power Station within the industrial zone of Mayak. The plan has been postponed due to the growing anti-nuclear social movement in Chelyabinsk oblast (see Mironova et al., this volume).

These continuing operations and planned expansions suggest that Mayak threatens to be a further source of ecological pollution and high environmental risks. It is little wonder that the Eastern Ural State Sanitation zone is often described as a zone of ecological disaster.

CASE STUDY: THE EASTERN URAL STATE SANCTUARY[8]

One positive outcome of the East Ural Radioactive Footprint was the creation of the Eastern Ural State Sanctuary in 1966 (About Mayak, 2003a).

The purpose was to prevent the removal of radioactive substances from the territory of the Mayak factory and to prevent access by the public to the territory that became radioactively polluted during the Mayak disaster. In addition, the sanctuary would serve as a site for research on radio-nuclide behavior in natural conditions and an evaluation of the state of water and soil ecosystems under the influence of ionized radiation.

Mayak had formed an Experimental Research Investigation Station in 1958 to investigate the influence of ionizing radiation on natural objects, agricultural plants, animals, migration of radioactive isotopes in food chains and to develop methods for using polluted land for agriculture. By 1967, this research program had returned much of the least contaminated land to human use. The remaining 16,616 hectares of highly contaminated land was formally turned over to Mayak in 1968 to manage as the Eastern Ural State Sanctuary (Kyshtymsky Okrug, 2003).

Despite its unfortunate basis for designation, the Eastern Ural State Sanctuary has emerged as a vital resource for maintaining the biodiversity of the forest steppes beyond the Urals. Species diversity quickly surpassed that in a longer established sanctuary not located on contaminated land. Some 186 bird species have been identified, including four types included in the Redbook. The Eastern Ural State Sanctuary is the only nesting place known for these four species. Because it was withdrawn from plowing after the accident, the sanctuary emerged as an important location for studying the restoration of the steppe and grassland ecosystem of the region from agricultural impacts. The sanctuary has also emerged as a foremost location for the study of the self purification and restoration of the biosphere in places of radioactive pollution (Evseev, 2003).

This specially protected area is under the control of Mayak. In other words, it is administered by the Ministry of Atomic Energy, not the Ministry of Natural Resources. Due to secretiveness, the sanctuary is not listed by the state as a protected territory and the state plays no role in maintaining it. And, with the end of research funding in 1991, the entire sanctuary is cared for by only one employee, who also does fire suppression work. Annual spring fires occur in the sanctuary. In 1996 a fire in the sanctuary engulfed 25% of the landmass and was stopped only with difficulty.

The end of the research is also of concern given that a continuous record of data had been collected from 1957 through 1995 on radio-nuclides in the environment. There is now an irreparable loss of information on the fate of Strontium-90 and Cesium-137 in the soils and in reservoirs and ecosystems.

Of further concern, given the lack of a buffer zone around the sanctuary and the potential for migration of animals with radioactive contamination beyond the sanctuary, there is a probability that people may shoot contaminated animals and eat them. The annual limit for Strontium-90 can be reached by eating only 6–10 kg of meat from the contaminated zone. In 1988 a hunting lodge was located in the region. Subsequently, there was a significant decrease in animal populations within the sanctuary. Furthermore, hundreds of hectares of the sanctuary are now used to produce animal fodder. This activity was begun by the experiential station to do practical testing of agriculture in radioactive zones. But its continuation is now illegal.

Mayak's plans for the sanctuary focus less on nature protection than production. It sought to use 450 hectares as the site for the proposed South Ural Atom Station and another 450 hectares is logged. Both uses violate Federal laws on protected natural areas. Activists seek to have the sanctuary reassigned as a separate unit of the Russian Ministry of Natural Resources.

THE DECLINE AND RECOVERY OF THE ZATO

By the 1990s, the ZATO confronted the new economic reality of the post-Cold War and the Russian military industrial complex faced the closure of nuclear facilities and a general slowdown. It was a very hard situation. There was a sharp decline in production, high unemployment levels, a "dequalification" for many highly educated people whose skills were now rendered irrelevant and a sudden loss of scientific potential and culture in the Closed Cities. The quality of life was dramatically affected. Social programs originally created to assist those impacted by industrial hazards now had to address the problems of people facing sudden poverty. A brain drain occurred as the most qualified individuals migrated from the ZATO to big Russian cities and even abroad (Tikhonov, 1996, p. 48).

The economic situation was stabilized to some degree by the end of the 1990s because the ZATO was highly competitive in the unregulated market when it came to such ecologically risky enterprises as the reprocessing of nuclear waste. However, a little bit of stabilization cost a great deal of added ecological risk. Being part of a completely rigid state system of supply for so many years, the ZATO now appeared to lack the flexibility needed to participate in the politics of a market economy.

This period of slow stabilization bought time for a new strategy of economic revitalization to emerge. Experts working in the ZATO began to see their high degree of specialization as an asset, offering a unique potential for the Closed Cities to take the lead in making their regions and all of Russia into a modern global technological force. However, for this dream to be realized, the Closed Cities would have to become open, even if strategic sites within them remained under restriction. The potential emerged for the ZATO to break away from their historic ties to the nuclear industry and establish a new economy based upon ecologically clean businesses.

In establishing a role in this new open competitive economy, Ozersk and other ZATO have distinct advantages but also have some major liabilities. On the positive side, the population is highly educated and the city offers vast opportunities for further education and training. People are highly entrepreneurial. The society is stable with little crime. Unused or underutilized industrial facilities abound. And social infrastructure and communications are highly developed. On the negative side, the city is closed, is far from regional centers and has an extraordinary legacy of environmental and technological hazards. The population also has liabilities. It comes disproportionately from the public rather than the private sector. People suffer from the psychology of having had the state provide for their needs for free. Many residents reflect an aging population. And they tend to be both overspecialized and overqualified, on one hand, yet ill-prepared for the contemporary challenges on the other.

LIFE IN THE ZATO

What forges the key elements of a ZATO?

First, Closed Cities have existed under top-secret conditions since their inception. They house the nation's most hazardous facilities. Their actions have been historically destructive of the environment and surrounding population. And some of these activities are still conducted.

Second, ZATO were established by Minatom and Minoborona (the Ministry of Defense) under the leadership of security head Beria, under conditions of the highest state secrecy. Their location was not even shown on maps. Although such external secrecy has almost been eliminated, the population remains afraid of persecution by the FSB and other state security authorities for any indications of public activity or dissidence. Such authorities remain visible and powerful in ZATO. Soviet-era mechanisms to threaten people are still very much alive in minds of the people who live in the Closed Cities. This fear, combined with their isolation from the world, has made the inhabitants of the ZATO a population that is easily managed.

Third, Federal and local mass media have no access to Closed Cities. Internal media is minimally developed and usually controlled by the agencies that they belong to. Nearly all information flowing in and out of the ZATO is censored.

Fourth, the special mentality of Closed Cities mixes fear with the power that the ZATO holds over the rest of the world. The community is closed and intimidated, yet its members hold a conscious superiority arising from their ability to develop and produce weapons of mass destruction. The main ideological postulate of the ZATO population thus reads, "No one can threaten us, but instead, everyone should be afraid of us because we handle something terrible for the whole world."

Fifth, Closed Cities develop a system of "mutual cover-up" required to hide crimes that would threaten the stability of the collective ZATO's existence. The ZATO are associated with serious ecological crimes and violations of basic human rights and freedoms. Yet, a Closed City is protected by a complicated network of kinship and interpersonal relations. In case of an emergency or illegal situation, a person is less motivated to follow the law than the orders issued at work and the rules established by the local community. Violators of this system of protection face various forms of punishment, including ostracism. To become a social outcast in a closed community is the ultimate threat against individual noncompliance.

Sixth, Closed Cities offer few possibilities of outward migration and severe limitations on introducing ideas that differ from official ones. There is no plurality of thought cycled through the community.

Seventh, there is historically no concept of Civil Rights among residents of ZATO. Whatever rights might accord those on the outside are surrendered at the gate. This lack of a concept of Civil Rights cuts in two directions. Not only do people not expect to be accorded rights, but they also do not understand the criminality of violating the rights of others. They consider the status quo of their life to be the norm because they do not know another life, and they do not know the norms held by outsiders.

In sum, the Closed City is a paradoxical environment. On one hand, the ZATO are known throughout Russia as "the Chocolates" because their residents hold a privileged status. On the other hand, Russians generally consider ZATO residents to live a life in treacle and strict discipline.

In reality Closed Cities are communities of fear, where residents must contend with arbitrary rules and a life without rights. Residents have been subject to a mixture of threat and the promise of good money for their silence and their health. People eat well and today live a quality of life equal to other provincial cities, but live in fear and remain silent while they live in a regime of strict secret rigidity. They cannot be condemned because they were physically threatened. Yet, their acceptance of having no rights translates to tolerance for breaking the world's rules and conventions. The conclusion is that people from the ZATO have formed an outwardly flourishing but inwardly manipulated society where civic activities are suppressed and no democratic conditions exist. Such a society is potentially dangerous for the world, creating a condition with unpredictable consequences. The population lacks an ability to change its own conditions without support from the outside. Thus, every ZATO is a small-scaled Soviet Union.

A further result of this closed social system is evidence of extreme stress. Alcoholism and mental disorders are wide spread among employees of the nuclear facilities. Criminality levels in Closed Cities assigned to Minatom are much higher than those found on average in cities of similar population. The number of AIDS patients and drug addicts equal the number of infected people found in much larger Russian cities.

ISSUES WITH HUMAN RIGHTS IN ZATO

The closed nature of the ZATO is fundamentally in conflict with the demands for freedom and change evident in the new market. As a result, the Closed Cities are easily susceptible to human rights violations, rousing public fervor aimed at protecting and restoring threatened and lost rights. It may be of interest to examine some recent examples of human rights violations in ZATO.[9]

Failure to Disclose Past Exposures

Beyond the background issues of secrecy, archives documenting environmental and safety violations at Mayak have been destroyed. As a result, possible victims of exposure lack information about the hazards of industrial toxins and radioactive exposures potentially affecting their health. Their lack of proof of exposure means that they do not qualify for a social status necessary to receive compensation.

Illegal Organ and Tissue Harvesting from Nuclear Workers

Special medical records of current and former staff of the nuclear enterprises or people working in hazardous conditions are kept. Such procedures include taking and monitoring of blood and tissue samples and collecting of organs from these workers throughout their lives and even after death.

People are never informed as to why these procedures take place, they are not asked for consent and results are are not told to them.

Misrepresentation of Birth Place

Reflecting the particular regime of secrecy that governed the Closed City, citizens born in ZATO prior to the 1990s are incorrectly listed on their birth certificates as having been born in an open city. This is particularly alienating to Russians. Place of birth has particular significance in Russian culture because of the close attachment to it, as well as the associated rights that relate to the birthplace.

Limitations of the Freedom to Chose Where to Live

While mobility in the USSR and even in post-Soviet society is substantially limited for all citizens, ZATO residents have particularly limited rights of choice about where to live in large part because it is very hard to get out of the Closed City. Promised subsidies to residents wishing to relocate fail to cover even a tenth of the real cost of an apartment in other open regions of the country, so even a permitted move may be unaffordable.

Limited Legal Rights

Russians have a right to select a lawyer if they must appear in court. However, ZATO residents are forced to choose among the relatively few lawyers living in their territory. These lawyers tend not to be the cream of the crop; moreover they are burdened with heavy loads of 200 or more cases. A resident can petition for counsel from outside the territory. Even if given permission, however, the lawyer is so delayed getting permission to enter the ZATO that there is no time to prepare a defense. These problems of access to legal help particularly affect those under 18 who are entitled to free representation but too often never receive it.

Unjust Denial of Entry

An outsider seeking entry to a ZATO must be invited by an individual or organization and must apply for permission to a special commission having the right to accept or deny the application. In Ozersk, for example, this commission includes the administration and lawyers of Mayak. They review applications, discuss them and make decisions about entry petitions with no obligation to explain or document their decisions or even to inform the applicant.

The consequence of such persistent violations and the lack of recourse is that, against a generally passive national population, citizens of ZATO are particularly passive in protecting their own rights. People do not believe in the possibility of living according to the laws that exist for all other citizens of Russia. As Closed City residents, they view their rights as forfeit.

Another cultural norm evident in the ZATO involves workers at the nuclear enterprises. The violation of environmental and safety laws there is so frequent and routine that it ceases to be viewed as a violation. It is taken for granted. It is possible to speak of this normative system as a "culture of hiding technological violations which cause ecological violations." The importance of this culture is magnified by the imperfections of the aging enterprises and the ease of making errors in technological processes which incrementally cause only small environmental impacts. Any responsibility felt by workers and decision makers is overwhelmed by the overall sense of power and lawlessness and the lack of punishment.

The public does not reject this culture of irresponsibility. In fact, too often, both the public and the judges punish the whistleblowers rather than those acting improperly. The exercise of democracy within a ZATO continues to be significantly limited by the interests of Minatom and the particular nuclear enterprises. Not only do they ignore the interests and rights of the larger population, but also manipulate the power structure to sanction citizens and limit their rights.

CONCLUSION

Closed nuclear cities in Russia continue to be zones of unsolved ecological and socio-cultural problems. As these territories engage in the new economic relations based on the market that characterize contemporary Russia, the potential for ecological harm expands rather than being addressed. Raising stakes through the permanent expansion of technological risks in ZATO threatens not only the ecological security of Russia, but also the whole of humanity. There may be no more urgent locus for the establishment of social-ecological rights and for redirecting economic activity towards sustainable development. These topics remain our interest for continuing research.

NOTES

1. These words were used by the first author's mother to warn her as she was leaving her Closed City to visit relatives on the outside. During the reign of terror conducted by Internal Security Chief Beria in the 1950s, the "Black Raven" was a euphemism for the black cars used by the security service to take people into custody. Many never returned.

2. NKVD – the People's Commissariat of Internal Affairs, KGB – Committee of State Security.

3. That is, a special document was issued to residents and workers necessary for entry and departure from the ZATO.

4. Initial reprocessing of the radiated nuclear fuel RT-1 was expanded to reprocessing fuel from reactor types VVER-440, breeder reactors BN-350 and BN-600, and composite fuel from the reactors of nuclear submarines, research reactors, and energy-producing facilities (Larin, 2001).

5. Mayak currently accepts nuclear wastes from reactors of the type VVER-440. To expand, it is rebuilding its RT-1 reprocessing capacity so that it can also reprocess spent nuclear fuel from reactors of the type VVER-1000. VVER reactors are Soviet Pressurized Water reactors. VVER-440 series reactors have six reactor coolant loops, most featuring a confinement design employing accident localization vaults within which leaks can be isolated and contained. In contrast, VVER-1000 reactors have four coolant loops and feature western style containment (http://www.nucleartourist.com/).

6. The waste water was composed of the following radioactive substances in these approximate percentages: 9% strontium-89, 12% strontium-90, 12% cesium-136, 27% isotopes of "rare earth elements", 14% zirconium-95 and niobium-95 and 26% ruthenium-103 and ruthenium-106 (About Mayak, 2003c, p. 7).

7. The "Joint Convention on Security and Secure Management of Spent Nuclear Fuel and Secure Management of Radioactive Waste" and the "Principals of How to Deal with Radioactive Waste" (International Atomic Energy Agency, 2002)

8. This case study was developed as the result of an investigation done in Ozersk by the public organization Planet of Hope with which the lead author is affiliated.

9. The examples given below coordinate with the Convention on Human Rights and other basic documents on human rights.

REFERENCES

About Mayak. (2003a). Nuclear Turnover, 163 (36), September 12, p. 2.

About Mayak. (2003b). Nuclear Shield of Russia. 151(24), p. 2.

About Mayak. (2003c). Pioneer Builders. 159(32), August 15, p. 7.

Evseev, B. (2003). The two-faced light of Mayak. *Ural Federal Okrug*(1), 78–80.

Glavolenko, U. V., Drozhko, E. G., Mokros, Y. G., Stukalov, P. M., Ivanov, I. A., & Alexachin, A. I. (2003). Contemporary conditions and how to avoid exploitation of water bodies with middle-activity waste: Lake Karachai and the Old Wetland. *Issues of Radiological Security*, *1*(29), 14–20.

Gubariev, V. (2001). White arkhipelag. *Literaturnaya gazeta*, *46*(5855), 14–20.

International Atomic Energy Agency. (2002) *About maintenance of ecological security of Techa cascade of water and basins of the watershed Techa-Isit-Tobol-Irtish-Ob*. Roundtable Report. Committee of State Duma on Ecology, Moscow, November 28.

Kyshtymsky Okrug. (2003). 63–64 (9–10), July, p. 22.

Kyshtimsky Okrug. (2003). Yesterday, Today, Tomorrow. October 28, p. 5.

Lappo, G., & Polyan, P. (1998). *Magazin, Sociological Research* (Vol. 2, pp. 46–48). Moscow.

Larin, V. (2001) Combine Mayak: Problem of the Century. Moscow, pp. 16–69.

Tikhonov, V. (1996). *Magazin, Problems of Prognosis* (Vol. 6, pp. 77–96).

Wolfston, D., Dougherty, E., & Astred-Chuen, K. (2002). *Nuclear Report* (Vol. 6, pp. 112–114). Moscow-Carnegie Center.

THE MOST CONTAMINATED PLACE ON EARTH: COMMUNITY RESPONSE TO LONG-TERM RADIOLOGICAL DISASTER IN RUSSIA'S SOUTHERN URALS

Natalia Mironova, Maria Tysiachniouk and Jonathan Reisman

INTRODUCTION

In this chapter we will look at some of the consequences of a sequence of nuclear disasters that occurred in the Southern Ural region of Russia beginning in the 1940s. Drawing upon the historical record, we document the steady increase in radiological contamination that resulted from a combination of accidents and a nuclear naivete that took nearly 60 years to outgrow. We will then analyze the dynamics of response to this contamination and health catastrophe. We will look at the population's reaction over the years, as well as the government's policy, or lack thereof, toward containing pollution, improving safety management, and protecting the health and environmental rights of the region's citizens. We will also compare the coping mechanisms of two different Russian cultures – that under the Soviet regime and that after perestroika – as a young democracy. Finally, we will examine

Cultures of Contamination: Legacies of Pollution in Russia and the U.S.
Research in Social Problems and Public Policy, Volume 14, 165–183

ISSN: 0196-1152/doi:10.1016/S0196-1152(06)14008-9

the effects of social movements and community action, issues of community conflict, and the phenomenon of ecodisaster tourism.[1]

A LAND OF LAKES

The extraordinarily scenic lake region of Chelabynsk Oblast is located in the southern Ural Mountains just at the continental divide between Europe and Asia. The region's major river, the Techa, is a tributary of the Ob River, its waters eventually flowing north into the Arctic Ocean.

A number of small villages formed along the Techa River in early times and are populated by ethnic minorities, mostly Tartar and Bashkir. Historically, the region was an early center for metalworking. The villages of Kasli and Kyshtim were organized around metallurgical facilities created in the 17th and 18th centuries under Peter the Great. Later, in the Soviet times, Kasli hosted industrial facilities producing radio and military equipment. Kyshtym added a military automobile factory. A third village, Muslumovo, consists of a very old settlement inhabited largely by Tatar people, as well as a newer section, inhabited mostly by Russians and better educated local people, which was built to serve the Trans-Siberian railroad that came through in Soviet times. Despite their industrial foundations, these villages had remained largely rural. Most of the population lived subsistence lifestyles. People raised animals, fished in the river, kept gardens, and gathered mushrooms and berries all summer. These populations were particularly vulnerable to the later radioactive pollution of the water and soil due to their direct dependence on the land.

The region's capitol, the city of Chelyabinsk, holds slightly less than half the oblast's population. Chelyabinsk was heavily industrialized by Stalin. The huge steel and arms plants and a major tractor factory were developed here because of the region's copious natural resources and distance from likely invaders. During the Second World War, the city's production of half of Soviet tanks earned it the nickname "Tankograd" (Whiteley, 1999).

THE USSR'S NUCLEAR PROGRAM COMES TO CHELYABINSK OBLAST

The USSR began its rushed entry into the field of nuclear development in 1944, eager to contend for creation of the world's first atomic bomb and address the eventual nuclear armament of the country.[2] The remote

Chelyabinsk region of the southern Urals was selected as an ideal center for nuclear weapons' development for several primary reasons. Because the oblast had been a center of metallurgy since the late 17th century and had become a major industrial center during the war, it already possessed a highly developed industrial infrastructure. The region's many lakes provided a perfect cooling source for nuclear reactors. And the local population of Muslim pastoralists from Central Russia provided a ready source of labor. The nuclear plant Mayak opened in Chelyabinsk Oblast in December 1948 as a production center for weapons' grade plutonium.

To facilitate the development of its military reactor facilities, the Communist party built "closed cities." These were "fenced-in" communities constructed near large nuclear plants for the sole purpose of housing plant scientists and workers, their families, and required service personnel (see also Kutepola and Tsepilova, this volume). The Ministry of Nuclear Energy (MinAtom) and the KGB (State Security Committee) administered these cities from distant Moscow.

Life in closed cities had its drawbacks. Long and drawn-out bureaucratic procedures and interrogation was required for everyday operations, including the entry and departure of citizens. Populations of closed cities were additionally severely limited in speech, movement, and rights to congregate. Because the nuclear project in Chelybanisk Oblast was a state secret until the glasnost period, the closed cities were not even shown on maps.

As compensation, workers were paid top salaries and given ready access to luxuries. As a result, the quality of life in the closed nuclear cities under the Soviet regime was reportedly even better than in the USSR's major cities, notably St. Petersburg and Moscow. Lifestyles were so comfortable for their workers that the closed nuclear cities were widely nicknamed the "Chocolates."

For the Mayak plant, the two "chocolates" Ozersk and Snezhinsk housed, respectively, workers at the plant and scientists conducting nuclear science and weapons research in nearby laboratories. The closed cities were located to the west and upstream of the Mayak facility. Mayak, in turn, drains to the Techa River as it flows east. Thus, while the Chocolates were relatively protected from the environmental consequences of the facility, the traditional villages of the region were directly in harm's way.

A SERIES OF NUCLEAR DISASTERS

From its beginning, the Mayak plant became a source of radioactive contamination for a large portion of the surrounding territories. Once mined, uranium metal is transformed there into plutonium. The process of dissolving and separating the various resulting transuranic elements results in a highly radioactive liquid byproduct. For its first two years of operation, Mayak discharged these byproducts directly into the nearby Techa River. Under the prevailing state-of-the-art of the time, reflected also in the practices at Hanford, Washington, dilution was the acceptable solution for pollution. Nevertheless, the riparian zones became extremely contaminated, adversely affecting the 8,000 people living downstream in villages directly on the Techa River, as well as another 20,000 living within the contaminated region.

By 1951, the water and banks of the Techa River were contaminated with extremely high levels of radioactivity, and the population living on the banks was becoming increasingly ill. As a result, a scientific expedition was sent to research the situation in the region and determined that the accumulation of radioactive nuclides in soil and organic tissue was, in fact, dangerous. They concluded that the water resources of the Techa River were ruined and were now totally unusable for drinking, bathing, or irrigation. The Mayak facility subsequently altered its system of liquid nuclear waste management, burying high-level waste in barrels, dumping middle-level liquid waste into Lake Karachai and other swamps in an attempt to use the water filtration system of natural wetlands, and continuing to release low-level waste directly into the Techa River.

The following decade saw many accidents and technical blunders at the Mayak plant, and pollution levels climbed. From 1952–1955, the site of the nuclear plant was contaminated on several occasions. The cause of these incidents was frequently the result of worker error resulting from a lack of technical training. Involved workers received immediate doses of extraordinarily high levels of radiation. Clean-up crews comprised of soldiers and prisoners were also put at risk.

The combined disasters also resulted in the relocation of residents from nearly all the small villages along the Techa River downstream from Mayak. Between 1952 and 1956, all 13 villages upstream of Muslymovo along the Techa and four villages downstream were permanently evacuated and their residents relocated (Whiteley, 1999).

These chronic problems were overshadowed by the acute event often referred to as "the Kyshtym disaster" that represented the world's worst nuclear release until the 1986 Chernobyl disaster. On September 29, 1957, a failure in the Mayak plant's cooling system allowed a tank of liquid radioactive waste to reach critical mass and temperature, resulting in a nuclear reaction. The detailed effects of this event remain unclear to this day. We do know that the resulting explosion blew into the regional atmosphere 70–80 tons of high-level nuclear waste containing 20 million curie of radioactivity (compared to 50 million curie at Chernobyl, three decades later). A territory of 23,000 square kilometers was contaminated, mainly with strontium-90. It has been estimated that as many as 373,000 people occupied the area receiving significant radiation. As many as 11,000 inhabitants were eventually evacuated, although most were not moved until long after the blast, some waiting almost two years for relocation. MinAtom continues to claim that only a small percentage of the radioactivity actually escaped the site of Mayak (see Medvedev, 1979; Whiteley, 1999).

Despite the soaring levels of radioactive pollution in Chelyabinsk Oblast, in 1964 MinAtom concluded that the use of Lake Karachai as a treatment lagoon was an acceptable solution for the site and ended research into waste management. However, three years later, yet another incident occurred; 600 curie of mostly cesium-137 was released from Lake Karachai in the form of wind-borne aerosols. According to the government, a dry summer left the banks of Karachai dry and exposed, allowing a subsequent tornado to lift radionuclide-impregnated dust into the air. The government, accordingly, called the release a naturally-caused event. Skeptics reject this characterization.[3]

These past events have produced a present-day territory with extremely high levels of radioactive pollution and widespread health problems. The territory of Chelyabinsk Oblast now contains high amounts of strontium-90, cesium-137, and plutonium-238, plutonium-239 and plutonium-240, as well as various other pollutants produced during the breakdown of these isotopes. In many areas, the amounts of radioactivity per square kilometer exceed the levels under Russian law that would trigger an emergency ecological situation. Long-lasting radionuclides continue to accumulate in the bodies of the oblast's 3.2 million citizens. In Ozersk, the closed city serving the Mayak plant, blood levels are 46 times the normal, global amount.[4] Common disorders involve those of the nervous system, digestive tract, respiratory system, heart, and blood.

THE CULTURE OF KEEPING QUIET

An important characteristic of this event is that the population remained virtually unaware of the cumulative disaster. News of the Kyshtym disaster first leaked out in 1976 due to the efforts of a dissident biologist, Zhores Medvedev, but the government did not admit to the accident until perestroika in 1989, after which it became widely known.

Although radiation affects the growth and health of organic tissue, it is invisible, has no smell, and no taste. Likewise, the release events were largely unobservable. Even the 1957 explosion, which occurred at 4:00 p.m., was obscured by the daylight. Villagers living along the Techa at the time subsequently reported seeing and hearing nothing of the explosion or its aftereffects, although one recalled seeing the northern lights, rarely visible in Chelyabinsk Oblast.

Anomalies began to occur as radionuclides flowed through the Techa and other rivers and built up in the soil. These symptoms of disaster, however, were few and tempered by seemingly beneficial changes in the environment. On one hand, human health declined and occasional sightings of fish with skin disorders or no eyes were reported. On the other hand, the intense radiation produced abnormally large mushrooms and berries for locals to gather. Forests were greener and more luscious. Although later these abnormalities might be cited as clues of contamination, nothing at the time exposed the truth of the disaster for the communities surrounding the Mayak plant.

To the extent that people were suspicious that something was wrong, there were strong reasons why they may have remained quiet and refrained from activism or questioning. The majority of Mayak's radioactive emissions took place in the 1950s when Stalin's recent reign of repression was firmly planted in people's minds. These repressions prevented open forms of criticism toward government policy. People had not yet regained the courage to complain or speak out.

Of similar effect was the widespread Soviet mentality of heroism. During the Second World War, the government of USSR used the mass media to establish the idea that sacrifice and deprivation should be tolerated for the good of the "motherland." One well-known Soviet film from the 1960s entitled "Nine Days of One Year" depicted a loyal nuclear scientist risking the loss of his family and exposure to lethal doses of radiation in order to further his experiments and science for national interest. The furtherance of Soviet power and the nation's stability was portrayed as an end that justified all means.[5] Such "nuclear sacrificing" propaganda combined with a strong system of secrecy enforced by the KGB and strengthened by Stalin's legacy of state security was used to fuel a cover up of the Mayak releases. When ill people from the Techa River communities visited doctors, diagnoses and data on illness and mortality were kept secret and vague so that no one would suspect radiation as the culprit.

A further factor in the affected region influenced perception of the disaster. The predominantly rural Tatar and Bashkir populations of the region are devout Muslims, whose faith pointed to the workings of divine will, not the incompetent MinAtom, for the cause of suffering. Many Muslim women in the region saw the high mortality and disease rates as punishment for their sins on earth. Rather than questioning or reacting to the unfolding disaster, local villagers referred to the heightened illness simply as "river disease" and took no action whatsoever.

THE GOVERNMENT'S REMEDIES

Until perestroika, remediation of the continuing disasters at Mayak was an internal concern within MinAtom. The response of the government was basically a series of band aid solutions that largely failed, coupled with total and complete secrecy. The plant's scientists tested the Techa's water in 1951, after two years of direct dumping of all liquid waste from the plant, and found the waters unsuitable for public use. That year, the government discussed a resettlement program for residents living on the river water. In place of this protective response, however, Mayak officials substituted an ineffectual plan to deliver clean drinking water to the villages. It would be several years before resettlement again resurfaced as a response.

While Mayak altered its operations to lower contamination in 1951, real protective measures for downstream residents did not commence until 1954, when a decree was issued for the restoration of over 60 existing wells in the region and the building of over 100 new wells. At the same time, representatives of MinAtom and the Ministry of Health met with household heads from many villages along the Techa and had them sign papers promising to stop all use of the river water, using the wells instead (Akleyev & Kisselyov, 2000). This decree was the beginning of a series of government restrictions issued over the next four years to protect citizens from the contaminated water. In June of that year, the government prohibited use of the Techa River's water for drinking, fishing, and agricultural needs. In February 1955 another decree applied the same restrictions to the nearby river Iset. Later that year a decree completely changed the status of the Techa River, officially removing it from all public use and contact. In 1957 the Techa River was fenced along its most polluted stretch, and the government issued an order to build bridges over it to avoid human contact with the water.

What is striking in retrospect is that these steps were taken in the absence of any information or explanation being given to residents of the region. In no small part, the void of information undermined the effectiveness of the mitigations. Residents had no appreciation of the grave basis for the restrictions. The government's intentions were unclear and at times contradictory. For example, in 1951, the government founded a settlement of geologists with a population of 1,600, across the Techa River from Muslumovo. Subsequent prohibitions on river use made little sense in light of this new settlement. Things could not be so bad if a new settlement was just built there. With the help of such contradictions and being uninformed, villagers continued to swim and fish in the Techa River and, although less than before, they continued using the river waters for irrigation.

Given the climate of secrecy, in which only the KGB and MinAtom knew about the radioactive releases, government policy was vague, contradictory and, therefore, ineffective. For example, agricultural officials continued to promote food production and sales from the contaminated area. As a result, throughout the 1950s, Muslumovo and other villages along the river continued selling their products in Chelyabinsk city markets. In 1953, as a case in point, 250,000 liters of milk, 1,400 kilograms of butter and cream, 30,000 kilograms of meat, and 300,000 kilograms of vegetables were marketed. As a consequence, contaminants were allowed to spread across the oblast.

In 1958, the government established sanitarian-protection zones around several rivers and lakes in the region. The new regime also hired security guards from the local population to guard the polluted waters and restrict access. This system also proved to be ineffective in protecting local people because the guards, being uninformed, did not take their work seriously. For example, in interviews we conducted, we had been told by residents that children were allowed to continue to swim in the Techa and other restrictions were not rigorously upheld. The guards were unwittingly put at risk, as well. The father of one of our Muslumovo respondents was a river guard. He died 10 years after assuming the position from "chronic radiation disease." This imperfect protective regime stayed in place for the next 15 years. A government reassessment of the situation in 1973 revealed that wells had deteriorated and become polluted. Moreover, the production of contaminated food was increasing.

Around this time, the government also reinstated the resettlement program, which ran a slow and unsteady course over the next two decades. In 1957, a plan to resettle the occupants of 18 villages along the Techa River was scaled back by scientists from the Mayak enterprise. Some contemporary observers believe that residents were kept in their villages as unwitting subjects in a study on the effects of radioactive exposure. By March 1957, only six villages had been evacuated by the time the significant explosion and radioactive release occurred. Thereafter, government resettled residents from the fallout zone. Another wave of resettlement occurred after the 1967 accident. By the mid-1970s, some 217 villages had been resettled from the region. However, some villages, notably Muslumovo, remained along the Techa. Although closer to Mayak than many resettled villages, Kasli and Kyshtim were not evacuated because government data showed them to be outside the plume of radioactive contaminants released in the 1957 and 1967 incidents.

The remaining residents continued to live unprotected from radioactive pollution. Their fate is suggested by the Muslumovo family of the deceased river security guard noted earlier. After her husband, the guard, died, the mother was given a job so she could care for her seven children. Put in charge of water sampling on the Techa River and lacking information and protective clothing, her job was to collect water, stones, and soil from the bank of the river for testing. To accomplish this task, she had her children help. For lack of a better place, they stored the samples underneath their beds throughout the winter. Five out of these seven children grew up to be sterile adults.

PERESTROIKA AND THE REBIRTH OF SOCIAL MOVEMENTS IN RUSSIA

In the late-1980s, the social and political structure of Russia and the USSR experienced profound changes. Gorbachev initiated perestroika ("reconstruction") and glasnost (free speech) in an attempt to open up the tightly closed and controlled society, while at the same time maintaining Soviet power. Finally afforded the right for social mobilization, people throughout the country began meeting and criticizing the government. New movements fed off of concern over nuclear energy. This social action was soon legitimized by the Chernobyl disaster. The People's Front, a large organization that began to function in Chelyabinsk city in 1987, brought about an eclectic gathering of activists interested in various facets of democracy.

In 1987, the government opened Mayak's records to the public, including information on the 1957 explosion. Citizens of Chelyabinsk Oblast learned of the disasters of the previous four decades for the first time. Information on the health hazards of Chernobyl began appearing in medical journals. Simultaneously, well-known Russian dissident (Medvedev, 1979) published detailed accounts of the Mayak disasters of 1957 (henceforth called "Kyshtim-57") and 1967.

Subsequently, in 1988, mobilized citizens in the city of Chelyabinsk from the amorphous People's Front coalesced into an active anti-nuclear movement, the Movement for Nuclear Safety. The catalyst was public disclosure of a proposal to build a new nuclear facility called the South Ural Nuclear Power Plant, secretly planned for years inside MinAtom. The resulting controversy about this project almost immediately drew attention to the region's existing health and environmental problems. The Movement for Nuclear Safety opposed the new plant, its protests citing the prior contamination. The Movement went on to demand compensation for victims of the Soviet nuclear program. People from contaminated villages along the Techa also participated in these campaigns. For the first time, radiation victims were given voice and were widely heard.

This campaign prompted the heads of Mayak and MinAtom to release information about the region's nuclear history and its present-day consequences that contradicted the anti-nuclear movement's findings and account. At the same time, the opening of democracy during perestroika not only liberated and empowered the public, but it also freed local heads of industry from the top-down system of the past to become advocates for new industrial directions. The industrialists coalesced as advocates for the South Ural Nuclear Power Plant.

The period between 1990 and 1992 saw dramatic changes in Russian politics. It was a time of political debate. Boris Yeltsin, as the chairman of the Supreme Council,[6] promoted liberal democratic reforms, while Gorbachev, as president of the USSR, pushed for a new socialism. At least rhetorically, government was oriented towards change and solving the problems created by past Soviet times (Mironova, 2005). In March 1990, a quintessential set of elections were held for federal and local Councils of People's Deputies. More than 2,000 rallies and marches occurred across Russia that year, with some 7.5 million participants. The elections brought democratic and liberal deputies into government at all levels, resulting in the birth of political pluralism in Russia. This was an unparalleled context favorable to Non-Governmental Organization (NGO) Activity.

Government policy continued to warm up environmental and health concerns during this period. In April 1990, MinAtom and the Ministry of Health pulled back some of the veil of secrecy regarding the Mayak enterprise's environmental and health impacts. On the federal level, new environmental legislation was passed, including the law "On Environmental Protection" that prohibited permanent disposal of foreign spent nuclear fuel in the territory of Russia. Furthermore, the government created and financed a rehabilitation plan for the entire Ural industrial region, and Gorbachev ordered a comprehensive investigation into the conditions of Chelyabinsk Oblast, resulting in a four-volume report. The Movement for Nuclear Safety was at the forefront of this research effort during the period 1990–1992, working closely with the deputies of the Supreme Council of the USSR for its formulation. A list of over 86,000 local victims of radioactive pollution was compiled and presented to the government.

Several activists from the Democratic People's Front were elected to the Council of People's Deputies on both the regional (oblast) level and for the city of Chelyabinsk. The Democratic Party of Greens was established at this time and received support from those in power. As a result, Natalia Mironova from the Movement for Nuclear Safety was invited to chair a new Commission on Radiological Security under the oblast administration. This action in Chelyabinsk marked a major policy departure for Russia as a whole. Until this point, radiological security was the exclusive province of the federal government. The creation of an oblast commission thus signaled a favorable localization in Russia's political system.

The oblast commission allowed the People's Front activists to participate in the expert assessment of the Ministry of Economy's plan to build the proposed South Ural Nuclear Power Plant. There were several significant outcomes. One was the 1991 publication of an ecological assessment of the plant's construction and operation by the Movement for Nuclear Safety. Under the egis of the commission, the movement held an international conference entitled "The Consequences of the Military – Nuclear Facility in the Urals." The conference included the participation of Mayak representatives and 500 representatives of nine countries, including 50 NGO activists. Finally, the movement received support from the Chelyabinsk city government to conduct a public referendum in 1991 on the plant. After the referendum resulted in opposition to plant construction, an agreement was reached with city leaders not to import or bury spent nuclear fuel within the oblast.

BACK TO THE COMMUNITY

Democratization continued to evolve at the federal level. On September 1993, President Yeltsin used his presidential authority, many say unconstitutionally, to issue Presidential decree 1400, dispersing the Supreme Council and the Councils of People's Deputies on oblast and local levels throughout the country. Yeltsin's goal was to save the federal government from instability by ousting his opponents, namely Vice President A. Rutskoy and Rutskoy's fellow communists in the Supreme Council. He also wanted Russia to secede from the USSR by replacing the Supreme Council with a two-house parliament called the Federal Council of the Russian Federation. After a dramatic shoot-out at the White House (seat of the Supreme

Council) in Moscow, Rutskoy was arrested and Yeltsin succeeded in disbanding the Soviet.

Inadvertently, the success of liberal democrats at the federal level had almost the opposite effect in Chelyabinsk Oblast. During the period 1993–1996, conservative communist leaders were successful in replacing the democratic and independent politicians and deputies. The communists then reestablished a strain of Russian politics unfavorable for NGOs, in general, and the environmental and human rights movements specifically. The Council of People's Deputies was disbanded, removing from office all democratic and independent representatives, including Natalia Mironova from the Movement for Nuclear Safety. In its place, the communists created a series of small councils one-fifth the size of the original Council.

The strong belief in democracy that was evident after 1990 began to waiver with the "wild market" (Yanitsky, 2000, p. 101), a turbulent period of fast social, economic, and political reforms leading to an economic crisis in 1992 (Yanitsky, 2000). People lost faith in the democrats. When elections were held for the new small Councils, the democrats lost to enterprise heads and former communist party members.

Activists in the Movement for Nuclear Safety now neither held positions nor influence with the government. Their relationship with the new oblast power structures degraded. The Commission on Radiological Security was closed in 1992 due to a lack of finance and government interest. Activists found themselves targeted by the Federal Security Bureau or FSB (the post-Soviet version of the KGB) that accused them of espionage and mounted a smear campaign against them in the newspapers. Involvement in the movement dropped off.

The Chelyabinsk movement now shifted to nongovernmental activities. The Movement for Nuclear Safety registered and received official NGO status in December 1993. Over the next three years, members of the movement helped form and register local NGOs, many in Muslumovo. The names of the Muslumovo organizations speak volumes: Kyshtim-57, Nuclear Hostages, White Mice, Techa, and the Muslumovo Committee for Self-Government. Temporary nonregistered NGOs also appeared in the small cities Kasli and Kyshtim for such specific purposes as direct actions or signature collection. Two things are particularly striking about these efforts to organize. First was the activism of rural Muslims just emerging from Soviet control. Second, Western funding was not an impetus, as it has customarily been in modern Russian social protest. Among the direct actions in Muslumovo, a local activist took a boat out onto the Techa River despite access restrictions, gaining media attention that reminded the nation of the village's plight.

Post-perestroika changes in government were accompanied by a new desire to compensate for mistakes under the Soviet regime. As noted, Gorbachev ordered an extensive research project, orchestrated by the Movement for Nuclear Safety, to ascertain the precise amounts of pollution in the territory and determine the state of victims' health. President Yeltsin visited the village of Muslumovo, and his administration initiated three programs to help victims of past nuclear disasters. Symbolically choosing to name its program "Kyshtim-57" as a signal that it sought to right past wrongs, government provided for the building of roads, schools, and hospitals, sought to address social problems, and provided financial compensation to all residents who remained along the Techa. In the end, the new government's programs proved just as ineffective as those implemented by the Soviet government in the 1950s and 1960s. Compensation, for example, was too low to keep people living in the Techa River villages. This program readily is abused by victims. One respondent to our interviews now lives in Chelyabinsk but maintains an official address in Muslumovo in order to receive compensation. She was unremorseful, explaining to our field researcher of her desire to "milk the government as much as possible for what they did."

INTERNATIONAL MOVEMENTS COME TO CHELYABINSK OBLAST

As perestroika opened up Russia to the world, an influx of national- and international-scale environmental organizations became involved in the anti-nuclear and human rights movement in Chelyabinsk Oblast. ISAR (Resources for Environmental Activists, Institute for Soviet and American Relations) began conducting workshops to train villagers in grant writing.[7] In the mid-1990s, Greenpeace and the Social Ecological Union joined the effort and headed several direct actions against the government and the nuclear industry. Greenpeace also published a volume in English and Russian entitled "Half Life: Living with the Effects of Nuclear Waste." The book presents the story of Muslumovo along with a photo album showing the pitiful and dire situation of the contaminated village. Interviews with disabled residents underscore a sense of hopelessness there (Raphael, 1986).[8,9] The book served as a showpiece supporting Greenpeace's efforts to bring the Muslumovo victims' case to the International Court of Justice in

Strausburg. Pictures from the book were exhibited at a Moscow show (Babakian, 2002) to coincide with a visit from President George W. Bush for nuclear disarmament talks .

INTRACOMMUNITY CONFLICT

Although villagers in the Techa region share a common victimization, over time several factors have corroded community. Much of this intracommunity conflict surrounds funding brought to the area by international movements for environmental and human rights and given to the small Russian NGOs surrounding the Mayak plant, especially in Muslumovo, with its preponderance of victims' civic initiatives. Both the Movement for Nuclear Safety and ISAR have conducted workshops on writing grant applications in Muslumovo. Yet some people do not apply. And some applications are rated more positively than others. As a result, grants are not distributed to all people equally, resulting in jealousy and a sense of injustice. Within the local culture, many Muslumovo residents do not fully understand the source or purpose of such grants and believe that grants should be given as rewards or honor to people who are important or have worked long in the community. As this is not the case, resentment builds against those who learn how to work within this new system of rules.

The jealousy is understandable given the extent to which some community members have benefited from grants, receiving not just money, but opportunities leading to a better life. The transforming nature of such opportunities was brought home during one interview, where a Muslumovo victim glorified her present life and gushed with gratitude for the international foundations and organizations that supported her. Before joining the movement, she was uneducated and lived a stagnant life selling cigarettes and food at a street stand. Now she runs her own NGO and travels extensively to Europe, Japan, and the United States attending meetings and conferences concerning the nuclear industry. She is understandably proud of her accomplishments. She was able to use opportunities created for victims by international humanitarians to escape from inertia, depression, and blame and to achieve genuine personal growth. This transformation is not easily understood by her neighbors who see only an unfair inequality in compensation.

Another source of community conflict has been MinAtom's propaganda campaign, which began when villagers learned of Mayak's history. MinAtom has repeatedly used the media to claim that large environmental organizations exaggerate the danger of living along the Techa River for their own benefit. MinAtom charged that, by promoting "radiophobia" in the region, these organizations justify their operations and maintain Western funding. Such activists, according to MinAtom, make money off the plight of local residents while doing nothing to actually help. When the chairman of Mayak was elected to the oblast government in 1993, the propaganda campaign intensified. Much of the propaganda was targeted on the cities of Kasli and Kyshtim, which have much larger populations than the small village Muslumovo. Yet, despite its relative isolation, the impact of infighting in Muslumovo is more acute. As an example, take the case of an activist who had created a museum of Muslumovo's history several years earlier, doing heartfelt work without pay. Under the influence of MinAtom, she felt hurt and cheated. In response, she began charging a twenty-dollar entrance fee to the museum aimed at tourists and visitors coming to learn of the community. Under the circumstances, everyone wants to get their share. This is one example of the pain and jealousy now floating around the community in various ways.

DISASTER TOURISM

Residents of contaminated communities always confront a contradiction of interests. On one hand, publicity is required if they are ever to succeed in getting action. On the other hand, with publicity comes disclosure, an invasion of privacy, and the potential for environmental stigma (Edelstein, 2004). This contradiction also has contributed to community conflict in Muslumovo. As with the museum described above, some local people have oriented themselves outwardly to serve an international audience. One of the more active villagers is a former journalist who constantly tries to publicize the village and its history. He gives tours of Muslumovo to Russian and international journalists and others, telling and retelling the story of past nuclear disasters. This benefits the community in some very clear ways. At the same time, it represents an adaptation to life in a contaminated community. Such tour guides now earn their livelihood because of the disaster. Many members of the community, however, are not ready or willing to participate in these activities. They desire a private life without publicity, and feel like zoo exhibits when visitors come. These latter residents are more likely to question the worth of the anti-nuclear movement because it brings unwanted media exposure. They are swayed by MinAtom's arguments.

Such "disaster-tourism" is also occurring at Chernobyl, where some residents who returned to the evacuation zone now earn their livelihood by showing visitors the key sites (Yanitsky, 2000). However, the situation is different at Chernobyl because of the clear government policies acknowledging the danger and the relocation effort. In Muslumovo, however, the population was never relocated. Tour guides here are not merely finding a way to cope with their victimization by devising a livelihood from the situation; they are also making a political statement in favor of relocating the population. This activism confronts those who are reconciled to living their lives in Muslumovo and who may see little protective gain in moving given their past exposures. For them, the most disruptive future would be one where they were forced to move to a new place with no home, with their community lost, without their land and farms, and possibly among those who would ridicule them for the stigma of who they are.

It should be noted that disaster tourism has thrived in Hiroshima and Nagasaki and in some former Nazi concentration camps. In these instances, tourism is helpful, allowing victims to tell their story and informing and educating listeners. In Japan, for example, the anniversary of the dropping of the atomic bombs allows survivors to impress thousands of young people with their experiences. Likewise, opening up the concentration camps allows for people to confront what really occurred, refuting some who wish to pretend that the holocaust never occurred. In both cases, there is the hope that awareness will help prevent repetition. It becomes the duty of the survivor to inform, to bear witness, and to serve the memory of those lost. Such sharing is encouraged and is cathartic for surviving victims.

Russia has a long tradition of celebrating its heroes in elaborate war memorials. For example, in the former Stalingrad (Volgograd), the immense statue of Mother Russia wields her sword over the vanquished Germans. The visitor confronts acres of exhibits that recall the horror of war and emphasize what brave Russians forbore. Yet, Russia has hardly encouraged its toxic and radioactive victims to express themselves. To do so is to expose tragic failures of government policy. Even though the government has changed, there is little interest in acknowledging hazards that remain and in legitimating claims for relocation and assistance that might be awakened and emboldened by the granting of legitimacy.

NEW THREATS OF CONTAMINATION: THE IMPORT OF SPENT NUCLEAR FUEL

The focus of the anti-nuclear movement shifted on December 21, 2000, when the Russian state Duma publicly disclosed three additions to Russian legislation: an amendment to the Environmental Protection Law lifted the ban on permanently storing foreign spent nuclear fuel, profits from trade in spent nuclear fuel were allocated to fund special ecological programs for radiologically polluted regions of the Russian Federation, and the federal law on the use of nuclear energy was updated (Spent Nuclear Fuel: Not a Valuable Resource but Dangerous Nuclear Waste, 2001).

Under these changes, MinAtom of the Russian Federation proposed the transport, reprocessing, and permanent burial of foreign spent nuclear fuel within Russia through 2010 (Ministry of Russian Federation on Nuclear Energy, 2001). One of the two main storage and reprocessing locations was to be sited in Chelyabinsk Oblast within the territory of Mayak. A U.S.-funded nuclear plant would be built there in order to offer "safe and secure" storage for the spent nuclear fuel of Russian-type power plants and reactors. In addition to reconstruction of the existing site, construction of two additional big storage units was contemplated. As a result, Russia would have the capacity to accept 20,000 tons of either Russian-type or non-Russian-type spent nuclear fuel from foreign plants while continuing to handle domestic spent fuel, the volume of which has diminished as plants operate at 35–40% of their normal capacity.

The anti-nuclear movement interpreted inconsistencies in MinAtom's argument for nuclear waste import to signal a further agenda beyond profit, namely the revitalization of nuclear energy in the new Russian economy (Center of Environmental Policy in Russia and Greenpeace Russia, 2001). Russia would then be promoting development of plutonium-fueled nuclear power plants, in line with waste import, at the very time when most other industrialized countries seek to phase out nuclear energy and plutonium fuel in particular.

One of the largest and most unified movements since perestroika formed against the Russian government's proposal and continued during the ratification process. This major mobilization reflected opposition coming from environmental organizations operating in Russia, as well as the Russian population itself. Chelyabinsk Oblast was an epicenter for this opposition. The region's importance to the issue once again piqued the interest of national and international NGOs and activists in Mayak's history and its victims.

Activists often partnered with victims to conduct direct action. For instance, on October 9, 2000, just a week after the new legislation's first draft was submitted to the Duma, Greenpeace joined villagers from Mulsumovo to place barrels of radioactive soil on the steps of the Duma building in Moscow. The soil came from farms and gardens affected by the Mayak nuclear site. To protest against adding more radioactive materials to the site, Greenpeace activists used a Geiger counter to show that this soil already had 30 times the normal background radiation levels (see Whiteley, 1999). Alexei Yablokov, a renowned environmental activist in Russia, brought lawyers specializing in human rights cases to the Chelyabinsk movement. His interest in the region emphasizes the linkage between environmental and human rights that appears throughout the issue of nuclear power.

Despite its collaborative protests and extensive mobilization, the opposition movement failed to influence the Russian government. This failure underscores that Russia has yet only immature structures for balancing power, allowing MinAtom to use its money and influence to manipulate the Committee on ecology to permit the new activities. What appeared to be legitimate decision-making procedures, hearings, and expert commissions in the review of the proposed legislation about importation of foreign spent fuel in reality was blatant promotion of the power structure's unchecked policies in a way reminiscent of Soviet times. Perhaps the message of Muslumova is that nothing really changes. Or at least it does not change easily. Perhaps we need to think about social half-lives as an accurate measure of change rather than revolution.

NOTES

1. Natalia Mironova, a thermo-electric engineer, is a native of Cheliabynsk and is renowned for her efforts to organize opposition to nuclear projects and for a cleanup of the Mayak reactor. Unless noted, this article draws upon the local knowledge and experiences of Dr. Mironova. The other authors, as well as Michael Edelstein, William Hallman, Olga Tsepilova, and Nadezda Kutepova visited Cheliabynsk in August 2002 and participated in a seminar with activists throughout the oblast convened by Dr. Mironova.

2. "Oblast" is an administrative unit in Russia, similar to a "state" in the United States.

3. Whiteley, 1999 reports a subsequent effort to fill Lake Karachai with debris to prevent a ubiquitous problem of radioactive mists lifting from the water surface to be carried out over the region. The task confronts the challenge that radiation levels around the lake can be lethal to individuals within an hour. He further indicates that earthen dams built along diverted portions of the Techa River remain in danger of a failure that would flood Muslyumovo with fast moving and contaminated waters at a height of over 12 meters.

4. Plutonium in the environment. Collection of materials for project *Plutonium Economics: A way out or a dead end?* Chelyabinsk, 1998.

5. Americans of this period were asked by President Eisenhower to not "rock the boat."

6. The Supreme Council was the federal legislature of the USSR.

7. ISAR sponsored an exchange of American and Russian nuclear activists that visited Muslumova. Their journal is available online at http://www.isar.org/docs/Nuclear.pdf

8. Now online at http://archive.greenpeace.org/mayak/

9. "Russian villagers deliver radioactive soil to Duma." Available on http://www.ecocities.net/Article11.html, Moscow, October 9, 2000.

REFERENCES

Akleyev, A., & Kisselyov, M. (Eds) (2000). *Medical, biological and ecological impacts of radioactive contamination of the Techa River.* Moscow: Urals Research Center for Radiation Medicine.

Babakian, G. (2002). A cold war legacy in black and white. Moscow Times, May 24. Page V.

Center of Environmental Policy in Russia and Greenpeace Russia. (2001). Press release of the Center for Nuclear Ecology and Energy Policy at the Socio-Ecological Union, Center of Environmental Policy in Russia and Greenpeace Russia, April 17. *Energy and the Environment, 28*, 4–5.

Edelstein, M. R. (2004). *Contaminated communities: Coping with residential toxic exposure.* London: Westview Press.

Medvedev, Z. A. (1979). *Nuclear disaster in the Urals.* New York: W.W. Norton.

Ministry of Russian Federation on Nuclear Energy. (2001). Technical-economic basis of the laws related to widening Russian participation in world market of spent nuclear fuel. *Energy and Environment, 28*, 27.

Mironova, N. (2005). *Social dynamics: Metamorphoses of selforganization and management.* Chelyabynsk: Chelyabynsk Publishing House.

Raphael, B. (1986). Nature of disaster. In: *When disaster strikes: How individuals copen and communities with catastrophes.* New York, NY: Basic Books.

Spent Nuclear Fuel: Not a Valuable Resource but Dangerous Nuclear Waste. (2001). Press Release of Joint Press Conference at Center of Nuclear Ecology and Energy Policy of the Socio-Ecological Union, Center for Environmental Policy Russia and Greenpeace Russia. *Energy and the Environment, 28*, 4.

Whiteley, J. M. (1999). The compelling realities of Mayak. In: R. J. Dalton, P. Garb, N. P. Lovrich, J. C. Pierce & J. M. Whiteley (Eds), *Critical masses: Citizens, nuclear weapons production, and environmental destruction in the United States and Russia* (pp. 59–96). London: MIT Press.

Yanitsky, O. (2000). *Russian greens in a risk society.* Helsinki: Kikimora.

PSYCHO-SOCIAL CONSEQUENCES DUE TO RADIOACTIVE CONTAMINATION IN THE TECHA RIVER REGION OF RUSSIA

Michael R. Edelstein and Maria Tysiachniouk

> I use the example from a nature film I saw of the African impala that go to the river and some get eaten by crocodiles but the others keep coming back to the river anyway. This is how we are. (Commentary of a Techa river activist)

INTRODUCTION

The Ural Mountain region is a remarkably beautiful landscape of forests and lakes. Here is the continental divide between Europe and Asia. One of the rivers originating here to eventually feed feeding the Artic Sea is the Techa river. But the verdant greenery that characterizes the region does not disclose the hidden dangers of plutonium and other radioactive materials either downstream or downwind of the Mayak Nuclear Complex. Major areas of the Techa river corridor and downwind areas were permanently evacuated after radioactive releases from the Mayak Nuclear Complex from the 1950s through the 1970s. Although acute and chronic water releases encompass this period, the so-called Kyshtim 57 accident, an air release

Cultures of Contamination: Legacies of Pollution in Russia and the U.S.
Research in Social Problems and Public Policy, Volume 14, 185–204

ISSN: 0196-1152/doi:10.1016/S0196-1152(06)14009-0

from Mayak in 1957, was the word's worst nuclear accident until surpassed by Chernobyl (see Mironova et al. and Kutepova and Tsepilova, this volume). But what of the inhabitants who remain? In this chapter, we explore some of the psycho-social impacts of living in contaminated areas, drawing primarily upon interviews with residents of the region. In doing this, we give voice to their perspective and views on each other and life in this contaminated region.[1]

Edelstein (2004) has suggested a framework for describing the psycho-social impacts of environmental contamination events based upon his own ongoing research program, as well as the overall literature on the subject. The resulting theory of environmental turbulence charts the disruption and forced change brought to peoples' lives by the discovery that they are living in a contaminated environment. These changes are described as behavioral, cognitive, and emotional.

Behavioral alterations occur for victims as "lifestyle changes," or adjustments in the patterns of everyday activities necessitated by adaptations to living in a contaminated environment or the new social rules constructed to manage that environment. Lifestyle changes include resettlement and evacuation, loss of the ability to use residential water, placing the yard or garden or parts of one's home off limits, having to use equipment to render the indoor air breathable, and changing one's pattern of outdoor activity to avoid exposures. Lifestyle changes can be temporary or permanent, may involve one area of a person's life (home, work, and community) or encompass its entirety, and may be relatively inconsequential or signify a fundamental and significant change in the meaning of life.

Cognitive alterations occur for victims as "lifescape changes," or adjustments in how people think about themselves, their health, their homes, the environment, and those whom they rely upon for help during trying situations. Lifescape changes frequently include the development of a pessimism about health, a personal sense of loss of control over life, an inversion of feelings of attachment and security toward one's home and community, a newfound fear of the environment, and a profound loss of trust for others, particularly government.

Emotional consequences include stress due to unwanted environmental change, the efforts to make adjustments, and failures to adequately cope with changed circumstances. Unsuccessful coping may result in psychopathology. Examples include posttraumatic stress disorder, depression, health obsessiveness, and other subclinical or clinically significant changes. People experience victimization due to human-caused changes; as a result, anger and blame are frequent consequences. The results of emotional changes and extreme efforts to adjust to new circumstances can also cause interpersonal strains on families and relationships. The required social costs and adjustments may also bring about community conflict or conflict between citizens and government or between polluters and victims. These conflicts may take the form of direct conflicts, grassroots actions, legal actions, and/or governmental actions.

Victims are frequently "disabled" by their loss of control over their lives and the capture of important decisions by experts and officials. The creation of local grassroots organizations and networking between such organizations can result in a countervailing empowerment or "enabling." Frequently, environmental stigma occurs as a response to contamination exposure, suggesting a prejudice or changed valuation of contaminated places and the people and products associated with them. Toxic victims frequently engage in anticipatory fears based upon potential changes to health and the environment. While people may engage in significant growth as the result of their response to contaminating events, they almost certainly confront severe costs.

The purpose of a psycho-social impact study, therefore, is to catalog the impacts of events that change the environment in which people live. What then were the psycho-social costs of the string of accidents and environmental releases associated with the Mayak nuclear complex?

PSYCHO-SOCIAL CONSEQUENCES OF RADIOACTIVE RELEASES FROM MAYAK

The region affected by Mayak is vast and includes two overlapping but distinct regions of danger. First, there are the areas along the Techa river that have been continuously impacted by releases of radioactive contamination from various parts of the Mayak complex to the river system. Muslumovo is one of a few villages to be left in place along the banks of the river; the rest were relocated long ago. A local activist estimated in 2002 that 4,000 of the original 4,500 people remain in Muslumovo, held by jobs in agriculture and the railroad station. Most are of the Tartar minority. People continue to build houses there, even though they know this is officially discouraged. And he bemoans the fact that so many of these people are young.

> Children should not be here. Young people should be replaced from here. There are a lot of them here.

Bounding the Threat

At least Muslumovo is a place where there is a clear demarcation between radioactively dangerous and safe areas. He notes that such clarity is not found in the second zone of danger, areas of the region receiving radioactive fallout from the Kyshtim 57 disaster or other airborne radioactive releases, where a much greater population continues to suffer potential exposures.

> A half million are on the list of victims of the Mayak disasters, but another million were affected. There is plutonium on people's dachas dropped from Mayak operations. If I had to chose where to live, I would chose Muslumovo because at least there you would know where to go, you could avoid the Techa. The zone of danger is clear. On one side of the road, it is okay and on the other, it is off bounds. In Muslumovo all people know they have to behave. Downwind there is increased uncertainty, and they do not know what to do.

This informant sited another village far from the Techa as a stark contrast to Muslumovo.

> There is air pollution and soil contamination, but no river exposure. It is like Hiroshima. The deposition (from Kyshtim 57) was near there. There were 4,000 people. There are now 750 people left. It is a dying village.

Point of Awareness of the Danger

Edelstein normally measures these changes against a precontamination baseline. In the downstream and downwind areas from Mayak; however, the precontamination baseline would stretch back to the period before the Second World War. A point of comparison of greater utility contrasts the time before and after the period of first real awareness of contamination. Local residents from throughout the region grew up in close contact with nature largely unaware of the hazards. They did not learn of their exposure to radioactivity until Perestroika began in the later 1980s. As a result, current generations all suffered environmental exposures through their everyday activities.

Of course, it was evident that something bad had previously occurred. Most villages in the region had been forcibly evacuated without explanation. In Muslumovo, as our respondents recalled, the Techa river was fenced off and guarded from the later 1950s. Officially, it was decreed a danger to health. Yet, it is instructive that residents apparently neither understood nor perceived any such danger. One respondent interviewed in 2000 recalled that he loved to fish there and was routinely caught by the police guarding the river. He told of mushrooms and berries picked. The biggest berries were right on the bank of the river. They were not allowed to take them, but nevertheless they would catch fish and pick the forbidden fruit.

Another informant, a mature woman from a Bashkir village on the Techa, recalled that

> When Mayak started working, we knew there was secrecy but not danger. Land was taken from people and a lot of construction occurred. It became like a foreign land. As a child I recall the river. Mushrooms and big berries. We would sneak there to eat the berries. We would feel like heroes. There was thyroid disease in all the people who did this. They had said not to go there, but they never explained the danger. There were military everywhere. As a result there was no competition for the berries. So the same people would go there. All the time there were discharges, and the people did not know it and continued to live there.

Another informant, a young Baskir woman, grew up in a village 3 miles from the Techa river and, although her family moved when she a young girl, she frequently returned to her family dacha and to see relatives. In 2002, she described her village as similar sociologically to Muslumovo except that the majority of its residents are Bashkir people instead of Tartar, both villages with some Russian families mixed in. Here the acceptance of the reality of the contamination did not come until much later.

> Many people before Perestroika did not know about the zone of pollution and they picked berries and everything. I picked mushrooms – huge big mushrooms near the river. There were no signs of the river dangers. They [the officials] tried to put up signs but people took them down all the time. We did not find out about the hazard until a few years ago. Before that I was still picking berries and selling them with my mom to others. There was a bridge across the river and a big road and transportation would cross. Drivers would stop on the bridge and take water to wash their cars. There are still people who cut hay. They pick berries and sell them to the drivers.

Lifestyle Change and Fear of the Environment

Once people knew about the danger, food became an interesting indicator of local peoples' fear of radioactivity. Food is probably the major pathway exposing people to radiation. And the last quotation alludes to two aspects of the food exposure issue, local consumption of local food and the sale of local food to others from the outside. A double standard became evident during our interviews whereby some local villagers along the Techa no longer eat at least some local food products, but, given the absence of other ways to make a living, they continue to sell products from the Techa area alongside products from throughout the oblast in the region's markets, principally the one in the major regional city of Chelabynsk. They catch fish or slaughter animals and sell them in the city without disclosing the origin. In fact, as we shall see, to identify the origin would potentially render the food unsalable.

In the view of a key Muslumovo activist, interviewed in 2002, residents there have learned to cope effectively in the way they manage the complexities of their situation. This informant reports that he is not obsessed with anticipatory fears that the Russians call radiophobia. Nevertheless, in his family, precautionary or vigilant behaviors are practiced.

> We eat by the dosimeter. We calculate our meal. We are afraid but not sharply afraid.

He believes that everyone should know the conditions in which they live. But while people have changed their eating habits at a great cost, he reported that parents are angry at him because they do not know how to feed their children. People from the village sell their radioactive hay and milk to the government. But lacking salaries and enough government aid to survive comfortably, they continue to sell their contaminated meat in Chelabynsk.

> People just know that everything from the river is dangerous. They pick far away for their family unless they are going to sell. I taught people that the hay and grass are dangerous. Tartars understood this completely. They sell their meat and hay in Chelabynsk. And they buy their own meat from Kazakstan.

This observer drew a distinction between the Tartar behavior in Muslumovo and that of Baskir people living in the contaminated region. He believed that the difference was less cultural than based upon differential circumstances. While the Tartar receive aid for living in the contaminated zone along the river, the Bashkir area lies within the comparatively gray fallout zone which earns no government subsidy.

> The Bashkir have no other source of support so they just use what is around them. And their lands were downwind. The mushrooms are really contaminated. But they all just eat them.

The mature Bashkir woman, also an activist, confirmed the account, citing that her people do not get compensation like the people of Muslumovo.

> Now everybody knows. They think about it. They don't necessarily talk. Today if you have a headache, maybe it is from the radiation. There is contamination of the fish and mushrooms. But it is not possible to not eat fish and mushrooms because people are hungry.

Our young Bashkir respondent, also interviewed in 2002, confirmed that local food was still consumed and sold by her fellow villagers.

> They sold berries because they needed money. My cousin who picks and sells berries does not eat them himself. But most of the people do eat them.

Beyond the need for money, she observed that most villagers still do not accept the hazard.[2]

> A lot of my relatives died from radio-illness. This is a continuing thing. All the people knew in the village it was dangerous, but they did not connect it to the illness. They did not believe it was a real hazard or that the berries that they sold were harmful.

A factor in this response is the ambiguity in drawing boundaries around the area of danger.

> Besides, the radiation in some places is really dangerous and in some places less. Of course, a low dose of radiation is also harmful in the long term.

As a believer in the reality of the hazard, our informant now personally takes protective measures with Techa food and tries to instruct others to do the same.

> Last year my mother went and picked mushrooms, but I refused to eat them when I realized the mushrooms were from there. I said I would not eat them, and I advised her not to either.

But it is hard for her to get this hazard message through to the villagers, even her mother.

> I told all my relatives. Most sell the food they pick right away for cash, but they eat it also. They say "I ate them my whole life and thought it was okay." I told my mother, but she also said all the people from the village have been doing this their whole life and it did not make sense to stop.

The compunction to eat food from nature is itself part of the Bashkir culture, as the mature informant noted.

> Here we cannot grow gardens. Potatoes and bread but no gardens. Even if we were permanently resettled, we would still act like nomads. We are not agricultural people, but gatherers. We do not adjust to modern circumstances. We grow food and pick food.... In Bashkir culture, people were very connected to the land. It was mother nature. We did not know what else to do but to live on the land....

The idea of precautionary selective consumption represents a real hardship for residents of the young Bashkir woman's village, who lack any other livelihood than what they can gather from the land.

> People have no jobs. My relatives have only a small salary, enough just for bread. They live from the land. They also have gardens. They eat vegetables. Eggplant. Potatoes. They have cattle. But they fertilize the soil with cow manure from cattle kept by the river. The milk is bad only if they take hay from near the Techa.

Her relatives earn additional cash by buying vodka in Chelabynsk and reselling it in the village, where the stores have closed. Vodka is an important folk curative potion in Russia, even if it also contributes to a baseline health threat in the form of alcoholism. Our informant said that she was fortunate to grow up without a father in the household, and thus less alcohol abuse than the norm. She stressed that vodka is viewed as a protective elixir, a view confirmed by the mature Bashkir woman from the Techa village.

> The opinion is that if I drink vodka, the radiation will go away.

Another traditional curative custom cited by our Muslumova informant is the banya, a hot steam sauna, long used by Russians to cleanse their bodies.

> People do nothing to improve their health. They do not exercise. They just work hard all day. They use banya and vodka for prevention. At Chernobyl, people took medicine and went to the banya twice a day. No one died.

While Muslumova residents have substantially cut their exposures through food consumption, some tolerate other environmental exposures. After all, as our 2002 informant indicated, they feel that, having lived on contaminated land for so many years, more time here would be like a drop in the bucket. Our Muslumova activist echoed this view:

> I continue to swim in the Techa river, any new exposures are minimal against my life history of radioactive contamination. Nothing frightens me now after I learned I had been irradiated.

In a 2000 interview, this informant reported that he still drinks from springs in the woods, but that he had started boiling the village water before consumption. Two years later he reported that the municipal water service had been cut off in the village, and that drinking water was being transported in. But the fact that people were no longer drinking local water did not end their habitual use of the river for other lifestyle purposes.

> In Muslumovo, all the people are from the Techa. People have a long time use of the river. Even though they transport in drinking water, we still use the river for bathing and gardening.

And people continue to depend upon the Techa for watering their farm animals.

> I try to keep horses from being taken to the river Techa to drink. But people ask, "where else will we get water for our horses?"

Cattle represent a further hardship, as he elaborated.

> But we are aware that if we are near the Techa the grass is wonderful for hay. You have a choice. Do you use the hay near the Techa or the bad grass not near the Techa? And if you take your cows to the Techa, you can watch with one dog from the bank while they get water and grass. You are in control. Or you can take the cows to a field and watch them scatter.

This Muslumova activist still plants potatoes near the river, even though he stopped fishing after he tested high for radiation. He told our interviewer in 2000 that he also still hunts ducks, even inviting her to hunt with him. In one week, he said he can get food for the year. But he is not totally careless. He measures the radioactivity in the ducks before eating them.

Our young informant on the Bashkir village suggested in her 2002 interview that class differences affect people's response to the radioactive hazards.

> If you look at different people, some are less lost in life. They have work. Their lives are more normal. Some have big houses with four cattle and some money. Others have no job and money and they do not care, they swim in the Techa. The better educated people may rationalize grazing cattle but might not take fish or berries for their own food. They care about their health more. They think access to the berries and mushrooms does not hurt, but they would not eat them every day.

Having only found out about the radioactivity in 1999 after she began training to become an environmental activist, she reported developing immediate fears of her newly understood contaminated environment.

> I found out by accident from the Social Ecological Union. They had a direct action camp for students. I went there and learned about the issue and the pollution. I realized that I did not want to eat the berries any more. And then there was an expedition of scientists to measure the radiation in the Techa. I accompanied one on a boat. I kept thinking that, all the time I was a child, I would run to this river. And now I realized that I did not want to go with the scientist. So I stayed on the bridge and he went alone. Now I fear to go to the river. I used to do it all the time.

She had always dreamt of living in her family Dacha along the Techa, but not now.

> Before I always planned to make our dacha my home. But maybe not. Maybe I will choose a cleaner place, my grandmother's house where I am registered. It is far from the Techa.

Now a city person, she also makes protective food choices at the market place, illustrating how knowledge of the contamination potentially impacts people living in the large regional center of Chelabynsk, seemingly far from the radioactive danger.

> I had no fear to eat the mushroom and berries. I will eat berries from other places. In the Chelabynsk market, people sell berries and mushrooms. I will buy them, but I try to get a sense of where they are from. I now prefer to buy factory mushrooms that are not from this area. Just at the railroad there is a market, and people still sell big fish from the Techa.

Even at this distance, the situation can be frightening even for those with no direct ties to the contaminated regions, as another woman living in Chelyabinsk explained to our interviewer in 2000. She was obviously mindful that this city has serious pollution problems of its own.

> Sometimes I am scared to live in this city. I open the water tap and before drinking I think "oh Gosh, what could be in this water?" I have become more and more aware and scared since I learned about Mayak.

Health Pessimism

Kept in the dark about the consequences of the Mayak accidents for so long, area residents were certainly aware of the spate of health problems affecting their communities. They came up with their own generic folk term for the rampant local illnesses—"river disease." Our Muslumovo informant described the local understanding of the cause of river disease.

> Regarding river disease, people knew that Mayak was a nuclear facility; many worked there. They knew there was an explosion in 1957, but they thought it was underground. They never knew about the deliberate early discharges to the Techa, and they never knew that accidental releases continued.

Such knowledge did not influence doctors' diagnoses of illness as they are forbidden to tie diseases to a cause related to Mayak. Whether or not people thought there might be a relationship between illness and the facility, they remained quiet, as our Muslumova informant suggested in his 2002 interview.

> Our parents' generation born under Stalin knew not to talk too much. They did not talk about it.

This informant, a Tartar, had reported in 2000 that the older generation had thrived much better than his generation. His relatives lived a long time. His grandmothers were at the time 96 years old and 100 years in age. His grandfather was a champion wrestler who had won horses in Afghanistan through the sport. They were Muslims and had multiple marriages; there were four wives in his father's family. The family was the epitome of health and strength.

But things have changed for the next generation. Four of his brothers have radiation disease, and his wife was the only of one of her siblings to be spared cancer. He was an energetic man in his forties who wanted to have children. But both he and his wife were sterile from radiation sickness. Of everything about the radiation exposures, he is most disturbed by being sterile. He thinks he could give birth to a child. He is strong. He brags that Tartars are stronger than Russians. Tartars never experience absolute poverty, only Russians. But now his roots are lost; he has no child.

The mature Bashkir woman told a similar story about her people.

> After Perestroika we learned that something was wrong. River disease. The Bashkir people are pretty healthy in the nation and before that. They are agrarian. All the families had sick children. At first we believed that god was punishing us, so we did not talk about it. In the peasant life you need healthy people to survive on the rations.

She later elaborated:

> Illnesses include cancer, leukemia and a lot of psychological problems—nerves. Some people have no feelings in their periphery or they have deformed bones. There is alcoholism. There was a legend that if you would drink, you could make the radiation go away. It is very hard to separate. They have lower immunity....Now everybody knows. They think about it. They don't necessarily talk. Today if you have a headache, maybe it is from the radiation. There is contamination of the fish and mushrooms. But it is not possible to not eat fish and mushrooms because people are hungry. People are very sick. With their health, they can't care for their children. They are badly dressed. They can't talk well.

Our Muslumovo activist knows that he has been affected by radioactivity. During Perestroika he went to Moscow and had genetic and blood tests. He was told that his radiation levels were 100 times normal levels. The doctors were surprised he was alive. In 2002, he described more recent tests done in Muslumova.

> They did genetic analysis of 15–20 families. Every child over 3 months old has radioactive substances in their blood, even if they never went in the river. They had plutonium in their teeth. The third generation suffers in Muslumova. They have loss of memory from 12 years on.

Reflecting on the causality for early death, he commented on the tendency to blame the persons' weakness, not an environmental cause, for such maladies.

> Many people say it is better to die old and sick than to be young and hungry. When young people die of cancer they say it is bad luck or they had a poor immune system.

Our young respondent from the Bashkir village made a similar point, but she also alluded to another dimension of their illnesses, the likelihood of multiple causality.

> The illness is due to the food. There are very many cancers. They are mostly among the people who were resettled. There is a lot of sickness but no evidence of radiation. There is a lot of poverty. People eat once a day. They consume a lot of alcohol. It is hard to assess whether they die from radiation, alcohol or hunger.

Even in thinking about her own health, this multiple causality is a problem.

> I have one major disease. I have few white blood cells. It might be due to the radiation. I am concerned. Hunger and stress during Perestroika could have caused anemia, so I am not sure if this is the cause.

A shroud of secrecy surrounds a special experimental health-monitoring program developed for residents continuing to live in the contaminated zone, described by the mature Bashkir woman.

> If you say you are from this region, you were put in the Biophysical Institute in Chelyabinsk and studied. Everyone is on their list.

The Muslumovo activist, independently, during his 2002 interview noted that the effect of the program is to turn residents' exposure to radiation into a long-term natural experiment.

> Now there is big belief that I agree with that Muslumovo and Karabyk are watched by the doctors to see what happens to people affected by radiation who have not been treated for exposures as they would be in Ozersk.[3] Neither village was resettled, although some Muslumova residents chose to resettle. So it is an experiment. You can look at the comparable genealogy trees of all the people. Doctors have it all on computers. Every person from Muslumovo or the other contaminated territories is invited to stay in the special hospital near here. The regional doctor just looks at your symptoms, but here you get a holistic approach to health. They take blood four times a day. They use new drugs and try cures. People want to go there. I have a pain in my stomach and bones, so I agreed to go there. The research institute studies rats, cats, dogs and us.

For this latter respondent, however, there is a major downside to being involved in this program of study, i.e. the loss of mobility.

> To resettle Muslumvo would cost one million dollars but they spent 10 million on the institute. They want people to stay there to be monitored. They would lose their experiment if people moved. So the government has not forced us to move. And we lose money if we move. The powers are interested in people staying.

At stake, moreover, is the perceived benefit of the research.

> If we resettled, we would bias natural selection and how our bodies fight radiation. A St. Petersburg institute found that there are special antibodies in the blood of people in

> Muslumovo. We don't react to radiation in the normal way. I believe we can save humanity with an antibody against cancer and radiation. People who are not sick have this in their blood. There is natural selection; some live and do not get sick.

Nonetheless, this activist wants to resettle those who want to. He is working with the European parliament to get the resettlement even if the Russian government opposes it.

> It is genocide. It is a concentration camp. They are research fascists. It is not ethical what they do now.

RELOCATION

But what of those that have already been relocated? The disruption of resettlement has affected most of the region's residents over the years. The mature Bashkir woman from the Techa village, herself resettled, gave this account.

> In 1957, there was the explosion. Some people saw a yellow cloud that dropped particles on the village. We were immediately resettled. It was a rude resettlement. Soldiers came in special suits. People were told to undress and put on other costumes. They did not explain. All the houses were immediately destroyed. All the cattle were put into pits and shot. People were put in cars and driven away. It was never explained that it was unsafe. 20 villages were resettled. Some people where separated and others put together. When the 1967 drought and Lake Kariachi dried up and its shores blew in the wind and spread radioactivity. It followed the same wind direction as the 1957 accident.

Our young informant from the Baskir village gave a very similar explanation of the experience of those living near the Techa river who were long ago resettled to her village.

> They made a separate street for the people being resettled. They integrated them with the village. New houses are built by the street. They left their cattle and things in the old house. They had to rebuild from scratch. They were never told why they had to leave. The military people just came and said they needed to leave everything and get to the other village. Their house and buildings are still there, falling down. People sneak back to the old house and carry things off to their new house. The cattle were killed and burned. People cried very much.

She explained that those resettled from the Techa are the beneficiaries of government assistance. They get $3 per month per person, by itself only enough to purchase one bottle of vodka. But, because families are large, the total sum for most families is enough to make them dependent upon remaining in the village. Regardless of the hazard, people do not wish to move. Moreover, this sum, as little as it seems to American eyes, is enough to cause local resentment. Our young Bashkir informant added:

> People in my village did not receive anything.

Another 2002 informant, an activist from Chelabynsk, noted that her mother came from one of Russia's most polluted cities to teach through the 1970s in Trudovoe, a Techa region village. There she met her husband-to-be, who grew up in a nearby village. Both places were covered by the Eastern Ural radioactive tail, the path of radioactive fallout resulting from the Kyshtim 57 accident. Nevertheless, their villages were not considered to be dangerously contaminated and so were selected to house many residents relocated from Techa river communities. She echoed the point about differential benefits.

> In places that were contaminated, there is food and clothes from Bulgaria and Holland. And there is state support in Ozersk. People who came to work in Ozersk are called "chocolate people." But in places with no pollution, there are no such good things.

In Muslumovo, the situation was different. The place was highly contaminated. During Perestroika, according to the Muslumovo activist during his 2002 interview, Gorbechev opened up the flow of information.

> We learned 100% of the truth about our exposure. His government promised to resettle the residents of old Muslumovo immediately and promised to build a new Muslumovo. But then they decided to split the people up in Chelabynsk and other places. Some people bargained and waited. The laws called for voluntary resettlement, but under the international radiation standards, we all should have been resettled. But later they explained that they only found high radiation near the river. So we lost the ability to resettle by law. International and Russian money was available afterwards. But people were picky about where to go. So the only people who resettled were those got an apartment immediately. Those people who were not picky were in a better situation. They just went. The others are still there.

The informant lamented that had Gorbechev stayed in power, he would have solved the problem, driven as he was by guilt over Chernobyl.

> Yeltson promised to resettle the people. But there was fast economic reform and no money to start over. Nobody has had any money since Perestroika. We can't buy a new house. We do not want to go to a worse place. We are more trapped by the circumstances of life.

In the ensuing economy, most people in Muslumovo have become dependent upon government assistance. They have no way to get the funds they would need to relocate themselves.

> The economy is tied to the zone. No money is coming in and only benefits help them. In Muslumovo you get 300 per month if you work and 30 per month if you don't. You pay less for gas and have a free education. There is an allowance for free medicine. If you move, you lose the benefits. So people are economically interested in staying. The incentives should instead encourage people to leave. If they can get 5000 rubles they can get a place in Chelabynsk. We believe that the entire village should be evacuated to clean territory. The administration did not evacuate people far enough. 10 k from the area is not enough. Many houses are on the banks of the river or just back. If people can get money, they can go where they want.

ENVIRONMENTAL STIGMA

Residents of affected areas find that their homes, their community, their lands, their products, and they themselves are feared, devalued, or discriminated against because of the contamination.

All of our respondents equate themselves to guinea pigs in an experiment, as suggested by the name of the local NGO "White Mice." Being victims of nuclear accidents, these people are looked at the same way by outsiders to the community. This is a common problem faced by contamination victims. Our informants explained the double bind into which they have been placed. On one hand, a local education program informs residents of the village that marrying within the community can be dangerous and raises the chances of bearing a sick child. On the other hand, marrying outside presents another problem, in that people know of Muslumovo's history and so worry about marrying a victim and having sick children. A woman from Muslumovo explained to our interviewer,

> If you tell the woman you love you are from here, you are turned away. Even if you fall in love at the University.

Likewise, Techa region residents find stigma when they take their products to market.

> You can eat the mushrooms if they are not from the river. For the Chelabynsk population, our people say they pick mushrooms in another region because, if people say the mushrooms are from Muslumovo, no one will buy them.

Issues of environmental justice are also evident. Residents of the region around Mayak are largely minority Tartar and Bashkir people who practice Islam. Other local residents tend to be descendants of people repressed and exiled under Stalin. It is often said that one of the major reasons that the Mayak complex was built where it was in the first place was the presence of a local labor pool. However, locals have a different explanation. Our activist from Mulumovo said in his 2002 interview that Mayak would never have been placed on Tartar lands because the Tartar are too independent. Instead, perhaps expressing some Tartar chauvinism, he said, lands of the much more passive Bashkir were selected for the Mayak complex.

> Tartars would have blown up, but the Bashkir do not respond. They are not involved and interested to take responsibility. They are passive. The Communists built Mayak on Baskir territory. It would have been impossible to build it in Tartar territory.

But, he noted, Mayak came after the Tartar anyway.

> But from the contamination of the Techa, the struggle moved to the Tartar. They resettled Russians and Bashkir, but Tartar did not want to resettle. So they did not. Tartar demanded to be all together.

Nevertheless, Tartar perceive that discrimination is evident in the postdisaster resettlement programs, as our Muslumovo activist noted.

> One part of Muslumovo is Russian and the other is Tartar. The Russians were evacuated and the Tartars were not. Russians are closer to the river but Tartars think it is discrimination.

ACTIVISM

Perhaps because they view their situation as a group threat, residents of Muslumovo have made the village a hot bed of activism. Our field interviewer had been deeply impressed by the many women activists. She found them organized, networked, and very educated. And they were engaged in long-range thinking. They were interested in sustainability as a vision for the future. Around 1995, they began a program of sustainable development for the region supported by the government only to have it squashed when the communists were subsequently elected.

Our Muslumovo informant also reported a cultural dimension to this activism.

> Muslumovo Tartars are very active; there is a big movement here. A high concentration of active people.... It is unusual. The big difference is that Tartars are fighters, aggressive.

Although, he also acknowledged a downside to these traits.

> There is a risky personality. It is like a casino. Some people can stop and others just play and play. In Muslumovo they will talk and bargain but not go. They will be resentful but they will stay in Muslumovo, even if only on paper, to get money.

His own activism is based upon his belief that people must have informed choice. He brought back equipment and trained school children how to measure for radioactivity in the milk and soil and at school. He told the farmers when it was safe to hay their fields. He then set up a second monitoring station. With other activists, he works to educate people in the village on how to decrease their radioactive exposures, what to eat and so forth. Using international funds, activists have developed programs with names like "To know is to survive" and "Life under the fence of Mayak."

This same activist won a grant to assist Baskir communities to deal with radioactive contamination. It gave him the opportunity to compare the Bashkir to the fiercely independent Tartar.

> The Baskir listen to their headman. They are under "the boss." They listen to commands and are disempowered. They have less a culture of independence. For the Bashkir, the risk message must come through their boss. If he says go, they will go.

A similar story came from our mature Bashkir informant.

> In Bashkir culture.... We had an eastern mentality. We lived in a community. If the boss said it was fine, we believed him. Those at the head did not know about the problems. Doctors who came were afraid to say what they thought. Muslim traditions also contributed. We were the most uneducated people in the oblast. In this culture, they live in a community. The head of the community makes decisions. It is like the security council. They came and said it was all wonderful. The leaders were chosen by the Communist Party. Since Perestroika, we can chose our own leaders.

Both of our Bashkir informants had become activists, despite the culture. But as activists, they had to confront the culture, as the mature informant explained.

> Religion is an impact? We look at life with fate. Therefore, you do not need to take an active position to change it. They continue to be passive ... and poorly educated.... Now we try to bring people knowledge and educate them about the situation. The government has never acknowledged the situation even though they give money.... Now we try to organize a local movement there, but it is very hard. Everyone is related. You can't say something against a relative or a friend. It is very hard to find someone in a small community who will speak out and against people. Some progovernment organizations were created by the government. People lose belief and self confidence.

The young Bashkir informant agrees of her village that

> The people are not motivated to be activists. There are no education programs. Some people know how to read the newspaper. There are no programs to explain what the dangers are and aren't. Or how to behave. In my village, no one is educated.

CONCLUSION

The impacts to Muslumovo and neighboring communities are best represented by the concept of cumulative impacts. Although several of the accidents were of massive or large scale, it is the combination of these with chronic smaller releases and smaller accidents that cumulatively produces risk in the region.

The communities that persist in the wake of radioactive releases from Mayak demonstrate significant psycho-social impacts. Illustrated are the complex issues of ecocide that destroy natural systems as well as the lives of those who depend upon them. Such natural resource communities have been well documented in the case of the Exxon Valdex accident (Dyer, Gill, & Picou, 1992). Russians traditionally take considerable food from nature through gathering, fishing, and hunting. When this bounty is then augmented by herds and by produce from their dachas, or garden cottages, a basis for subsistence living is provided, as evident in Techa river region. And it is possibly even truer for native groups living in such settings. But contamination makes unsustainable that which once was. The mixed natural resource/subsistence farming way of life found in this region was decimated by the cumulative disasters. Now the choice is to eat local foods at ones' peril, to sell local food for cash at the peril of unsuspecting customers or to lose the subsistence foundation for life, depending instead on government subsidy or facing poverty.

Furthermore, as established by my collaboration with Hastrup, (Chapter 5 this volume), a close historical relationship to place in combination with resource dependency maximizes the adverse psycho-social impact to toxic victims. Moreover, health issues, such as fear of cancer (or "river disease") combined with other prevalent health conditions, contributes to a substantial loss of normalcy. As we see in both instances, even the act of producing a healthy child, or any child, may be confounded by the conditions of contamination.

We get a taste here of the issues of forced relocation that accompany some toxic events, including the Love Canal case discussed in the first chapter. Here we see the marginalization of those left behind, who must then advocate for their own relocation – an issue found at the heart of the Love Canal event, as well. We also get a sense of how those relocated are also marginalized through environmental stigma. We also see the issues of delayed response. In Edelstein and McVay Hughs's Chapter 18 (this volume) on the World Trade Center Disaster, we learn that residents are exasperated by the five year delay that has occurred there in addressing community impacts. In Muslumovo, in contrast, we see the impacts of a 50-year-delay.

The Muslumovo case also well illustrates how government's response to an event, its short- and long-term assistance, its information policies and practice of informed consent, all contribute to victims' response to the disaster. Here government also caused the problems, as well as exhibiting many failures in its response. Even the act of receiving health care is viewed, in the account of one of our observers, as capture by those studying residents as experimental subjects, the "white mice" referred to in the name of one of the local grassroots groups.

The subject of the local grassroots movement is also interesting. As fully explored in Chapter 8, events along the Techa not only illustrate how people are disabled by contaminating events and government responses, but how they rise above their conditions by organizing and engaging in personal and collective growth in their knowledge of contamination, government policy, health and ecosystem impacts, and alternative courses of action. By the time of this writing knowledge and organizing has spread throughout the region, whereas at the time of our interviews it was more evident in Muslumovo than in other areas. As a result of grassroots clout combined with alliances with outside NGO supporters, progress is also being made toward the goal of voluntary relocation.

This case also raises questions of cultural impact. Here we rely entirely upon the constructed realities of our respondents who carry the baggage of built in historical biases but also the weight of observations garnered personally over time. Suggested are some interesting cultural differences in response to contamination and government efforts to address it. Moreover, an evident environmental injustice has occurred here, exploiting the marginalization of the region's population. It is not just that local Tartar and Baskir people were victimized because the Mayak facility was present, but the plant's location itself reflected the presence of marginalized peoples in the region, considered expendable by the dominating group in control.

In sum, the Mayak contamination has created a condition of Environmental Turbulence for affected populations. This case illustrates that theory related to toxic victimization developed for U.S. applications works well as a structure for understanding psycho-social impacts in an FSU/RF location far away geographically, culturally, and historically.

NOTES

1. As an outgrowth of a Fulbright Scholarship spent with Edelstein in 1999–2000, Tyschiachniouk translated into Russian Edelstein's interview format used in his

contamination research. She then trained an interviewer to undertake a study in Muslumovo, completed in early 2000. Although only a few of the interviews were translated to English and forwarded before a computer crash destroyed the Russian files, Edelstein was briefed by the interviewer on the overall results during his visit to Russia in 2000 and retained his notes. Subsequently, he conducted several additional interviews during his visit to the Cheliabynsk region in 2002. References to interviews in this chapter are drawn from these sources.

2. Our respondents hold thoughts about differences between Bashkir and Tartar peoples that are reflected in our account. As with many of their other statements, we did not have the opportunity to independently confirm these characteristics and differences. Accordingly, please consider such views as those of our informants rather than as our own independent conclusions. They are of interest in no small part as reflections of the social constructions of local people attempting to understand and address the issues that they confront.

3. Ozersk is a ZATO or closed city associated with the Mayak facility (see Kutepova and Tsepilova, this volume).

REFERENCES

Dyer, C., Gill, D., & Picou, S. J. (1992). Social disruption and the Valdez oil spill: Alaskan natives and a natural resource community. *Sociological Spectrum, 12*(2), 105–126.

Edelstein, M. R. (2004). *Contaminated communities: Coping with residential toxic exposure*. London: Westview Press.

ENVIRONMENTAL RISK PERCEPTION AFTER CHERNOBYL

Irina A. Zykova

INTRODUCTION

Personal life experience is not sufficient for an adequate environmental risk evaluation. People cannot understand environmental danger without having necessary information. Once established, however, environmental awareness has a direct influence on people's evaluations and, consequently, on their lifestyles (Sjoberg, 1996).

In the case of radiation risk, the danger is intangible and cannot be felt. There is, therefore, an obvious influence of the knowledge of exposure on the risk's evaluation. In this chapter, I explore radiation risk in the context of the case study of Chernobyl, drawing some conclusions that may be applicable to the broader discussion of environmental risk perception.

Hundreds of thousands of people were exposed to excess radiation as the result of the Chernobyl accident 20 years ago. Russia's central computer database alone contains information on 600,000 people who received radiation doses from the accident (Tzyb, 1998). This population includes those who were engaged in addressing the consequences of the accident as well as the population of the radioactively contaminated territories.

The whole of Russian society was unprepared for the Chernobyl radiation accident. Professionals, central and local public officials and the general public were unable to evaluate the extent and consequences of this occurrence.

Cultures of Contamination: Legacies of Pollution in Russia and the U.S.
Research in Social Problems and Public Policy, Volume 14, 205–221

ISSN: 0196-1152/doi:10.1016/S0196-1152(06)14010-7

As a result, necessary protection measures, such as providing preventive iodine treatment during the first days after the accident, were not taken immediately after the accident (Zykova, Arkhanguelskaya, & Zvonoa, 2001).

The Chernobyl accident caused emotional stress (Liberman, 2002; Shubic, 1996) and gave rise to high general anxiety and a special psychological condition called "radioanxiety" that persists for many years (Zykova et al., 2001). Local people had no adequate concept of radiation and no knowledge of radiation safety. In fact, Soviet propaganda created an idea of the "safe peaceful atom" that operated in the cultural background and, against which, the antithetical radioanxiety developed. Once established, however, radioanxiety introduced expectations of inevitable grave consequences of even small radiation doses affecting one's own health, that of one's children and all future generations.

The accident at the Chernobyl Nuclear Power Station is an example of an environmental disaster in which hundreds of thousands of people are suddenly and violently exposed to a global environmental danger. Protective measures were recommended for all town and village residents. Yet, people did not see the fire at the nuclear power station with their own eyes and the danger was not clear and visible. The disaster was at once widespread and invisible; information and advise about protective measures given to the public afterward caused the population of the contaminated territories to subjectively assess that there was a high level of radiation risk.

For more than 20 years, I have worked for the St. Petersburg (or Leningrad) Institute of Radiation Hygiene. Right after the Chernobyl accident, our staff was deployed to the contaminated Russian territories. We took measurements to evaluate the degree and extent of contamination, to quantify individual radioactive loads and calculate radiation doses received by exposed residents, and we developed radiation protection strategies and behavioral recommendations (Leningrad Institute of Radiation Hygiene, 1990).

From the first months of our work in the contaminated territories, Institute researchers began to understand that social and psychological consequences of the accident (i.e., emotional stress and radioanxiety) were more dangerous to health than the small radiation doses that people received. As a result, it became necessary to study this new condition of radioanxiety in order to understand its main causes. This fascinating issue has remained the focus of my research over the intervening 20 years. These studies were conducted by a team from the Institute of Radiation Hygiene, first under the guidance of Doctor of Medical Sciences, Professor A.N. Liberman (1986–1998) and afterwards under the guidance of and jointly with Doctor of Medical Sciences G.V. Arkhanguelskaya (1988–2006).

Drawing upon studies conducted between 1986 and 2006, this chapter provides an objective evaluation of the Chernobyl accident, as well as evaluations of subjective understandings of radiation danger, the issues of radioanxiety and the print media coverage of radiation.

THE CHERNOBYL ACCIDENT

The Chernobyl Nuclear Power Station is located in the Ukraine close to the border between the Gomel Region of Belarus and the Bryansk Region of Russia. Geographically, it lies almost at the center of Europe. In 1986, all of these territories were part of the USSR.

The Chernobyl accident occurred on the night of April 26, 1986, as a result of human error. The operators switched off the accident protection system while testing maximum reactor power, stopping the flow of water in pipes in the outer reactor wall. During normal operation, this water takes heat from the reactor and uses the heated water to power generators producing electrical power. At maximum power the reactor overheated, boiling away the water in the outer casing pipes, resulting in a thermal steam explosion that destroyed the reactor. This official version of the accident remains controversial; many Russian scientists do not rule out that the occurrence of an uncontrolled chain reaction that caused a real atomic explosion.

The developments that took place after the destruction of the reactor are well known and cannot be disputed. The reactor was filled with graphite, which burned at a high temperature, the fire lasting nine days through May 6. The hot air rising off this hot fire lofted radioactive particles into the upper atmosphere where they were carried to areas very far from Chernobyl. As the air cooled, radioactive particles precipitated out, either in raindrops or as dust particles. To a greater or lesser extent, the contamination affected 101 countries worldwide.

As a result, the accident had unexpectedly grave consequences. The severity of these consequences were determined by two factors, the long period (nine days) during which radioactive particles were launched into the atmosphere and the fact that radioactive particles fell far beyond the theoretically calculated 30 km zone around the Chernobyl plant. This situation was not foreseen in any regulations, instructions or scientific forecasts. The worst contamination was concentrated in three areas:

- directly around the Chernobyl station (the so-called Ukrainian–Belarusian contamination area);
- 200 km to the northwest of Chernobyl (the so-called Bryansk–Belarusian contamination area);
- 500 km to the northeast of Chernobyl (the so-called Kaluga–Tula–Orel contamination area).

In the first days after the accident, the town of Pripyat, only several kilometers from the plant, was evacuated. Evacuation also occurred in the most contaminated areas around Chernobyl and in some villages in the remote Bryansk–Belarusian contamination area.

Unfortunately, contamination of distant areas was not confirmed until several weeks after the accident. All through June and July 1986, radiation hygienists could not believe their eyes as they measured contamination in the landscape and in food products. It took an extended time to evaluate the extent of contamination across vast territories and several years to fully identify the most contaminated areas.

Identification of contaminated Russian territories extended into 1989. The extended period of evaluation protracted the stress of residents waiting to learn if they had to relocate. Afterwards, residents of some areas were offered the opportunity to move to uncontaminated areas of their home region. Decisions about such relocation continued through 1993. In Belarus, the relocation program was terminated as late as in 1998. So, this extended period of evaluation and partial relocation protracted the period of high emotional stress far beyond the accident itself.

As many as 500,000 people still live in radioactively contaminated areas in Russia and they are forced to take protective measures in order to reduce their exposure to radiation.

WHAT WAS THE RADIATION DANGER OF THE CHERNOBYL ACCIDENT?

The main threat to the population of contaminated areas was ingestion or inhalation of radioactive particles with food, water and air. Right after the accident, radioactive iodine posed the greatest radiation danger, but this element decays naturally in 2–3 weeks. Delays in admitting the accident and identifying its extent meant that iodine tablets, capable of blocking the effect of the radioactive iodine, were not distributed until nine days after exposures began. Thus, this protective step was largely ineffectual by the time it was implemented. Of course, radioactive iodine was not the only concern. With a half-life of 30 years, the danger from radioactive cesium (Cs) will persist for many years.

There were also delays in prohibiting consumption of food products grown in the contaminated areas. In any case, such prohibitions have not necessarily deterred consumption of local products. The residents of contaminated areas used (and still use) local food products, milk from their own cows and goats, eggs and meat from their hens and livestock fed primarily on what grew in the fields.

In Russia's Bryansk–Belarusian contamination area, radiation conditions began to stabilize after 1991. Additional radiation exposure was at much lower levels than that encountered in the first days, months and years after the accident. Twenty years after the accident, the population in some territories still needs to use protective measures to reduce doses of radiation to "normative" levels.

SOCIAL AND PSYCHOLOGICAL CONSEQUENCES OF THE CHERNOBYL ACCIDENT

In this section, I draw upon my own experiences in the contaminated territories during the first months after the accident. I witnessed first hand the great psychological difficulties experienced by the residents as they sought to evaluate the radiation danger, understand the necessity of protective measures, visualize the danger and seek somebody to blame.

A participant observation was conducted between May and July 1986 in the Bryansk Region (Bryansk–Belarusian contamination area). By the time of the accident, at the end of April, families had already used up supplies of the past year's crop and were completely dependent upon what they were growing themselves. Food products were not routinely delivered to local stores because there was no demand for them. The incomes of local residents were so low that they usually purchased only bread, cereals, salt and sugar. Uncontaminated products from other regions were very scarce.

Protection measures such as the prohibition of local product consumption were explained by distributing leaflets and in meetings with government officials. Newspapers and radio did not provide such information before mid-summer 1986. Nobody – neither the population, nor the authorities – knew in advance what protection measures would be needed in the case of such an accident. People hardly understood the meaning of the prohibitions recommended for their protection.

Nevertheless, it was evident that such protection measures as the prohibition on using local food products could be implemented only with the personal consent and participation of those who lived in the radioactively contaminated areas. The greatest psychological problem was that people did not understand what had happened. Not having seen the explosion and fire at the Chernobyl Nuclear Power Station with their own eyes, local residents did not understand why they were being told to eat store-bought products. Media coverage of the accident was very limited given suppression of information about the tragic events. Many people merely did not believe that they were facing any danger. A lively and optimistic middle-aged man explained that it took a longtime before he came to believe that there truly was danger.

> It was not until the deputy health minister of the USSR came to our region that I understood that something had happened! We had never before seen a minister here.

The local residents evaluated the danger by such indirect signs as the arrival of commissions and government officials or evacuation of big bosses' children from the contaminated areas.

There was official resistance to some confirming occurrences that might cause panic. After local doctors and officials evacuated their children, Communist Party bosses demanded that their children return home under the threat of expulsion from the party. Party expulsion meant either a demotion with a decrease in salary or dismissal from their high-paying job. So most of the children returned to the hazardous area. As illustrated, it was easier for officials to restrict information about radiation danger than to face a panicked and stressed population.

For similar reasons, the protective prohibitions on consumption of local food products were not officially recommended until 2–3 weeks after the accident. After mid-May, local residents were asked to take some measures for decontaminating food products from radiation. Meat was to be boiled for a longer time, and the broth was to be thrown away or given to farm animals. Villagers made cottage cheese from the milk of their cows more often than before because water-soluble cesium remained in the liquid whey that was then used as a fodder. Milk plants began to accept private milk for processing. All schoolchildren were taken to recreation camps for the summer. Pure products were delivered to pregnant women through children's milk kitchens. Efforts were made to take pregnant women to sanatoriums at the expense of the government, trade unions or enterprises.

Anxiety kept growing as bosses most often did not tell people about the dangers yet demanded the observance of measures including not drinking milk from one's personal cows, delivering milk to state-owned milk plants for processing and even having livestock taken from the owners by force. And, there were instances of contradictory or inconsistent demands. While the pregnant women and older children were evacuated, women with young children were forced to stay behind in contaminated areas due to the shortage of family rest facilities in uncontaminated areas offering the special food and living conditions required by young children.

Psychologically, local people took the emergency situation very hard. Local people asked lots of questions about what was and what was not safe to eat. All residents of these areas lived in villages and small towns. Their diet consisted of the produce of their gardens, forest mushrooms and berries, milk from their cows and goats, eggs and meat from their hens and geese. According to our surveys, 70% of these residents kept eating their products the whole summer after the accident despite the prohibition on eating local products.

Some people tried to comply with the prohibitive measures without understanding the full nature of the risks. In one case, a mother was mindful of the prohibition on eating local meat, milk and pot herbs. She explained how she adjusted the feeding of her three-year-old child.

> The child did not eat the "prohibited" poultry meat. As fresh milk is also prohibited and the child needs some fresh products, we gave him chicken broth.

Unfortunately, this mother was unaware that when she cooked her hens, cesium dissolved easily in the water and contaminated the broth while leaving the boiled meat cesium-free. By complying with the prohibition, she failed rather than succeeded in protecting her child.

In a second case, parents fed their one-year-old child milk from a goat kept in its shed. Even though the animal was fed with fresh grass cut everyday in the field, the parents thought that they were fully complying with the prohibition on allowing livestock to graze in the field. Fed with contaminated grass, the goat produced contaminated milk. When these young children were tested in July 1986, both had very high cesium levels. Their levels were even higher than those found in their parents.

Another woman revealed in private conversation that her mother-in-law delivered only a portion of milk to the government plants, as required. She kept the rest of her cow's milk for her grandchildren to drink. She did not understand that the milk was dangerous. Instead, as an aged woman who had survived the war and many agricultural reforms, she understood the introduction of a mandatory collection of milk by the "government" as a confiscation of her products. She was not about to give away her milk and deny her grandchildren.

The new rules also contradicted the basic customs of the Russian household. In response to the proposal that all food had to be purchased at the store, a young woman with tears in her eyes recounted her concern that she would be considered a bad housekeeper by everyone, including herself.

> What kind of housekeeper will I be, if everything grows around me and I go to the store for every crumb? I'll be a good-for-nothing housekeeper!

At present, village residents in the contaminated areas continue to eat the produce of their farms and get paid for their work in the form of local grain, hay, combined fodder and food products. Yet, at the same time, they are recommended to observe protective measures according to the generally accepted radiation safety principle ALARA ("As Low As Really Achievable").

The prohibitions on consuming local food underscored a basic contradiction in their situation. On the one hand, they were told there was not much danger and that the dangerously contaminated areas had been evacuated immediately after the accident. On the other hand, they were requested to observe very difficult or unfeasible protective measures. Another contradiction occurred with the planned relocation of people from the most contaminated areas begun 3 years after the accident, in 1988 and 1989. The message carried by this relocation was clearly that the prior 3 years had been spent in a dangerous situation.

Relocation, whether immediate or delayed, was a turning point in the life of residents. One elderly woman who had survived the Second World War commented of the evacuation.

> It was worse than during the war.... During the war we did not leave our homeland, we did not leave our places, the graves of our relatives and our homes....

Residents of the contaminated areas circulated rumors, created myths, composed short humoristic songs ("chastooshki") and even wrote long folkloric poems about their experiences. For example, a poem was written about a high-ranking gray-haired professor who came, gathered all young women and undressed them for examination. The story was told as a parable of real events, but with a lewd implication intended as a distraction from the reality. In the summer of 1986, experts and scientists often gathered young women with small children who stayed in the contaminated areas for a lecture on protective measures. In a separate vein, dosimetrists often asked people (both men and women) to take off their working garments during measurements of bodily uptake of radioactive cesium. Their goal was to avoid confounding tests because the clothing was significantly contaminated by radioactive dust. This folkloric composition thus unified two utterly unrelated yet simultaneous events, which appeared to be an open breach of the principles of everyday village life. Moreover, people perceived them as "permitted" by the authority of a highly educated professor. The story thus takes the perceived crash of everyday lifestyle and associates it with "the face" of the guilty, the authorities who allowed this to occur.

People seemed to have an easier time recounting such tales or spreading rumors than coming to grips with the facts of the nuclear accident. A rumor circulated for a rather longtime that the director of the Chernobyl Nuclear Power Station had committed suicide. While it was untrue, it signified that the director was blamed for the accident without any proof of actual culpability.

To understand the danger of radiation, people had to feel it as something visible and tangible. Thus, their stories often made the invisible visible. We were often told about an extraordinary yellow "rain" with bubbles that had fallen after the accident. For several years after the accident, many people spoke about a colored rain that they associated with unusual behavior of their animals. We also heard stories that radiation had fallen as a special rain even before the 1986 accident. Others claimed that they had seen radiation hiding somewhere as if it were a house-spirit.

I share the opinion that the accumulated radiation doses resulted in an insignificant growth in the incidence of thyroid gland cancer (mostly in children) and in a greater incidence of all somatic diseases. However, as is generally accepted, emotional stress and high radioanxiety of local residents are even more important causes of health problems than was the additional radiation exposure.

Researchers have not yet adequately quantified the social and psychological consequences of the accident. There are no generally accepted methodological guidelines for such evaluation and the consequences are very difficult to analyze and forecast. The Chernobyl accident remains in the memory of all Russians as the "people's disaster."

The population of contaminated areas continues to decline. Local medical institutions are 50% understaffed; two hospitals and a maternity department were closed. A 2005 law abolished the free medication previously provided. At the same gradual pace, new houses, townships, specialized medical and social-psychological centers have been built in uncontaminated areas.

Efforts to offer rehabilitation programs for both the population and contaminated areas on a large-scale notwithstanding, problems of individual psychological adjustment often persist. The radiation situation improves, but the psycho-emotional condition of people living in the contaminated areas does not. People view themselves as "victims." Adding to the other aspects of this victimization is an evident dependency. Residents of contaminated areas came to rely on monthly government payments as their main source of income. During the painful revolutionary economic reforms affecting the whole society, this income made for a comparatively successful life. Residents' desire to retain these benefits maintains their mentality as victims, posing a further obstacle to their post-accident psychological rehabilitation.

PUBLIC OPINION ABOUT RADIATION DANGER

The second fragment of the chapter contains results of the author's own studies, defining and describing what constitutes the special condition of radioanxiety. Studies in the form of questionnaire surveys were conducted in three settings of varied radiation exposure: the territories known to have radioactive contamination (Bryansk Region of Russia and Gomel Region of Belarus). potentially contaminated territories (Murmansk, Russia) and St. Petersburg, a large industrial city far from the areas of contamination (Arkhanguelskaya & Zykova, 2000). More than 5,000 people were surveyed between 1988 and 2005.

Given their experience with radiation, one would hardly expect the Russian population to be naive about this hazard. Radiation is clearly not a "tangible" factor. Nevertheless, only a half of the surveyed population agreed with the statement, "It is impossible to define radiation without special devices." The other half were certain that they could detect ionizing radiation from its negative influence over their general state.

During the period of turmoil caused by the Chernobyl disaster combined with the ecological awakening of the Perestroika period, concerns about radiation and contamination dominated public concern. In our early surveys, Russians rated radiation as their greatest risk. Profession and level of education also influenced subjective risk assessment. By 2000, our surveys show that ecological and radiation danger had slipped to second or third place in rankings of everyday danger in recent years in Russia (see Table 1).

Table 1. Rankings of Greatest Contemporary Dangers for Russians 2000.

School Teachers of Ecology ($N = 200$)	Rank	Specialists and Journalists ($N_s = 57/56$)
Bad economic situation	1	Bad economic situation
Ecological radioactive and non-radioactive (chemical) contamination of air, water and food.	2	Lack of legal defence for people
Bad medical care	3	Bad medical care
Lack of the legal defense for people	4	Non-radioactive (chemical) pollution of air, water and foods
	5	Radioactive contamination of air, water and foods

It is interesting that experts and journalists ranked ecological and radioactive threats lower than ecology teachers.

A sample of 200 third year medical university students was next asked to delineate and rank the various risks of radiation. They ranked first danger of radiation from an accident and second the danger from radionuclides in food and from environmental contamination. Their third highest risk was professionally encountered radiation danger. They ranked everyday radiation danger fourth and radiation danger due to roentgen-radiological medical procedures fifth.

These rankings match those found by Paul Slovic and his colleagues over several decades of studying risk perception in Americans (for example, Slovic et al., 1982). Slovic has consistently found that nuclear weapons, nuclear reactor accidents and nuclear weapons fallout are viewed with dread. These threats are seen as catastrophic, involuntary risks with major significance for future generations. In contrast, Americans saw risks from X-ray diagnostics as controllable, easily reduced and voluntary. X-ray risk was not viewed with dread and posed a perceived low risk for future generations. These results appear to reflect personal familiarity and perceived benefit found for medical diagnostic and treatment procedures employing radioactivity or radioactive substances.

However, the public, generally, views radiation danger through a different lens than do scientists employing objective analysis. The latter uses toxicological data to establish proven dose-response relationships. These relationships can then be modeled for the level of exposure, individual doses and collective doses to suggest strategies for radiation protection. The scientist is less interested in the exposure situation than in this objective model.

Without any doubt, the public employs other criteria for evaluating radiation danger, focusing instead on radiation in the context of their lives. Thus, what appears to have been a modest radioactive exposure to me emerges as a major source of stress for exposed residents living in the territories contaminated by Chernobyl.

THE CONCEPT OF RADIOANXIETY

The foundation of "radioanxiety" is the belief by residents of the contaminated territories that radiation is dangerous to their health (see a parallel discussion of "chemophobia" in Filatov et al., this volume). This belief has persisted in 70% of this population since 1986 after the Chernobyl accident. I use the term "radioanxiety" as a substitute for the more commonly encountered term "radiophobia." One finds "radiophobia" used in the Russian mass media since Chernobyl, in Russian school manuals and even in some science publications. However, in 1993 and 1994, scientists discussed terminology and distinguished between "radioanxiety" as indicating stress within an otherwise healthy psychological status and "radiophobia" as indicating psychological disease (Sjoberg, 1996).

Monitoring of radioanxiety was undertaken between 1993 and 2005 in radiation-contaminated and non-contaminated territories as already described. Respondents were asked to rate their level of radioanxiety using a 5-point scale. They also compared the health danger of radiation to everyday health dangers. Not unexpectedly, respondents living in the radiation-contaminated areas revealed much greater radioanxiety, giving a higher evaluation to radiation health risk based upon their personal exposures to radiation. Their evaluations exceeded expert estimates of the objective risk from radiation.

The ratings were also influenced by level of education and professional knowledge. Thus, medical students, other specialists, journalists and physicians assessed the radiation danger for health lower than others did in the radiation-contaminated region. Their subjective ratings had been influenced by the objective assessment of risk.

In these studies, it was interesting to observe that all respondents estimated radiation danger for themselves lower than for the general population of contaminated areas and for professionals dealing with hazards. Respondents saw their children, relatives and neighbors as being in more danger from radiation than themselves. This view feeds the rationalization "I am not in the danger situation, and I don't need in to do anything to protect myself." This finding may also reflect Russian traits of psychological passivity.

Ratings were also made showing preference for measures to increase radiation safety. Here laypeople gave the highest ranking to three steps for which government would be responsible: "regular control of radiation background and environment," "implementation of programs for diminishing radiation doses to the population" and "new legislative decisions." Last on their list was a choice that would transfer responsibility to them for protective actions: "maintenance of elaborated radiation protection precautions by the population themselves."

In contrast to these general ratings, professionals preferred radiation control measures and implementation of state programs to aid affected populations. They evidently understood that it was a hard burden for the population to carry out protective measures, especially in the long run, and that maintenance of protective behaviors would require ongoing intervention by professionals. While a continuation of monthly support to residents of the contaminated areas was favored over Western style insurance, few journalists and specialists found such payments to be suitable. Instead, they preferred support for specific local assistance programs in areas of healthcare, job assistance, social support and economic benefits.

MASS MEDIA AND THE POPULAR UNDERSTANDING OF RISK

In this third section of the chapter, I analyze the Russian print media coverage of radiation danger. The importance of such coverage stems from the fact that radiation danger awareness must be conceptual because radiation is "intangible." Awareness rests mostly on existing knowledge. For local people, the evaluation of radiation danger is determined by the available information about radiation levels, radiation danger and the consequences of radiation risk.

While public opinion may influence mass media in an open society, in a country with hard censorship, information in the mass media reflects the official point of view and government position and not judgements by the greater part of the population. It was not until after 1990 that Russians could find an open sharing of information about hazards in their place of residence. The Chernobyl incident was an influential social experience for subsequent ecological risk communication.

My methodological approach to the study of the process of risk communication involved monitoring the radiation information in the mass media. It was possible to contrast coverage in the radiation-contaminated territories (Gomel and Belarus) with coverage from St. Petersburg, a large industrial city, and Murmansk, an ecologically clean area in spite of the danger from its atomic submarine fleet.

The study analyzed newspapers with the largest subscription, two in Gomel along with a special weekly ecology newspaper; two in St. Petersburg and two in Murmansk. A content analysis was carried out for the years 1996 and 1997 considering the annual number of ecological and radioecological articles, the emotional tone (neutral, intimidating or soothing), the author and the content.

The largest frequency of articles about ecological problems was found in Murmansk, averaging more then one per issue particularly in the summer. Gomel offered half this level of information and St. Petersburg only a quarter. Everywhere, coverage of radioecological problems was one-tenth that of other ecological information. Radiation coverage was greatest in the radiation-contaminated territory (Gomel and Belarus) but nevertheless rare and neutral in tone. Coverage tended to focus on updating the situation rather than discussing public health and protective measures. By contrast, across Russia at large, 70% of the publications discussed radioecological problems, often emotionally appealing for active action.

In Murmansk, radiation information tended to involve short information notes about science conferences and expeditions. Mention of radiological disaster focused on the loss of a submarine a decade earlier, the "Komsomots," and the activity of the organization Bellona, which is concerned with the North Sea environment. Publications on radiation tended to focus on the story of Alexander Nikitin, a military officer who leaked damaging information to the press about the military radioactive storage and contamination from Russian submarines. There was also coverage of international efforts to resolve nuclear safety in the region, including the problems of radiation protection in the atomic industry and the military fleet and radioactive waste storage.

The emotional tone was mainly but not entirely neutral. The coverage did not tend to deal with radioecological problems in Mirmansk. Only coverage of Gomel and Belarus, in the region contaminated by Chernobyl, openly discussed issues of contamination. But here, too, coverage was neutral in tone, omitting both descriptions of the public troubles about radiation contamination and protective measures and also propaganda presenting the official state position on ecological radiation risk assessment.

An important test of post-Chernobyl government openness occurred on May 5 1998, when what I will call the "radiation-fire" incident began. In the face of official silence, information about the incident came from indirect sources. First, "radio-amateurs" overheard naval personnel communicating over non-secure lines and, based upon what they overheard, reported potential danger from a radiation incident involving a fire on a Russian atomic submarine in the Kola Peninsula near Murmansk. Local people began to phone one another, advising each other to stay at home, close windows and evacuate little children. Local radio and the BBC ran radio programs about personal protective measures, recommending that people stay at home, close windows and take iodine.

Official word did not come until some days later when the Russian Navy informed the public that a training accident had occurred. The incident was first reported by a Murmansk newspaper on May 9. On May 16, a Norwegian newspaper published a photo of the burning submarine. Local print journalists knew about the incident from the local radio station and from family and friends. Nevertheless, of six articles in May 1998 on radiation in the local newspaper "Murmanski Vestnik," only one mentioned "the rumor" of a fire on an atomic submarine on May 5 and the fact that people were preparing to evacuate children and that some were taking iodine pills. The paper based its information on May 8 interviews with the regional governor and two senior naval officers. All denied that there had been an accident, describing events as special training. The authorities sought to pass the entire incident off as a mistake of the North Fleet press-service for not clarifying the training mission. The topic was not revisited in additional articles. And further information only came out as a result of an inquiry into an earlier incident on a nuclear sub where a sailor was killed. The navy took the position that information in that incident was secret. By extension, the media assumed the same was true of the fire and withheld coverage.

The second major newspaper in Murmansk, "Poliarnaia Pravda," published an article on May 12, reporting that panic had occurred because the public was unaware of the special training exercise. The article indicated that no radiation had been released, as the incident was not even real. In a later article on June 30, a picture was shown of a billboard on a Murmansk building that read: "We don't trust rumors, but we do sell iodine from sea-cabbage."

The following February, the paper printed an angry piece by a specialist in radiation safety questioning public distrust given that there had been no radiation release. The article failed to explain why the normative dose of background radioactivity had been changed around the time of the incident. Nor did the author use clear language understandable by the average member of the public. The article underscored the reasons for the very distrust that had prompted an article bemoaning the loss of trust.

This brief media review suggests that the lessons that might have been learned from the communication blackout that surrounded Chernobyl have not been learned. Rather, issues of nuclear risk have not reached a point for open disclosure, discussion or public involvement in Russia.

DISCUSSION OF RESULTS

The general public continues to give a high rank to environmental dangers. This concern is not mirrored in mass media coverage. A key reason evident from my research is that both professionals and journalists give a considerably lower estimate of the danger from environment pollution, especially radioactive contamination, than does the public.

The Chernobyl accident was viewed as a "public disaster" by the public and not only for the contaminated territories but for other state territories as well. Two decades after the disaster, when it can be put in better perspective, we see that a high level of radioanxiety is its most important social-psychological consequence. This result stems from the fact that the public understands radiation to be dangerous to their health. The belief that radioactive exposure has occurred and threatens health is well established in the contaminated areas. Everywhere, but particularly in these designated contamination zones, the public ranks the radiation risk for health higher then other health risks. Professional groups tended to downplay this risk, particularly physicians.

The level of radioanxiety contributes to the controversy that continues to surround atomic power energy development. The public's assessment of radiation risk varies greatly from the criteria used by specialists concerned with radiation protection. People examine risk in the context of their lives, not dose-response models. This difference is exaggerated by the fact that specialists refuse to translate their jargon into the vernacular. They seek to reduce social control over atomic projects precisely because they fear the public's radioanxiety. Security concerns are readily cited as an excuse for secrecy and misinformation. The process becomes circular when being "kept in the dark" contributes to this public anxiety.

In a democratic society, we would expect mass media to reflect public concern. That is not yet true in Russia. A reading of the Russian press provides unemotional and routine news and official pronouncements. But information about the consequences of a radiation accident, necessary protective measures or discussions of public opinion about radiation danger are rare. As a result, neither the press nor its official sources are widely trusted. In contrast, the Russian population places a great deal of trust in professionals as evidenced by the fact that between 70 and 90% of the respondents in our surveys gave them the highest rating. While Russians prefer personal meetings with specialists, TV has emerged as the most popular source of information.

The legacy of Chernobyl and the persistence of radioanxiety underscores core needs in an effort to democratize the nation. There is a need for effective two-way radiation risk communication if public opinion is to be heard. The traditional trust by Russians of specialists suggests that experts must be drawn into the everyday communication process using language that can be understood by all. The relationship between the public, specialists and journalists must be nurtured if common ethical positions, points of view and criteria for evaluating danger are to emerge. For now, we must conclude that radioanxiety reflects more than the fear of unseen radioactive dangers. It also reflects a society that does not disclose such hazards in a timely, clear and effective manner.

REFERENCES

Arkhanguelskaya, G., & Zykova, I. (2000). Risk perception of nuclear waste in Barents region. In: P. Axensten (Ed.), *Risk and nuclear waste* (pp. 43–50). CERUM: Umea University.

Leningrad Institute of Radiation Hygiene. (1990). *The main questions of radiation hygiene consequences of the Chernobyl accident*. Science Report number 02.10 037741.

Liberman, A. (2002). *Radiation and stress*. St. Petersburg: Novy Vek.

Shubic, V. (1996). *Immunology health after radiation and ecological accidents*. St. Petersburg.

Sjoberg, L. P. (1996). A discussion of the limitations of the psychometric and cultural theory approaches to risk perception. *Journal of Radiation Protection Dosimetry*, *68*(3/4), 219–225.

Slovic, P., Fischhoff, B., & Lichtenstein, S. (1982). Psychological aspects of risk perception. In: D. Sills, C. P. Wolf & V. Shelanski (Eds), *Accident at three mile island* (pp. 11–42). Boulder: Westview.

Tzyb, A. (1998). Medical consequences of the chernobyl accident. *Journal of Medical Radiology and Radiation Safety*, *43*(1), 18–23.

Zykova, I., Arkhanguelskaya, G., & Zvonoa, I. (2001). *Chernobyl and society: Risks assessments*. St. Petersburg: Russki Esfiqmen.

HABITUAL RISK TAKING IN DZERZHINSK: DAILY LIFE IN THE CAPITAL OF SOVIET CHEMISTRY☆

Alla Bolotova

> Two astronauts land on the moon. One of them has no space suit. "Are you crazy?" the other one asks him. He answers, "Nonsense! I am from Dzerzhinsk." (Anecdote from the 1970s.)

INTRODUCTION

In both natural sciences and social sciences, there is relative agreement about the fact that the 20th century saw great diminishment of the earth's natural resources. In addition to dwindling materials and space for human activities, our industrial mode of natural resource consumption brought various ecological problems, including waste, and pollution of water, soil, and air.[1] The specifics of any given social system influence an individual's perception of pollution of the surrounding environment and its consequences, and also influence the reaction of a society in general to ecological problems. In other words, different societies develop different collective and individual strategies for coping with problematic situations related to "technogenic"

☆Translated by Maria Tysiachniouk and Jonathan Reisman.

Cultures of Contamination: Legacies of Pollution in Russia and the U.S.
Research in Social Problems and Public Policy, Volume 14, 223–252

ISSN: 0196-1152/doi:10.1016/S0196-1152(06)14011-9

pollution of the environment. This article, based as it is on an in-depth case study, analyzes the peculiar relationship of people to ecological issues in Russian society. Research was carried out in the city Dzerzhinsk, which, throughout the Soviet period, was proudly called the "Capital of Soviet Chemistry." This city is thus a demonstrable example of the Soviet period, and the history of the city will serve as a lens through which we will analyze contemporary ecological problems of the city and the relations of its citizens to these problems. Dzerzhinsk was selected for study after it was described in newspapers as "the dirtiest (i.e., most polluted) city in Russia."

In contemporary Russian society, sensitivity to ecological problems is generally relatively low. During the late 1980s and early 1990s, a period of sharp societal change, the USSR saw a flash of interest in ecology. At that time, truly massive environmental movements appeared. These were some of the most powerful and influential protests and political movements of the Perestroika period. At the same time, there was a wave of media publications concerning environmental topics. After only a few years, the public's interest in these topics began to wane next to other issues. However, the environmental movement did not disappear completely in those foggy times of economic crisis. Activists managed to create sustainable NGO networks that remained connected to each other through informal contacts.[2] Currently, between 25,000 and 30,000 people are involved in some Russian environmental organization.[3]

During the Soviet period, closed, mono-industrial cities like Dzerzhinsk served the military-industrial complex.[4] According to experts' assessment, pollution of the environment often reached catastrophic levels (Fedorov, 1995). However, this legacy of contamination has not proven to be the subject of deep public concern. People try not to think about it. Similar denial characterizes the response of the city's power structure, which unanimously chose to ignore the problem, and, in general, keep quiet about ecological issues. Their inaction was supported by developments in the 1990s, principally the shutdown of many industrial plants due to the collapse of militarism and the economic downturn, with a resulting dramatic decline in air pollution. These conditions favored the prevailing and reassuring rationalization "nyet problem" (i.e., no problem).

There is a huge gap between the dire assessment of the situation in Dzerzhinsk made by outside (Russian or Western) experts and the daily awareness carried out by a majority of the city's citizenry. In Western Europe or the US, no doubt, available information suggesting even one tenth the pollution problems in a city similar to Dzerzhinsk would suffice to ignite a powerful environmental protest movement. In post-Perestroika Russia, environmental protest in such cities is rare. Ordinary citizens prefer not to know about the status of their environment. Explaining the public's "blind eye" to the environment, therefore, is the major research question for this investigation.

Material needs are currently the most poignant problems for many people in Dzerzhinsk. It is tempting to conclude that more abstract issues of environmental risk are ignored because peoples' consciousness is focused on survival, financial security, and dealing with materialistic needs. However, this is a simplistic conclusion. For one thing, it is notable that even affluent residents do not show concern about the environment. The Russian roots for ignoring risk are much deeper than simple "Maslovian" economic explanations.

The social history of Dzerzhinsk has played a significant role in the development of this risk acceptance. Of specific importance are the absence of any "public space" for discussing risk and the strategy of ignoring the problem adopted by the power structure. In these ways, Dzerzhinsk is representative of all of Soviet society, especially the formerly closed cities such as Dzerzhinsk, which served the military-industrial complex. The creation and functioning of these technological systems and the problems they caused resulted directly from the core characteristics of Soviet technocratic society, specifically the long-lasting atmosphere of secrecy, neglect of worker safety issues, and inattention to the needs of citizens. Dzerzhinsk demonstrates how 70 years of Soviet technological experimentation established a dangerous legacy for present and future citizens of the country – not just in creating environmental hazards but also in deadening people to their consequences.

Information was gathered using a qualitative approach. Field research occurred during two site visits, in 2002 and 2003. In total, 28 in-depth interviews were conducted with citizens of Dzerzhinsk selected through a snowball approach. The sample included five interviews with long-term residents, three with experts, two with environmental activists, six with other socially active residents, and the remaining 12 with ordinary Dzerzhinsk citizens of varied age. Overall the sample was biased in favor of older informants in order to assure a historical perspective. Our investigation also uses information from a roundtable discussion conducted in August 2002 as part of the Russian-American project "Contaminated Communities of Russia and the United States." In addition, we collected a large archive of public material about the city and its citizens.[5]

ECOLOGICAL RISKS AS COLLECTIVE CULTURAL CONSTRUCTS

Contamination of the natural environment is not an objective phenomenon, even though it is possible to study it using strict scientific methods. Modern society often creates risks that are invisible and cannot be assessed by the human senses (Beck, 2000). We accumulate information about these risks on the basis of expert assessment and interpretation. However factual in form this information appears to be, this expertise is itself very much the product of historical, political, and social process.

But is expert subjectivity better equipped to see invisible contaminants than is the public's perception? Here we return to our core questions. How is it that, in Russia, little public concern over environmental problems is manifested, even in those regions where pollution levels are viewed by experts as catastrophic? In such cases, the experts, who represent a scientific type of rationality, are more likely to be concerned than local citizens that deal with the risks everyday and who have an on-the-ground knowledge and rationality of a different type. How is such a discrepancy in risk perception possible? What are the social roots of different interpretations of risk, and, in particular, the tendency to ignore risk?

Risk is a term with somewhat varied meaning across different branches of science, engineering, psychology, economics, sociology, and anthropology. In the post-modern era, risk reflects the systematic interaction between society and the different threats and dangers caused by modernization and is magnified by uncertainty and fear (Beck, 1992). Widespread social mobilization in the west against environmental threats is explained by such perceived risks and fears. This does not mean that all fears are recognized and lead to action or that an ignored risk is an unfeared risk. Fear may itself result in the neglect of some danger (Beck, 2000).

There are many other possible reasons for inaction. People may know of a risk and yet not be conscious of it. This lack of active awareness may be due to the hazard's invisibility, a lack of reminders about the risk, a lack of any perceived ability to control it, and/or the fact that risks have become the realm of experts possessing tools for measurements and means of action (see Beck, 2000).

Because risks are collective cultural constructs, it follows that the risk that threatens one group might have a very different meaning and urgency for another. Societies, having different characteristics of social life, might adopt very different strategies for dealing with the same type of risk (Douglas &

Wildavsky, 1983). Thus, the comparison of varied societies offers a ripe potential for illuminating the subjectivity of risk.

One finds perfect material for comparative analysis of risk perception in the contrast of socialist and capitalist systems. Both societies contain large-scale, risky enterprises oriented toward industrial development. However, significant discrepancies are evident in the perception of pollution between socialist and capitalist countries. That these differences are due, in part, to social dynamics is suggested by the contrast of public perceptions of risk between citizens of the former West and East Germany. While technological risk played a major role in political discourse in the west, East Germans lacked any public forum for such debate and risks were seemingly unimportant to East Germans (Hunnius & Kliemt, 1993).

How then do Russians perceive risk? Yanitsky explains the tendency to appear apathetic to environmental risk by citing the demands on Russians living under the strain of material need to address their own security and survival. But his point is not just that they are distracted by more basic priorities. Rather, for Russians, all of life is a fight for survival. As a result, people accustomed to being continually forced to mobilize under extreme circumstances become desensitized to risk. The result is a society with an extraordinarily high tolerance of risk (Yanitsky, 2003).

Modernization brought with it a new class system largely unnoticed in a society focused upon removing all vestiges of traditional stratification. Yanitsky (2003) writes that this new type of social inequality was between the producers and consumers of risk. He might better have argued that it was between those that profit from risk production and those victimized by it (Edelstein, 2004).

Before shifting to our case study of Dzerzhinsk, it may be useful to also consider the phenomenon of Russian's acquiescence to risk. This acquiescence reflects basic personal and cultural tendencies for dealing with bad news. Environmental crises seem so difficult and complex and often so depressing and hopeless that it is easiest to adapt a strategy of ignoring unfavorable information. Then, there is the importance of Soviet history. During the entire existence of the USSR, the ruling Soviets paid very little attention to the issues of industry and its effect on the health of the population and the environment. The system as a whole ignored environmental risks, hiding them, keeping them quiet, and deeming information about environmental problems' secret. For an entire society, such issues came to be viewed as inappropriate for discussion and action. As a result, individuals were forced to develop their own strategies for adapting to the resulting problems, often with help from their relatives and friends.

The perpetual lack of public space for open discussions about problems clearly influenced the perception of environmental risks among Dzerzhinsk's citizens. Risk discussion appeared only after Perestroika, when the city changed its closed status and the regime of secrecy softened somewhat. In place of a severely limited and controlled press, a whole wave of environmental coverage appeared in newspapers. However, widespread interest in these topics waned, and the space for risk discussion once again shrunk. As media increasingly became the voice of the new power structures, it came to reflect powerful business interests. It ceased to serve the role in a free society of a major channel for spreading information openly, providing the public with access to critical expert systems, giving the public information about the state of the environment, and offering an arena for discussion of different assessments of risk. An informational vacuum was created.

The historic lack of opportunities for direct political action on behalf of communities and causes also served as a constraint. Lacking were the legal and political strategies of recourse upon which grassroots responses might mobilize.

In sum, in a society in which secrecy and maintaining a quiet passivity toward environmental problems is a phenomenon of the system, people incorporate the strategy of ignoring dangers into their individual lives, and so have little protection against these dangers.

AN ECO-HISTORY OF DZERZHINSK

Edelstein (2004) posits that the "eco-history" of a community is an important starting point for understanding how pollution problems are addressed. Dzerzhinsk is a city of a quarter million in Nizhny Novgorod Oblast, 30 kilometers from the city Nizhny Novgorod. Founded in 1930, Dzerzhinsk was built as an industrial city dedicated to chemical enterprises, which employed the majority of its citizens during Soviet times. Over much of its history, Dzerzhinsk produced chemical armaments and poisonous gases. Today, 3 of 15 enterprises involve chemical processing.

We identify three distinct historical stages in the development of Dzerzhinsk. These are the Stalin era of the 1930s and 1940s, the Cold War Period (particularly the 1960s and 1970s) when large-scale socialist industrial and social development occurred, and pre-Perestroika to the present (from the middle of 1980s) when chemical plants declined and social and economic crisis hit the city. These three stages will prove useful in understanding the foundations of environmental perception in Dzerzhinsk.

STRATEGIC BUILDING CONSTRUCTION OF THE SOCIALIST CITY OF DZERZHINSK

> Now, on the site of this abandoned place, the Soviet power has decided to build a big industrial city. In a number of years, you will not recognize your small settlement.
> M. I. Kalinin, "All-Union Governor," from his speech to the meeting of laborers of the future City of Dzerzhinsk (1927).

By the end of the 1920s, Russia had been weakened by the lengthy period of civil war, there was much destruction, and many citizens faced hunger. The government decided to force industrialization of the country. Chemical industry was given the highest priority. An infrastructure for industrial chemistry was viewed as necessary to defense production in order to prepare for another war and eliminate dependence on imports.

The "City of Big Chemistry" was to be located nearby pre-revolutionary workers' settlements, notably the local village Rastiapino. By 1929, the decision was reached to combine several of these working settlements into an industrial city named Dzerzhinsk.[6] This site offered several strategic advantages. First, transportation costs were reduced due to the proximity to rail, the river Oka, and to the regional capital, Nizhny Novogorod. Second, the site was far enough from national borders to allow defense in the case of war. Third, there was a pre-revolutionary base of chemical enterprises that could serve as a foundation for building new enterprises. Fourth, the surrounding villages became an important source of labor for the new proletariat as hungry peasants flocked to the city to earn money.[7]

Decision making in Stalin's times failed to take into account ecological factors that, in retrospect, suggest that the site was really not so ideal. The locality can be characterized by sand and limestone deposits with high levels of diffuseness. The sand layers are near the surface and serve to quickly disperse contaminated liquids through the water table, concentrating surface pollution in the Oka River.[8] At the same time, the dolomitic-limestone bedrock below the sand is Karst topography, riddled with solution cavities allowing pollution to move easily down into groundwater. In short, the environmental features of the site could not hold greater potential for environmental harm.

The new industrial city was to produce enamel for metals, phosphorous for matches, and poisonous gases for military weapons. Technology and equipment for the chemical enterprises were acquired from abroad. By the start of World War II, eight chemical production facilities operated in Dzerzhinsk. Because the main emphasis at that time was the production of poisonous substances, bombs, weapons, and other means of chemical defense, the city was veiled in an atmosphere of super secrecy.

The peasant workers relocating from failed collective farms came with no qualifications for their jobs. Already traumatized, they were willing to accept low wages and a harsh quality of life. If they were unable to walk to work from their own villages, they were forced to live nearby their factory in barracks housing 40–50 families and lacking indoor plumbing. Some workers lived in earthen huts, even in the wintertime (Kotlyar, 2001).

The intention was not to create hard living conditions for laborers. But promises of good housing, transportation, utilities, and cultural institutions were delayed by the state of emergency in the pre-war and the war years. All industrial enterprises were required to work at full capacity, as suggested by the slogan, "the plan must be completed at any cost." Not only were all other needs relegated to a minimum level, but worker safety was also ignored and many mishaps occurred.

The quality of life in the 1930s was further reflected in the high level of mortality from diseases and malnutrition. An engineer describes worker conditions in Dzerzhinsk:

> The tram going to the city around 8 o'clock in the morning comes to a stop at Igomnovo. It was half filled with workers returning from the night shifts, tired, with pale, overtired faces. Many of them were coughing the whole way. They were squinting from any light that shone upon their sick, red eyes. At the stop Kalininskaya, they were joined by another group of the same overtired tormented people. Many of them had faces and hands colored a bright yellow. It was possible only to guess that their skin is penetrated by nitrogen oxide. (Kotlyar, 2001, p. 10).

As better housing was finally built, "temporary" worker settlements were created in close proximity to the factories. This design allowed a longer workday by eliminating the inefficiencies of commuting. It is said that, in Russia, there is nothing more permanent than the temporary. As a result, residents today can still plainly see nearby factories from their orchards and homes.

Beyond the workers, other social groups included the administrative personnel of the enterprises and a class of technocrats called the "young specialists." These ranks were filled by young graduates of universities and middle level chemistry institutions from across Russia who were assigned mandatory jobs in Dzerzhinsk. A series of chemical, technical, professional education institutions was also created to help fill the need for specialists (Shalnov, 1999, p. 88).

Living conditions for qualified personnel was slightly better than for workers. They lived in hotels or in multi-level apartment houses with bathrooms and other standard luxuries. However, as one of the veterans of "Dzerzhinsk Chemistry" recalled:

> Life in the emergency conditions, where we were placed, could seem to one like heaven if he does not take into account the air in the settlement, filled with poisonous gases and steams discharged by the plants. In this "heaven," people were permanently sick, and some of them slowly died. (Kotlyar, 2001, p. 10)

But what forces make people tolerate such hardships of life? The majority of the workers and engineers were relatively young. The Soviet regime relied upon the stamina inherent in their youth to motivate them to work with enthusiasm. Propaganda was used to intensify their belief in the ultimate progress of the socialist system despite "temporary difficulties." Indeed, the myth of "the bright future" was one of the most influential ideas in the Soviet Union in the 1930s. In Dzerzhinsk, as in the rest of the country, worker motivation was further spurred by "the Stachanov Movement." Named after a record-setting laborer, this government designed and maintained productivity program proposed the need for workers to go beyond the formal plan of production and become "labor heroes," earning special rewards. Local officials often inflated their production numbers in order to show many "Stachanovsky workers" and thus acquire additional resources for their enterprises. This form of "paper heroism" encouraged violations of worker safety measures, as laborers sought to become labor heroes.

HARD TIMES: A PLAN AT ANY PRICE

By the mid-1930s, a more complex system of labor was established for chemical production facilities that dramatically increased the pressure on workers. Production goals were further separated from reality. Ideological propaganda was used to directly stimulate superhuman efforts. In combination with repression, there was an atmosphere of great fear.

Socialist economic planning was increasingly used to enhance the industrial power of the USSR. Industries were given quotas to meet. The major slogan of the time was "the Plan at any Cost." Failure to complete the plan could cause punishment, not only for the individual violator, but also for the whole work collective. The result was a "whole encyclopedia of falsifications" aimed at salvaging the reputation of an enterprise and the safety of its workers (Kotlyar, 2001, p. 28).

Workers at the chemical enterprises in Dzerzhinsk, as in the whole country, were exposed to large-scale repressions. Industrial sabotage was a major excuse for arrest, and failure to fulfill the plan was deemed equal to sabotage. The secrecy of military enterprises provided authorities with freedom to quickly "clean out" all "bad elements." People lived under the permanent fear of being arrested. These pressures were increased by the military context. Given the inexperienced and increasingly disillusioned work force and volatile work environment, actual completion of the plan was only possible with widespread violations of worker safety. Emergency situations and accidents were ordinary occurrences. The health of the worker and of the environment held a comparatively low value under the circumstances (Kotlyar, 2001, p. 28).

By the end of the 1930s, feelings of impending war permeated Soviet society, from the country's leadership to the everyday lives of ordinary citizens. Everyone lived with the thought constantly in mind "if there is a war tomorrow." The country's preparation for the war included emphasizing enterprises of chemical armaments. At this time in Dzerzhinsk, three enterprises for the production of chemical armaments were functioning. They produced tetraethyl lead for airplanes, explosives, and a big quantity of chemical defense supplies. Later, during the war the slogan "plan at any cost" was replaced with a new slogan, "everything for the front, everything for the victory." Now even more superhuman efforts were demanded of the industrial workers.

Spurred by the urgency of the war effort, the industrial environment was rife with environmental crime and worker injury. As an example, Kotlyar (2001, p. 15) writes of his experience working in Dzerzhinsk at a production unit for the poisonous gas Yperite:

> The production of Yperite where we were sent was pretty much primitive without modern technological measures of protection. The air in the enterprises was filled with Yperite fumes. Frequent discharges of Yperite were covered with saw dust and the floor was cleaned with Chlorine chalk. Masks against gas, and rubber suits, shoes, and gloves, failed to prevent harm to the skin, eyes, and trachea. That is why every shift had twice the number of people, half for working while the others were in the hospital or in profilactoreum.[9]

The double-staffed shift was a cynical solution to the problem of ever-present diseases among the staff. The scale of harm from poisonous substances at a Dzerzhinsk chemical industry at this time is suggested by 1942 data from enterprise number 96 (now known as Kaprolaktam). At just this one plant, there were 2,397 cases of poisoning by Yperite and another 89 by another poisonous gas, Lewisite (Fedorov, 1995).

Work in a special chemical weapons' production unit had the most serious consequences for health. Within a special unit, Yperite concentrations often exceeded current permissible standards. This air was ventilated directly into the atmosphere without any cleaning or treatment. There was no sanitarian protective zone around enterprises. Living quarters were adjacent to the plant. As a result, workers were exposed not only at work, but also with their families at home (Fedorov, 1995). Young engineers of the time later reported that they ignored safety instructions, not appreciating that their exposures were cumulative. "We understood that only much later when retired workers from our special unit began to die while still young" (Kotlyar, 2001, p. 11). Workers were denied any choice. They had to work where they were assigned. Given the intolerable conditions, it is no wonder that more than 500 people are recorded as having escaped from Dzerzhinsk during the war (Fedorov, 1995).

In the period after World War II, the revitalization of Russia's industry continued to hang like a knife above the heads of industrial laborers, along with severe military secrecy and threats of repression. Living conditions continued to be very hard, and the health of the workers of chemical enterprises continued to be damaged by exposure to different chemical substances released during permanent violations of worker safety and frequent emergency disasters. Beside the severe conditions of life, people were working very hard. It was necessary to rebuild the country after the war, and a majority of people continued to believe that the socialist system would reward their work and that the proverbial "bright future" would come soon. After the death of Stalin, the political situation in the country changed significantly. Life and working conditions also changed.

"BRIGHT PRESENT: IN DDTGRAD"

"You know, we were more working than living." (From an interview)

After the war, the production of chemical armaments in Dzerzhinsk was decreased and production of poisonous gasses such as Lewisite and later Yperite ceased. As the Cold War period commenced, the military-chemical orientation continued to dominate the industry of Dzerzhinsk even as some defense enterprises were reoriented to "peaceful production." New scientific research institutions and new industries were founded.

Here as elsewhere, an active new war was declared on pests as part of an effort to rebuild the agricultural productivity of the nation after the war. For instance, in 1965, two plants producing chemical armaments switched to producing the pesticide dichlorodiphenyltrichloroethane (DDT), referred to as "dust." Ironically, the environmental harm caused by production of DDT was comparable with the prior production of chemical armaments. And, instead of being known as the "Capital of Soviet Chemistry," Dzerzhinsk received a popular new nickname, "Dustograd" (i.e., "DDT City"). While Dzerzhinsk continued to be a secretive and closed city, the 1960s and 1970s were a time of pride in the city's status as a chemical center. Old-timers recall this as a period of stability and prosperity when the lives of citizens became significantly easier. The fear of repression disappeared and the system of social welfare improved. New technologies for the chemical enterprises appeared, as did better living conditions and cultural institutions.

In the 1950s, officials had developed a plan to reduce the harm caused by chemical exposures for workers by creating "sanitarian protection zones" around various plants. Implementation began a decade later. Later plans were even more protective. Worker settlements were to be relocated, and green buffer zones created around the enterprises. The result would be three rings around the plants. The sanitarian zone was not to be inhabited at all; the buffer zone was to be used only for recreation, and the settlement zone was for living.

In the late 1960s, Premier Khrushchov undertook the "habitat" program of intensive apartment construction to give all Russians housing. At that time the living conditions in Dzerzhinsk were a very pressing problem. For example, plant 96 (i.e., Kaprolaktam) was totally surrounded by settlements consisting mainly of barracks or private houses without bathrooms. Four settlements stood only 1–2 km from the plant. A kindergarten for workers' children stood only 500 m away. Data from the early 1960s revealed that the concentration of toxic substances in the sleeping rooms of this kindergarten equaled and at times exceeded concentrations in the plant itself (Fedorov, 1995). Extensive building under the Habitat Program allowed many of the workers living just "under the fence" of the chemical factories to move into the city, 7 km away. They moved into multiple-apartment houses with cramped, low-quality apartments derogatorily called "Khrushchovka."

This program of relocating industrial settlements continued until the end of the Soviet period but was never completed. Success was achieved with the primary objectives. A majority of workers in Dzerzhinsk's factories were resettled away from the industrial zones and chemical plants slightly decreased the levels of air pollution. However, in present-day Dzerzhinsk, citizens continue to live in three settlements abutting chemical plants and long designated for replacement. Creation of green zones was also never completed. Often trees could barely survive because the soil was so contaminated.

The post-war period saw some recompense for Dzerzhinsk's previous privation. Old-timers recall that there was a "good" supply of food in the city in the 1960s and 1970s, compared both to the previous wartime period and also to the perpetual food shortages faced by conventional "open" cities during the Cold War period. This "elite" food supply drew people from as far away as Gorky who came to make purchases. Such recognitions boosted civic pride and the resident's sense of well-being. Under the Soviet system of social protection, such benefits were to be permanent. Employees of chemical factories regularly received vacation tickets to the Black Sea with all expenses paid. And for health, they were routinely sent to "profilaktoreums," spas where they could eat, relax, and exercise, and "sanitariums," places in nature where people were sent to restore their health. By the 1980s, each big enterprise in Dzerzhinsk had its own health center where workers could periodically vacation free of charge.

In general the Cold War Soviet state became a "state of overall wealth." The state and the Communist party were major distributors of goods and services, paternalistically caring for the populous. Now, approaching 20 years after the end of the U.S.S.R., elderly citizens of Dzerzhinsk recall with a sense of nostalgia the time when the state was really taking care of them. But there was a downside to the era as well. The planned economy continued to force enterprise leaders and workers to violate safety rules in order to fulfill scheduled tasks. Negligence, a lack of security, and theft became systemic characteristics of Soviet industry (Graham, 2000). In the chemical factories, such lapses were quite dangerous. It was as if the good times invited even more risks. For example, one 65-year-old informant recalled her time working at the Sintez plant, where tetraethyl lead was previously produced:

> People themselves were negligent. On the production of ethyl, people began to get employment through personal contacts and friends. There were good salaries, a wonderful schedule, three days of work, and two days of rest. Their special profilactoreums were great, and there were pensions for men in their 50s and for women at 45 years old. And people were not paying attention to dangers.

In reality, people knew little about the potential harm to their health or about the polluted environment. After all, information on this topic was strictly closed. For example, evidence that workers at chemical armament factories had been exposed to contaminants was suppressed under a 1933 order claiming "absolute secrecy" over such information.[10] Information was also distorted. Officially, the harmful substances in the air were presented as occurring at lower concentrations than the norm, and, therefore, of no threat to workers at the enterprise. Evidence of the threat due to bioaccumulation of hazardous substances was suppressed.

In sum, the Cold War period saw improved living conditions for many citizens of Dzerzhinsk. The system of social protection guaranteed compensation for harm to workers' health in the chemical industry. These positive changes gave residents a psychological sense of realization of the long promised "bright future." But with 15 chemical facilities operating in the city, environmental problems were accumulating. Through the 1970s, the increased volumes of industrial production combined with the startup of new enterprises created a situation where smoke was so dense above the city that the green protection zones were insufficient for plumes to be dispersed and treatment plants were overwhelmed.

THE CRASH OF THE SOVIET CHEMICAL INDUSTRY

With the beginning of Perestroika in the Soviet Union, it became clear that the technical infrastructure of Soviet chemistry had become badly aged. Formerly powerful and long working chemical enterprises lost business. The near collapse of the Soviet Union's military-industrial complex brought down closely linked industries. Continuing state reforms further decreased production until many other enterprises stopped working altogether. While patriotic Dzerzhinsk citizens viewed this crash as the work of competitors trying to steal their place in the market, the fact was that Dzerzhinsk's enterprises were no longer competitive. Their technology was outmoded, and they were economically inefficient.

This abrupt downturn fomented a crisis in the city. In post-Perestroika years, the quality of life among citizens fell dramatically. Once again material needs came into the picture. Pensioners who gave their lives and health to what they termed their "native enterprises" were now forced to live with miserable pensions averaging \$30–\$90 per month. Many returned annually to the local disease center to renew their "sick" status, entitling them to such benefits as a higher pension and tickets to profilaktoreums. In this way, illness becomes a kind of symbolic capital. An individual's sickness can be converted to economic gain.

Youths also became severely disadvantaged. Young people sought to leave the city, partially because of its high level of unemployment and the lack of opportunities. Local universities were created to prepare students for chemical professions that were no longer in demand. Old people and other low-income groups focused on the challenges of survival in a time of hard economics. In the resulting atmosphere of social apathy, alcoholism and an acceptance of ill health prevailed.[11]

As the downturn spreads, the potential for catastrophe is never far from the surface. The current opportunity created by reduced factory operations, which limits new pollution production, does nothing to remove the huge repository of past contaminants vulnerably situated atop the Karst topography. Thus, there is an ever-present threat of disaster associated with past environmental sins that a period of abstinence cannot cure. The effect was seen, for example, a number of years ago when the Korund plant could not pay its energy bill. Facing the prospect that the plant's power would be shut off, environmental organizations were forced to beg the utilities and government to not throw the switch. As a former manufacturer of chemical arms and poisonous gasses, with huge remaining stockpiles that could not be controlled without inputs of energy, this shutdown would certainly have spelled disaster.[12]

In 2003, a special governmental commission was formed to assess the chemical enterprises of Dzerzhinsk and develop strategies and solutions. Options are limited by the antiquated technology and the large investment required for new equipment. Potential investors are additionally frightened off by the ecological problems that have accumulated over the years.

The positive side of the industrial crisis is the opportunity to create alternative types of development. Only leaner, smaller, and demand-oriented enterprises can survive the new conditions. As a result, a diversity of small businesses has blossomed, manufacturing products including domestic chemicals, detergents, and building materials.[13] These small businesses will never replace the economic base represented by the chemical industry. But they represent an important reorganization of the city's economic structure.

The civic life of Dzerzhinsk is also modest. Of the 200 registered NGOs, only one standing organization, Social-Legal Environmental Association (SPES), actively addresses ecological issues. The Regional History club is also of interest because it attracts seniors who reminisce about the area prior to industrialization, when it was a cottage district. The records of such clubs reflect a strong nostalgia for not only the long-lost ecological health of the region stretching back to the early 1900s, but also the material security of the late Soviet period.

Another observation about Dzerzhinsk applies nationwide. As industrial production dropped in the 1980s, concern over environmental problems eased. This relaxed concern directly reflects the fact that air quality visibly improved with plant slowdowns and shutdowns. On the other hand, the invisible legacy of soil and water contamination remained. But, lacking a concerned public constituency, it has not been in the interest of political and economic forces to recognize these problems. They are instead pushed under the rug. It remains for small groups of activists to make claims to largely indifferent audiences that the city is a chemical time bomb ready to explode. But is there such a time bomb?

A TICKING CHEMICAL TIME BOMB

> A group of children were taken to a summer camp to relax in a beautiful and clean place. When they started to get out of the bath, surrounded by beauty and nature, suddenly three boys lost consciousness. "What happened?" screamed one of the attendants. "Don't worry," said another, "they are from Dzerzhinsk. Throw them into the cesspool and they will recover." (Anecdote heard repeatedly in Dzerzhinsk).

During Soviet times, there were many environmental anecdotes about Dzerzhinsk. All focused on the challenges of surviving there under such extreme conditions with so little chance to breathe fresh air. The telling of such sad jokes by citizens of the city helped them live with the obvious pollution of the 1970s and 1980s, when Dzerzhinsk was practically engulfed in smoke. Citizens could tell which way the wind was blowing by identifying which plant's pollutants were in their nostrils. At this time, despite rhetoric about being the capital of Russian chemistry, officials of Dzerzhinsk fundamentally ignored the fact that the city was a mixing bowl for an unbelievably diverse chemical soup and did only the most limited chemical monitoring. As a result, only limited data now exists to characterize the harm to people and environment from 70 years of chemical production. Adding to the lack of pollution measurements was the legacy of arms development and testing, which demanded extreme discretion. Not only didn't people share what they knew, but they asked few questions and didn't expect to be informed. Thus, they often knew little about what even their own firm produced and its uses. Public information was not available, in many cases, until after 1997 when Russia ratified the international convention on prohibition of chemical armaments.[14] Most residents are still fundamentally uninformed.

The biggest unresolved issue in Dzerzhinsk is the liquidation of wastes from the chemical armament enterprises. Past efforts to dispose of these materials posed additional threats to residents and the environment alike. Beginning in the 1950s, waste from the production of Yperite and Lewisite was crudely burned in a large oven, exposing both workers and nearby residents to potential dangerous discharges (Fedorov, 1995). Unauthorized dumping occurred in the region's forests and settlements, with sacks of unknown substances, building debris and metal scrap still visible. Every hole in the ground was filled with wastes. Most often, wastes were also stored onsite at each chemical enterprise. At Kaprolaktam, for example, a huge lagoon was dubbed "the White Sea" by residents. The some 10 million tons of waste in this lagoon represents a "hellish mixture" of chemicals sitting in a hole in the sand.[15]

Predicting the fate of these wastes is complicated by the underlying Karst topography. A common feature of such dolomitic limestone formations are sink holes, caves, and solution cavities. Over the 70 years of industrial practice in Dzerzhinsk, some 5,000 sinkholes have been recorded, ranging from small to huge. In 1988, as a case in point, a large sinkhole opened suddenly, swallowing a newly built unit at the plant Kaprolaktam where 500 workers worked. In the absence of any plan for secure storage or disposal of wastes, these holes are invitations for dumping, many serving as industrial waste disposal sites. Neither covered to slow water infiltration, nor lined, these sites all represent sources for continuing groundwater contamination as liquids drain quickly though the Karstic rock formations. This very outcome was demonstrated after officials approved the dumping of the industrial waste ammonium sulfate. The compound quickly appeared in groundwater throughout the entire region.

Both activists and officials agree on the magnitude of this problem. A 1999 municipal report on the "state of the environment" does not mince words, concluding that the 15 local enterprises use powerful and active hazardous and poisonous substances. Of 190 chemicals discharged to the air, 40 percent were classified as highly toxic. Thirty-six of these compounds are described as being carcinogens or mutagens. Chlorinated compounds are found to be poisoning not only a large area of land surface but also to have penetrated the ground to a depth of 80 meters. Similar dispersal figures are reported for oxide ethylene, phosgene, hydrogen cyanide, acid nitrile, and ammonia. The waters and sediments of the Volosanicha river, a tributary of the Oka and Volga, are heavily contaminated with DDT, hexachlorobenzol, and trichlorophenol.[16]

A worst-case scenario constructed for the 15 plants anticipates a chemical disaster at one that spreads to others through a domino effect. In this instance, it is projected that 100,000 people would die and twice that number be injured. In fact, a series of smaller but significant industrial accidents occurred during the time of the chemical weapons' production.[17]

- In October 1941, an explosion occurred in the facility producing Yperite at plant number 96 (now Kaprolaktam), killing 24 people.
- Of seven people poisoned at plant number 148 (now Orgsteklo), two died.
- In July 1947, at Kaprolaktam, chlorine was discharged, poisoning 28 people.
- In June 1949, at plant number 148 (Orgsteklo), one person died of Phosgene gas poisoning.
- In February 1960, an explosion in the Yperite facility killed 24 people.
- In December 1967, a discharge of phosgene from the Chernorechensky Chemical Plant (now Korund) injured 78 people.
- In January 1974, a discharge of liquid chlorine at plant number 96 (Kaprolaktam) severely poisoned 11 people.
- In November 1987, a discharge of hydrocyanic acid at Orgsteklo killed one worker.

Less visible than such accidents is everyday consumption of chemically polluted products by citizens. Pensioners and others across Dzerzhinsk are forced to grow vegetables and build orchards for their subsistence. In fact, continuing the tradition of Soviet times, most families have dachas with orchards and gardens, many situated close to the Industrial zone and sometimes even inside the zone. Soils are polluted throughout the Dzerzhinsk region. Moreover, only contaminated water is available for irrigating plants.

The situation is most dire for the three settlements nearby the industrial zone that were never relocated. Carrots, beets, and potatoes grown in these settlements have been found to be contaminated with DDT, lead, and other pollutants. Despite the knowledge of these dangers, authorities have done nothing to prevent exposures, and the residents, themselves, evidence little concern. Local people continue to consume "gifts from nature." An estimated half engage in fishing, a third in gathering mushrooms, and many additionally collect berries. Not long ago, an English visitor took home a small gift package of harvested mushrooms from her hosts in Dzerzhinsk. Curious, she sent them out to be tested for dioxin, finding levels 556 times higher than permissible levels. Yet, residents continue to consume such foods over the course of their lives.

TERMINALLY ILL VILLAGE

> When people talk, they say "anyway we will not die, anyway we will not die!" (interview of a resident)

The most tragic situation in Dzerzhinsk involves residents of Babino, Igumnovo, and Petryaevka, the three industrial settlements that were never relocated. Residents not only live out their whole lives within sight of chemical plants, but their bodies are also perpetually exposed to the influence of multiple chemical compounds. These residents also grow gardens and keep orchards. An official of the Ministry of Natural Resources described the situation:

> A nightmare! Their water is impossible to use even for industrial purposes, yet they water plants with it. They water it and then eat their tomatoes. That's why the sickness there is high. These settlements are certainly terminally ill!

Water in the settlements was poisoned long ago. Efforts to pipe in water from new and safe sources largely failed, forcing people to rely on their polluted wells. An additional threat is the traditional reliance by residents of these villages on graphite blocks called "mazute" as a source of heating fuel. Mazute consists of blocks of industrial waste made available at low cost by the local firms. While this may have proven to be an efficient means for enterprises to get rid of the toxic wastes they generate, it is a major source of potentially dangerous environmental exposure. One source claimed that the 400 tons of graphite burned in the village of Igumnovo per year released 150 tons of dioxin, other chlorinated compounds, and other contaminants to the local air.[18]

Activists have recently succeeded in putting a stop to the burning of these graphite blocks. However, residents have not been relocated from the contaminated zones. Many factors continue to block action and contribute to a general lack of concern, even among residents of these villages. Two key factors from our prior discussion deserve further consideration as causal factors in this pollution acceptance: that risks in residents' everyday lives have become a routine, and thus mundane, and the influence of a culture of secrecy on the perception of environmental problems.

ROUTINIZATION OF RISKS

> We all are chemists here! (Interview with 65-year-old woman)

During the years when big chemistry flourished, citizens of Dzerzhinsk paid little attention to the smoke above the city because they were accustomed to it. Yet, one recalled, "*The smell was suffocating. The city was under a blanket of smoke.*" Citizens of the industrial settlements situated close to the enterprises drew particularly dangerous pictures. A 60-year-old man shared his recollection:

> I remember how it was. Here in the house, in the yard, there was a fog from discharges. Often you would enter the yard and nothing could be seen. Everyone was closing their windows. Everyone was covering their windows and covering themselves with wet towels for the night.

To most, the smoke actually signaled something positive, the stable functioning of the chemical enterprises that provided work for citizens, reflected their accomplishments, and spelled their contribution to the nation. This benefit seemed to outweigh any concerns. A number of factors contributed to this attitude. Most residents were rank workers or service employees for the polluting enterprises and loyal to their employers. A professional class of engineers, scientists, and chemists believed that the benefits of technological production outweighed risks, which were controllable, in any case. Some people expressed bravado; to work at the most dangerous plant meant that you had the greatest skills; you were also paid better. A local journalist, one of the informants, captured the local work spirit in these recollections about life in Dzerzhinsk:

> Workers at the enterprises were proud to work in the units where pollution levels reached 100 times the maximum permissible concentration. And there was such poison – God should not have let them. I had one friend, and his wife was pregnant. They lived in a dorm, in such a hole – in a wooden barrack. She was happy when she was assigned to work in a unit producing Cyanide. And she was pregnant! She was so happy because (given her work assignment) they put her on line to get an apartment.

Everyone in the city was aware of their dependence upon the monoculture of large chemical industries; there was social resistance to protest and even to discussion. The lack of information and the climate of secrecy reinforced these tendencies.

And then there was the comparison of the work and residential environments. For plant workers, often forced to work in gas masks, the city seemed comparatively clean. As one 50-year-old man commented, "it is heaven here in the center of the city, even if at night you feel like you cannot breathe." Dissent and concern only received expression through humor, evidenced in the nickname "Dustograd" and in the many anecdotes, and through gossip, evident in rumors and in circulating reports about sick children. But humor and gossip did not raise the issues to a level of discourse; rather they helped maintain denial.

Of course, some indications of danger were hard to ignore. People knew of accidents or discharges from the smell in the air, their difficulty breathing or by hearing the stories of their friends. A 65-year-old woman recalled, "All discharges were done at night. At night it was absolutely impossible to breathe." She shared the experience of her friends who lived behind the Sintez plant. Some days they would awaken in the morning to find their chickens suffocated on the ground. Thus, they knew both that the toxic discharges had occurred that night and that the gasses released were heavier than air. Had the gasses not tightly hugged the ground, the family, asleep in raised beds, would have also died. This was just a fact of life.

People adjusted their lives as best they could to the patterns of contamination. In the 1970s, when all the enterprises were discharging smoke, residents found ways to lower their exposures by modifying their activities. The worst releases of the week routinely occurred on weekends so residents tried to go to their dacha then. Discharges were also bad after dark, so they put wet gauze over their faces at night. If the smell was intolerable or a lot of gas was being released, they closed their windows.

Yet, the citizens of the "City of Chemists" never questioned such industrial release practices. This was normalcy. Residents never assessed whether it was right or wrong. Before they began to encounter pollution as an issue in the media during Perestroika, it was only salient for them when they could directly smell or see it. Risks were only what you might sense or perceive. Invisible risks did not exist in their consciousness because they had no information about them. As this woman reports, people detected pollution, but it was so common that they did not ponder its significance.

> In the morning, when we got up, the snow would be yellow. It would turn gray in the mouth or on the tongue. Fish were dying in the rivers. But maybe we were not really noticing that too much. I don't even recall that people would talk too much about all this.

After Perestroika, citizens suddenly confronted masses of information about the environment, their health, and the threat of toxic exposures. Suddenly they realized the dangers surrounding them. Now it became an issue that nobody had warned them not to swim here or to live there. But after a while, the novelty of even such information wore off. The knowledge that they were immersed in dangers either became too overwhelming or too oft repeated. People once again began to avoid the subject.

Nevertheless, a lesson remained for the workers' children, a generation who grew up covering themselves with wet towels to not smell the smoke. A woman of 20 reported:

> I would never spend my life working in chemistry. Both my parents worked in chemistry and both are dead. My mother died at 40, my father at 50. I don't want such a life for my child.

SECRET ZONE: DZERZHINSK – WHERE PEOPLE GET SICK AND DIE QUIETLY[19]

The less you know the deeper you sleep. (Russian proverb)

Extending from the rise to the fall of the Soviet Union, Dzerzhinsk was a closed city of extreme strategic importance to the military-industrial complex. Before 1991, foreigners could not go to Dzerinsk and Russians needed special permission. The only visitors allowed were involved in industrial chemistry or armaments, and they had to pledge to keep state secrets. Even to this day, former plant employees avoid talking about their work in Soviet times. In our field interviews, we encountered people who were circumspect in describing where they had worked or who would only talk off the record.

As a consequence of this strict secret regime, even residents of contemporary Dzerzhinsk are afraid to talk openly about environmental issues. Secrecy as a phenomenon is incorporated in both the thoughts and actions of the city's citizens. Reinforcing the effect of secrecy is the sense that individuals can do little to change the situation in a way that would reduce their risks. Risk reduction is greatly tied to changing where you live. But who can afford to move?

The natural way out of this uncomfortable situation, then, is a return to ignorance – pushing away information and not accepting it. Those in the local power structure who are responsible for solving the city's problems are instead focused on property and privatization issues with an eye to making themselves richer. All local newspapers have ties to the owners of the enterprises or to the power elite who have an interest in keeping issues quiet and dormant.

Despite the volatility in municipal leadership since the collapse of the Soviet Union, there has been no change in the environmental power structure. Observers claim that the approach has consistently been based upon lies, keeping things quiet, and doing nothing. Risks are controlled by limiting information and thus knowledge, not by addressing problems. One informant, a male of 55, charged that:

> This approach is useless. Everything is about keeping quiet. It is all based on lies. Given the ecological situation in Dzerzhinsk, what is killing people is the lying from every side of every administration. And even when the administration changes, the approach to the environment remains the same.

Given this normative structure, it is not surprising that those who try to call attention to the accumulated problems of the city are dismissed as "the voice of the enemy," as "competitors," as "spies," or as "antidevelopment and progress." Critics – eco-activists, critical publications in the Russian media, unfavorable city reports broadcast on central TV, and depictions by foreign journalists highlighting local ecological problems – are routinely dismissed as agents for competitors of the city's chemical enterprises. This post-Cold War version of the "enemy of the state" is used to neutralize negative ecological information. Citizens routinely accept the administration's characterization of journalists and activists as troublemakers.

These explanations were employed in the year 2000 after three reports on Russian national TV (NTV) highlighted the ecological problems of Dzerzhinsk. The city was depicted as in an "eco catastrophe" and its citizens as all sick from the long-lasting influence of the chemicals surrounding them. A counter attack by the local power structure charged that the involved reporters were working for unnamed competitors of the chemical firms in Dzerzhinsk who employed television reports and environmental activism as part of a campaign to discredit the city and its industries in order to capture the chemical production market. The mayor even sued NTV and local media outlets that followed its lead.

Dzerzhinsk's counterattack against defamation did not stop there. An environmental audit of the city and territory was ordered by the city administration to contradict the stigmatizing information and create positive publicity. The study was funded by $1,000 contributions from each of the local industries. A Moscow firm was hired to give an "independent" assessment of the safety of the local environment. The audit emphasized pollution indicators that were not significant (i.e. they were found at levels below health standards). These pollutants were examined individually, avoiding issues of cumulative effect. Declaring that the situation in Dzerzhinsk is much better than in other places, the report went to great lengths to compare the results in Dzerzhinsk to the environmental profiles of several other Russian and foreign cities. The audit emphasized higher levels of soil pollution in these cities chosen for comparison. And it stressed the sharp local decrease in Dzerzhinsk's air pollution over the past decade due to plant closures and the slowdown in the chemical industry. In this way, the audit was "spun" so as to refute the NTV coverage and to invite new economic development in Dzerzhinsk. Particular emphasis was placed upon the goal of publicizing the city as an ideal location for resort development.

Environmental activists and independent thinkers found this effort to pitch Dzerzhinsk as a tourist destination to be laughable. A staff member of the Ministry of Natural Resources offered this characterization of the city's environmental report:

> And they came up with this "independent" and "scientific" ecological audit which proved that Switzerland is a dirty place compared to Dzerzhinsk. Bremen is a dirty landfill, and Kaliningrad is a wasteland. But Dzerzhinsk is a place that people come to for recreation. You will come to the city and it will cure you! Everybody bugged their eyes. Such lies didn't exist before. This was called an "independent eco audit"!

In contrast to such skeptics, "partriotic citizens" who trusted the administration and were accustomed to accepting provided information as unquestionable truth found the audit sufficient to assuage their concerns.

A NEW BRIGHT FUTURE: BUT WHICH ONE?

These dynamics of Dzerzhinsk reflect more than adaptation to pollution and a culture of secrecy. They reflect the fact that there had been one consensual viewpoint governing the city for scores of years. A decade after the breakup of the U.S.S.R., Dzerzhinsk now evidenced at least four competing viewpoints reflecting some combination of focus on leftover problems from the past and the possibilities of a totally new future. It was clear that no consensual direction was at this time possible.

Leftover Legacies – Ecological Spin

The first of these viewpoints is focused on responding to the widespread environmental contamination that already exists. It is notable that this viewpoint is publicly represented by just one tiny organization, SPES, and its two staff members. As the only environmental NGO active in Dzerzhinsk, SPES has played a key role in educating the public and "burning the interest," as one staff member put it. They then added a cautionary note, "If there were no public organizations, then ecological issues would be absolutely hidden."

Leftover Legacies: Capitalist Spin

The contaminated legacy is less a problem than an opportunity in this second viewpoint. To the vestiges of the old military-industrial complex that must seek ways to survive now that there is no Cold War pumping millions into their budgets, buried contaminants appear as "gold in them over the hills." In Dzerzhinsk, as elsewhere in the former Cold War states, there is money to be made from environmental cleanup. The now downsized former defense enterprises can claim special expertise in this area given their direct role in creating the contaminants to begin with. As a result, Dzerzhinsk firms have actively sought project contracts from the European Union as part of the decommissioning of weapons' sites under the Convention on Killing Chemical Armaments.

New Future: The Industrial Spin

A third viewpoint consists of those seeking to improve the image of the city and boost the potential for investment in its industry. The leader of this development effort is the Mayor himself, the owner of one of the few large enterprises to survive the post-Perestroika period and become profitable. The power structure promoting new forms of economic development has generally sought to keep problems quiet.

New Future: Post Industrial Spin

A fourth viewpoint consists of those ready to break away from the contaminated industrial past and chart a new course for Dzerzhinsk. For this group, Dzerzhinsk no longer seeks to be "the city of chemists." Efforts to create a new strategic plan point away from redevelopment of the chemical industry. At the same time, there is little interest in dwelling on difficult environmental problems that present obstacles to this new direction.

Given these competing directions, the lack of a public space for discussing ecological problems has been a major constraint. The absence of discourse further helps to explain the low level of risk acceptance in Dzerzhinsk. The administration has recently formed a Public Environmental Council to involve interested citizens in decision making. It remains to be seen if this council will allow dialogue about contamination or if it is focused only on new development.

Despite such changes, the city evidences difficulty in overcoming the past legacy of the Soviet system. Journalists covering ecological problems continue to play a significant role in pushing the power structure. When environmental problems are revealed in the media, contrary to motivating responsiveness, publicity leads officials to be defensive, to keep the problem quiet, and to discount the views of critical actors. Dzerzhinsk remains burdened by old technology and the regime of secrecy. There is no indication that the problems of industrial pollution will ever be addressed.

Beck cautions that when serious environmental hazards do not generate simultaneous political activity, "the society of risks" has a tendency to become "a society of scapegoats." Blame is placed on those who spread concern rather than those who caused the danger (Beck, 2000, p. 91). This conceptualization is appropriate to Dzerzhinsk. As capitalism allows the private interests of industrialists or entrepreneurs to be reformulated under the guise of public benefit, this definition of public good competes with claims by SPES on behalf of the victims of environmental contamination. These competing definitions of public interest are contradictory and perhaps mutually exclusive. To recognize ecological problems is to discourage investment. And, investors are required to get the plants working again or build a tourist or alternative economic base, to generate jobs, improve quality of life, and improve material well-being. But is the health of people and environment contradictory to this path to a radiant future?

CONCLUSION

In 1997, Dzerzhinsk was placed in the Guinness Book of Records as the "Most Chemically Polluted Small City in the World."

Soviet history was one of high promise and major disappointment. In the end, that system lacked a method of decision making that allowed for the promised radiant future to be attained. Rather, the achievements of the society entailed extraordinary costs, including the sacrifice of people's health and the health of nature. In Dzerzhinsk, invisible risks remained invisible because public information was suppressed. The city was closed within itself, not just from the outside.

The consequence is tragic. The past system of production met extraordinary quotas, particularly when the record was exaggerated. But the stretch for productivity was an excuse for a lapse in safeguards. The heroism of workers was genuine. But their sacrifice was much deeper than realized.

The conundrum we have tried to examine is not that the land and waters were so polluted or the workers and their families put at risk. Instead, it is a seemingly more basic issue. *Once this havoc has been wreaked, can it be admitted?* Do we face our self-caused problems and address them as best we can? Or do we look the other way?

Dzerzhinsk's unwillingness to face the continuing problems from its past reflect many factors. Two stand out in importance:

- Secrecy, with its effect that most people lacked and lack any clear idea of just how bad things are. Moreover, in a closed society, there is a paucity of opportunities for sharing and discussing such information.
- Routinization, with the result that the everyday nature of risk exposure makes it mundane rather than frightening. We are already used to living with not having solved this problem. In fact, it is invisible and we want to keep it that way.

To these we have added several considerations. The health consequences of contamination are not easily proven and even an environmental audit can hide or distort such effects. Competing challenges, such as unemployment, and competing health threats, such as alcoholism, can readily distract the focus from environmental health issues. The economic crisis also deprives resources to address the problems and to modernize production. The social system makes citizen action and protest unlikely. Competing interests and concepts of development are threatened by too overt a focus on seemingly intractable problems of contamination. From this perspective, it is better kept quiet.

In sum, to create a society in Dzerzhinsk that would recognize its problems, a new set of political values are required. The leaders and citizens of Dzerzhinsk today face complex and difficult problems that prior leaders could not address. Instead, these problems were concentrated in space and time and left for the future. The question now is will current leaders do the same, hiding the problems for others to someday confront?

Clearly, there is an interest to not maintain a system that generates such insurmountable problems. The new strategic effort may open up new patterns for development of Dzerzhinsk. This pattern will certainly abandon the mono-industrialism of the chemical era. The real question is whether economic development is really possible without confronting past problems, or whether in the end a new radiant future is only possible by facing these issues.

NOTES

1. It is obvious that alterations of the environment from previous epochs were the "other side" of civilization's development. However, contemporary societies have reached a stage of reflection and critical assessment of the impacts of people on their environments. Putting aside the inherent contradictions between people and nature, which was basic to many philosophical systems, this reflection is a turning point in the progress of human ideas. "Dichotomizing society and nature is a construct of the 19th century that has worked toward two goals – conquering nature and ignoring nature." (Beck, 2000, p. 7)

2. For more details about Russia's environmental movement, see Yanitsky, 1995, 1996, 1999, 2000; Pickvance, 1998.

3. According to Yanitsky's assessment, 2000, p. 1.

4. The military industrial complex in Soviet times was related to production of rockets, nuclear weapons, chemical armaments, artillery, aviation, and strategic objects. "Closed" cities were literally off-limits to outsiders, in contrast to all other cities, which were comparatively "open."

5. This material included books on Dzerzhinsk, its history, its older citizens, on the history of certain enterprises, and published memoirs of citizens, local, national and international newspaper publications, video materials, a BBC TV show about Dzerzhinsk, and a film made by local activists about environmental issues in the city.

6. The city was named after the first chair of the ChK (All Russian Emergency Commission on Fighting Contrary-Revolution Speculation and Sabotage), known for instigating the first wave of arrests and repressions in the Soviet Union and as a feared interrogator of prisoners (Vert, 2002). A monument to this bloody revolutionary leader remains today in a key city square.

7. Shalnov, 1999; Also interview with the chair of the Club of Regional History.

8. From interview with geologist, staff person of Ministry of Natural Resources.

9. A spot for respite and recovery at the workplace that facilitated continuing labor.

10. Lev Fedorov, Expert on Issues of Chemical Armament. From the Interview Radio Liberty. 26.04.2002. http://www.svoboda.org/programs/eco/2002/eco042602.asp

11. Don Janishevsky (2001, December 16). Disregard for health is killing Russians.

12. http://www.Regnum.ru/allnews/159920.html

13. By way of example, one informant described how her husband's efforts to start an ice cream business in Moscow failed due to the competition. By contrast, he was so successful in Dzerzhinsk that he now exports ice cream to Moscow.

14. This convention has been ratified by 140 countries. Russia was one of the biggest producers of chemical armaments. In 1997 Russia acknowledged 24 chemical arms objects (i.e., facilities) as part of 6 major chemical complexes in 5 regions – Novocheboksas, Chuvash (5 objects), Volgograd (8), Chapayevsk, Samarskaya Oblast (3), Dzerzhinsk, Nizhny Novorgo Oblast (7), Berezniki, Perm Oblast (1). According to the agreement, six objects needed to be decommissioned, one in Volgograd, one in Chupayevka, and four in Dzerzhinsk. The objects in Dzerzhinsk included two units producing Yperite, used for military weapon material and hydrogen cyanide and one object at the firm Kaprolaktam involved in the production of

Lewisite. For part of this work, the EU gave Russia 7 million Euro in 2000 (Nezavisimaya gazeta 6.09.2002). See also the chapter by Filatov in this volume.

15. From an interview with the staff of Ministry of Natural Resources.

16. From a speech of the administrator of Dzerzhinsk cited in the electronic bulletin: Problemy Khimicheskoj bezopasnosti (Problems of chemical security). Announcement UCS, 4, November 1, 1999. http://www.ecology.iem.ac.ru/ucs/b507.txt

17. Compiled using materials of L. Fedorov: Fedorov, 1995; Issues of Electronic Bulletin Problemy Khimicheskoj bezopasnosti (Problems of Chemical Security); interview with Lev Fedorov on Radio Liberty (26.04.2002).

18. Ekologija i prava cheloveka (Ecology and Human Rights), *Bulletin*, 11, 2000.

19. "Ammonium and chlorine in milk of mothers." *Pravda*, 10.01.2001.

ACKNOWLEDGMENT

The field research upon which this article is based involved collaboration with two colleagues from the Center for Independent Research, St. Petersburg, Russia, Svetlana Pchelkina, and Dmitry Vorobyev, who also contributed to key concepts used in this chapter. Their involvement is gratefully acknowledged.

REFERENCES

Beck, U. (1992). *The risk society: Towards a new modernity*. London: Sage.

Beck, U. (2000). *The risk society: Towards a new modernity (Obschestvo riska: na puti k drugomu modernu)*. Moscow: "Progress-Tradicija." (Russian printing of the original).

Douglas, M., & Wildavsky, A. (1983). *Risk and culture. An essay on the selection of technical and environmental dangers*. Berkeley: University of California Press.

Edelstein, M. R. (2004). *Contaminated communities: Coping with residential toxic exposure*. Boulder, Co. and Oxford: Westview Press.

Fedorov, L. (1995). *The unannounced chemical war in Russia: Politics against ecology (Neob"javlennaja khimicheskaja vojna v Rossii: politika protiv ekologii)*. Moscow: Center for the Ecological Policy of Russia (Centr ekologicheskoj politiki Rossii).

Graham, L. (2000). *Ghost of a public engineer: Technology and the fall of the Soviet Union (Prizrak kaznennogo inzhenera. Tehnologija i padenie Sovetskogo Sojuza)*. St. Petersburg: Europian House (Evropejskij dom).

Hunnius, G., & Kliemt, J. (1993). Risk as a social construction: The perception and communication of risk in the Federal Republic of Germany and the German Democratic Republic – a comparison between systems. In: B. Rueck (Ed.), *Risk as construct: Perceptions of risk perception* (pp. 221–236). Munich: Knesebeck.

Kotlyar, I. (2001). *The city of big chemistry (V gorode "bol'shoj khimii")*. Moscow: Greenpeace.

Pickvance, K. (1998). *Democracy and environmental movements in Eastern Europe. A comparative study of Hungary and Russia*. Boulder, Co. and Oxford: Westview Press.

Shalnov, S. (1999). *Dzerzhinsk – Our home*. Nijhny Novgorod: "Litera".

Vert, N. (2002). *History of the Soviet state (Istorija sovetskogo gosudarstva)*. Moscow: "Ves' mir".

Yanitsky, O. (1995). Evolution of the ecological movement in contemporary Russia (Evoljucija ekologicheskogo dvizhenija v sovremennoj Rossii). Sociological investigations (Sociologicheskie issledovanija, p. 8).

Yanitsky, O. (1996). *The ecological movement in Russia (Ekologicheskoe dvizhenie v Rossii)*. Moscow: RAN.

Yanitsky, O. (1999). Transnationalization of civil society: A case study of non governmental organizations in three post Soviet states (Transnacionalizacija grazhdanskogo obschestva: na primere nepravitel'stvennyh organizacij v trekh postsovetskih stranah). *Mir Rossii, 1–2*, 135–150.

Yanitsky, O. (2000). *Russian greens in a risk society*. Helsinki: Kikimora.

Yanitsky, O. (2003). The sociology of risk: Key ideas (Sociologija riska: kljuchevye idei). *Mir Rossii, 11*(1), 3–35.

HANFORD: THE CLOSED CITY AND ITS DOWNWIND VICTIMS

Michael R. Edelstein

INTRODUCTION

The post-Cold War period allowed the U.S. nuclear legacy of ecocide to be declassified and made public. The policy of nuclear secrecy, evident in Russia (see Mironova et al., this volume), was not merely an eastern practice. Western nuclear releases were kept equally under wraps. In England, for example, the Windscale disaster was not fully disclosed until 1987.[1] Likewise, releases from the Hanford Nuclear Reservation, in Washington State, and other U.S. nuclear sites were kept undercover until the same period. The irony was that Americans learned of many of the nuclear skeletons in their closet around the time that Russians learned of theirs (see Mironova et al., this volume). It would appear that glasnost was contagious.

In 1986 and 1987, just at the cusp of public disclosure of these releases, the author was part of a team assigned the task of calculating the Social Impacts of siting a High Level Nuclear Waste Repository at Hanford. Of particular significance to this volume, from within the vast American military nuclear program, the Hanford Reservation is the site most often compared to the Mayak facility in Chelyabinsk (see Dalton et al., 1999; Edelstein & Tysiachniouk, this volume; Marshall, 1990b; Mironova et al., this volume). Both were the principal plutonium production sites within their respective contenders in the Cold War. For this reason, a return to this 20-year old research offers a fascinating point of comparison to the Mayak chapters.

Cultures of Contamination: Legacies of Pollution in Russia and the U.S.
Research in Social Problems and Public Policy, Volume 14, 253–303

ISSN: 0196-1152/doi:10.1016/S0196-1152(06)14012-0

BACKGROUND

The nearly 800 square mile Hanford Reservation in western Washington is located on an arid plain traversed by and bounded on the east by the Columbia River and bounded to the south by the Yakima River. Until 1943 the small, sleepy towns of Hanford, Richland and White Bluffs served an agricultural area known for the northwest's quickest growing season. In 1943, the area was abruptly transformed by the U.S. military to serve a central role in the wartime race to build the atomic bomb (Hanford would produce the bomb that destroyed Nagasaki, Japan). Hanford was chosen for its remote location, high background levels of radioactivity due to the presence of geologic radon, abundant water in the ground and the Columbia River and abundant electricity from the nearby Grand Coulee Dam. Within an 18-month period, residents of White Bluffs were ordered to leave and the city and its orchards were razed, the first reactor was built and 50,000 new residents arrived. Richland was selected for scientist housing and neighboring Kenewick, as the commercial center. These "Twin Cities" subsequently joined Pasco, the old agricultural and railroad center across the river, to form "Tri-Cities," home to nearly 150,000 people at the time of the research (GAO, 1986; Rhodes, 1986; Whiteley, 1999).[2]

The need for plutonium intensified during the Cold War. By mid-1955, eight production reactors lined the Columbia River. Two plutonium separation plants, Purex and Redox, standing toward the center of the tract, were involved in recovering, processing and shipping plutonium from irradiated reactor fuel. Several other specialized test facilities were subsequently located at Hanford, including the fast flux 400 megawatt thermal liquid cooled research reactor. The Atomic Energy Commission or AEC, later succeeded by the Department of Energy or "DOE," placed day-to-day operation of the facility in the hands of private contractors. Three nuclear reactors were later built by the Washington Public Power Supply System (WPPSS), only one of which ever operated, contributing to the largest municipal bond default in U.S. history (GAO, 1986; Rhodes, 1986; Whiteley, 1999).

The Green Run Test

Of many radioactive releases during the early days, one deliberate experimental release deserves special mention because of its Cold War connection to Mayak. Shortly after the USSR's first nuclear weapon test, the U.S. military undertook to test radioactive fallout at Hanford in December 1949.

To simulate Mayak's operations, a Hanford spent-fuel reprocessing plant was modified to use more radioactive "green" fuel cooled only a fraction of the normal time. Air scrubbers were disconnected and a cloud containing 27,800 curie of radioactive gases was deliberately allowed to escape. Nearly a quarter of the release consisted of hazardous Iodine-131, twice the anticipated level. When unexpected weather conditions limited dispersal, the cloud of radioactivity hugged the ground, giving significant exposures in excess of standards to the Hanford Reservation itself as well as highly populated regions of southern Washington and Oregon. Exposures were further maximized by holding the test on the weekend, when people were likely to be home, and by the lack of public warning. Dose reconstruction suggested that downwind infants were at risk from drinking cows' milk affected by fallout. The incident is not considered to be a major source of risk only because releases during wartime and immediate postwar Hanford operations were routinely considerably higher. In fact, the Green Run experiment accounted for only 1.1% of radioactive Iodine releases for 1949. A report on the Green Run remained classified until 1989 (GAO, 1993; AP, 1986). At the time of my interviews in 1986–87, the Green Run test was still a matter of rumor and vague recollection.

Contamination

The Hanford complex has proven to be a serious threat to the adjacent Columbia River either directly or via groundwater. In 1945, a large quantity of radioactive runoff was pumped down two deep wells directly into the aquifer. Additionally, various Hanford operations directly discharged 110 million curie of mostly short-lived radionuclides to the river and intentionally dumped some 450 billion gallons of liquid waste on the ground. Big spills occurred in 1956 (55,000 gallons of high level nuclear waste) and 1973 (115,000 gallons), releasing to the soil above the water table 40,000 curie of Cesium 137, 14,000 curie of Strontium 90 and 297,000 curie of other fission products. Beyond these releases, some 61.4 million gallons of high-level liquid waste from reprocessing spent (used) nuclear fuel has been stored in large underground tanks. The original 149 of these, buried between 1944 and 1964, were unlined; 26 began leaking within 10 years, releasing a million gallons containing 1,000,000 curie into the soil, and another 30 had "questionable integrity" according to a 1983 report. There were also 39 active and at least 337 inactive low-level waste disposal sites and 35 transuranic waste sites containing at least 710,000 cubic meters of materials generating an unknown amount of leachate. As a result of ongoing discharges, 80 square miles of groundwater and soils beneath the Reservation became significantly contaminated by chemical wastes and radionuclides, with plumes of contamination discharging along 10 miles of Columbia River shoreline and approaching the City of Richland's water intake. Strontium 90 and other radionuclides threatened important salmon nesting sites and were detected in river shellfish. Surface contamination was spread by such ecosystem processes as blowing tumbleweed and rabbits that had visited a badger hole. Air releases of significant amounts of Iodine-131 occurred between 1944 and 1972 at levels dwarfing the Three Mile Island nuclear disaster and dropped fallout over the Columbia River and lands downwind (D'Antonio, 1993; GAO, 1999, 2005; Marshall, 1987, 1990a; Whiteley, 1999).[3]

Cleanup

The cleanup of Hanford has been termed the largest public works project in history, estimated to take between a hundred and a thousand years to achieve at a cost approaching a trillion dollars. To date, a series of reports with names like "Train Wreck along the River of Money" suggest that the Hanford cleanup has been inefficient, mismanaged and wasteful (Lovrich & Whiteley, 1999). The General Accounting Office (now the Government Accountability Office) blamed the problem on what they termed the

> "Hanford Syndrome," a risk aversion approach common to large bureaucracies whose symptoms are an unwillingness to alter familiar behavior patterns, sticking with unproductive or failing procedures, to enhance tendencies for excessive resource allocation and regulation, and to oppose innovations. It is an important element in sustaining unproductive patterns of work (GAO, 2005, p. 29).

The cleanup has also been dramatically under funded, suggesting a cautious note.

> Careful observers of the Hanford case in Chelyabinsk might learn an important lesson from the events that transpired in this tragic set of circumstances all too similar to their own. Government promises of long term commitments to address the real costs of nuclear arms production are easily made, but they cannot really be counted on by the people most directly affected (Lovrich & Whiteley, 1999, p. 330).

The Repository Search

The 1982 Nuclear Waste Policy Act unleashed an effort to site a national high level nuclear waste repository (henceforth "the repository") in geologically stable bedrock where wastes could be segregated deep underground from earth processes for the many tens of thousands of years required for radioactive decay. After an initial screening, the list of candidate sites deemed worthy of "site characterization" was winnowed to three: Yucca Mountain, Nevada; Deaf Smith, Texas and Hanford, Washington. The process of selection was to be based on careful scientific studies and active federally funded participation by the affected states and tribal nations. In 1987, with the characterization studies in progress, Congress abrogated the act and chose Yucca Mountain for a full environmental impact study as the intended repository site.[4]

THE STUDY

The author served on a team assembled for the State of Washington by Impact Assessment, Inc. to conduct a social impact assessment of the proposed repository. My task was to examine and contrast the psycho-social issues for two distinct cultural settings, the relatively closed nuclear residential city of Richland associated with the Hanford Reservation and "Downwinders," residents of areas downwind from the facility who shared an identity as possible victims of radioactive releases from Hanford. Proximate Downwinders lived in the farming communities of Franklin County, just across the Columbia River from Hanford, and Distant Downwinders lived in the comparatively far off Spokane area. The focus of the two Downwinder populations would prove to diverge, with Franklin farmers fearing crop stigma more than they feared radiation and Spokane Downwinders focused on threats to their health. As a result, the first was highly constrained in their activism and the second lacked any deterrent to acting against the source of their threat.

Beyond document review and observation, the crux of the study was a series of interviews. In 1986, I began my work by participating in a group interview with twelve representatives of diverse community interests in the Tri-Cities area. The list included a nuclear safety expert at one of the operating plants; a nuclear weapons design consultant; and officials concerned with regional development, tourism, farming, wine production and the local science museum. The study of Franklin Downwinders involved interviews around three key informants.

In 1986 and 1987, I interviewed members of two extended farm families living in Franklin County, Washington, a temperate but arid area requiring irrigation to grow crops. The lead spokesperson for one was "the farmer/nurse," who for thirty years had grown crops on a 300 acre farm in Mesa with her husband and three now married children while also serving as the local home care and hospice nurse. Second was "the sheep farmer's wife," joined by her husband, mother-in-law, sister-in-law and friend. All five were present at an interview in August 1985 in a crowded, smoke filled kitchen with lots of kids and animals running around. I talked with the farmer's wife and mother again in October 1987 while they were inoculating their herd. The third informant was farmer and Downwind activist Tom Bailie. On October 10, 1985, Bailie played host to an extended conversation I held from 7:30 a.m. until 1:15 p.m. with 23 locals at the Mesa Cafe in Mesa, Washington. As people entered the café, Bailie informally introduced me as a researcher for the State of Washington studying the proposed high-level nuclear waste repository. This was their chance to give their views to the state about how it should respond to the repository. The flow of mostly male farmers functioned as an extended family who met every morning for breakfast and to exchange views. People came and went; some participating at breakfast, then departing to farm, only to return for lunch.

As a point of demographic similarity for comparing the Hanford culture of Richland and the anti-Hanford culture of Spokane, I chose the League of Women Voters, an organization dedicated to promoting an understanding of public policy issues through education. There was an active rivalry at the time between the strongly pro-nuclear and pro-Hanford Franklin-Benton League (here termed the "Richland Chapter") and the anti-Hanford Spokane League. While I never made it to Spokane before the outside-the-process Yuka Mountain selection ended the study, I did interview eight members of the Richland Chapter in one of their homes. Typical for the chapter, all were residents of Richland, college graduates and wives of workers at Hanford or other nuclear businesses. The informants stressed that their economic tie was to Hanford, not to nuclear issues per se. Let's take a closer look at the Hanford culture before returning to the proximate and distant Downwinders.

RICHLAND, AN AMERICAN CLOSED CITY

A useful point of departure before we venture into our American Nuclear City is Kutepova's (2002) poignant description of life in Ozersk, the ZATO associated with the Mayak nuclear facility, presented at our project conference in St. Petersburg on August 9, 2002 (see also Kutepova & Tsepilova, this volume). Kutepova's observations were based upon participatory observation derived from the multigenerational experience of her family working at Mayak. Ozersk is a closed community where people exert uncommonly little control over their lives.

> It is a community managed by outside forces.... The ZATO is a system of control from birth. ZATO are closed to the public and for information. It is a local community in a closed space. There are few immigrants.

The closed status magnified the traditional Russian importance of friends and family.

> There is a high level of mutual influence. The whole community is just penetrated by connections between friends and family. In Russia, relations and friends always know what they are doing, and they act through a relational network.

As at Hanford, residents were misled about environmental dangers due to releases.

> Mayak was built in 1946. Up until the 1990s, we were told that the winds kept radiation from the city. Recently we learned that this is not true.... There has apparently been several discharges to the cities. There is a significant difference in thyroid disease, but most people do not know this. The public attitude is that it is safe to live there.... At the official level, Ozersk is "a clean place" and "the streets are green."

However, there was a different understanding of risk in the private realm.

> There is a difference between the public and family discussion of the plant. People know there is terrible pollution. In private, if someone does not feel well or their health changes, they explain it by the influence of radiation.... The doctor says, "The husband works there, the mother lives here and bore her children here. What do you want?"

This private reality, however, did not influence the public culture of silence.

> It is the same as the population living in the contaminated territories, they keep silent. The effects of strict control linger in people's mind even if that control has been relaxed.

Moreover, close family and friendship ties act to constrain public speech and action.

> The situation is so stigmatizing in its consequences for people that they are afraid to act for fear of the negative consequences to follow. They have no civil initiatives on this issue.... People see civic action as dangerous.... If people want to act, they are afraid because of the responsibility.... They will become surrounded by a negative response from the community. They might lose their job.... If they act, their family and co-workers pressure them. Friends will be afraid to even interact with them.

Richland, Washington was never closed to the same degree as Ozersk. Nevertheless, several key elements mirror the Russian closed city: concealed hazards, enforced norms of secrecy, development of a distinctive local nuclear culture among a population of nuclear scientists and workers and domination of local culture by a belief in technology and expertise.

Concealed Hazards

The Hanford record is striking for its inattention to public health. During the first decade, scientists focused on worker exposures, although a 1945 campaign promoted community use of iodized table salt to prevent thyroid gland damage by radioactive Iodine.[5] It was recognized that the Green Run had harmed range animals, justifying compensation of farmers, but there was no concern over whether farmers had themselves been harmed. After the 1946 documentation of bioaccumulating radioactivity in fish, bans to fishing, hunting and ingestion of untreated river water were considered (Whiteley, 1999).[6] Similar actions were considered in the mid-1950s over exposure through milk (Thomas, 1990). In these instances, officials were deterred from acting by fear of alarming the public and disclosing security information.

Similarly, a mid-1970s report documenting groundwater contamination from Iodine-129 was withheld. The information only became public when, a decade later, a Hanford employee cited it in a conference presentation. The whistle blowing employee was punished. A review by the General Accounting Office later concluded that the DOE had deliberately "inhibited or suppressed the disclosure of information." Hanford contractors suppressed news of the iodine release even after the DOE declassified the information, at least partly to avoid arousing public fear and to avoid hurting the chance for Hanford to be selected for the repository (GAO, 1988; Bruscas & Wilson, 1987).[7]

A Culture of Secrecy

Despite a promise of "open disclosure" (Gerber, 2002, p. 69), Hanford operated under total secrecy in a manner parallel to the Russian program (see Kutepova & Tsepilova, this volume and Mironova et al., this volume). Even the reference to the ZATO as "Post Offices" has had its parallel, for example in the reference to the Los Alamos lab as Post Office 2663 (Rhodes, 1986). Even informal disclosure was limited. The director of the Hanford Science Center noted, "the security was so tight that people could not tell their wives." A Mesa farmer explained that, when Hanford workers retire, "they are required to sign a paper promising not to talk" (see also Gerber, 2002).

Secrecy was extended to the environmental and health impacts of Hanford's operations, reflecting, in part, the psychology of wartime and the Cold War (Whiteley, 1999). Residents downwind were never given results from numerous studies in which they consciously or unwittingly participated. One Downwinder did not need the results to interpret a study intended to see if gardens and milk were contaminated over a 10-mile radius. "The DOE wouldn't have done the study unless they believed there was exposure." Security breeches sometimes confirmed that there were lingering effects from such releases.

> We knew someone who worked at Hanford and who went fishing a lot. He had read a report. He told us not to east the fish, especially the bottom fish (cat, sturgeon, salmon).

Even community members abided by secrecy norms. Someone at the sheep farm interview mentioned a local farmer who had kept quiet after witnessing a release from the N reactor. Some progress was noted: "Now when there is a radioactive release they put it in the paper."

Richland's Nuclear Culture

"Uncle W.," the only surviving "leveler," had driven a caterpillar in the 1940s to level White Bluffs to make way for the nuclear plants. Announcing during the Mesa Café interview that he would hate to express his opinion about Hanford, he nevertheless proceeded to recall.

> There were farmers in Hanford and White Bluffs. They kicked them off. They didn't give them nothing. They couldn't take nothing. It was the biggest rip off in the State of Washington. It was against the constitution!

In its place, there arose a new community of scientists and workers embroiled in the Manhattan Project (Gerber, 2002). This new community adopted a "nuclear culture" (Loeb, 1986) that revealed public nuclear imagery. A Richland high school football team called "the bombers" displayed a mushroom cloud in its insignia. The bowling ally was called "the Atomic Lanes." There was an evident deep rooted faith in nuclear technology, a pride in the community's role in winning the war against Japan, a belief in the safety of nuclear projects and a social system normed by the expectations of secrecy, support for the nuclear project that was Hanford's reason for being and belief that one's neighbors ran the plant for the benefit of all.

This perspective was readily demonstrated during my interview with the Richland League of Women Voters. All of the women had husbands working at Hanford or a related enterprise. All played down the risk of harm from the Reservation. Not themselves experts, they were consumers of others' expertise; "We are influenced by the people around us, by what we see and read." But they drew their own conclusions, "We need to listen to what others say and draw our own opinions."

League members saw the Tri-Cities as "a typical community," attractive because of climate, small size and orientation to family activities. They were frustrated that people outside their community saw them as merely "...billboards," as one explained.

> We see ourselves as a community with a major industry. Outsiders see us in a one-sided way.

Indeed, some comments garnered from the blue color railroad and farming community of Pasco and from Franklin County revealed such bias. Richland residents were described as "Arrogant. Very pro nuclear. Very stuffy" and as a "technical community of well-educated mobile yuppies." On their part, Richland League members agreed when one suggested

> Most people outside think in either a pro or anti-nuclear way. Most people living here [Tri-Cities] see it as grey.

Another Richland resident was more blunt in describing Hanford culture, "Nuclear is a religion!" Faith in this religion fell off by distance from the Tri-Cities (Lovrich & Whiteley, 1999).

The Power of Expertise

Hanford was run by scientists who lived largely in Richland. It was, therefore, a community with uncanny nuclear knowledge and an unbridled faith in the knowledgeable. League respondents trusted the competency and responsibility of the scientific community running Hanford and reported no personal threat from the Hanford operations.

> We have no reasons to feel that we are at risk. We're not living next door to a reactor. If an accident happened at N [the N reactor], it is 30 miles away to the north. There are vast spaces out here. People don't realize this.

Several of their husbands wore dosimeters, so there was constant feedback about whether hazards were encountered and brought home. Their experiences were reassuring. When one League member's husband triggered a radioactive alarm at work, the cause turned out not to be anything at

Hanford, but rather the luminescent dial of his old army compass that sat on his dresser where he dropped his badge each night. She commented,

> We respect radiation. We know that it does not travel far. We take measures to protect ourselves. We know that there is monitoring here. It is something we are very careful with. Radiation can be detected – with a Geiger counter – other hazards can't be.
>
> We are not less fearful. We are just more clear about what we can do. I'm more afraid of sharks than radiation.... I don't know sharks.

Nuclear risk was garnering undue attention. Other issues cried for attention. Thus, they argued that teenage pregnancy "is a bigger threat than nuclear."

The Nuclear Imagery of Tri-Cities

I asked Richland League members to describe the place they live as if they were informing a new acquaintance. The adjectives they mentioned included agriculture, the mountains, climate, the desert, the river, good schools and the difference between eastern and western Washington. In the ensuing discussion, some indicated they would not mention nuclear industries; others would mention it only later in the conversation depending on "whom you are talking to." Previously, it was their experience that people had never even heard of their area.

> Until the repository issue, people did not know the Tri-Cities. Now, there is a good chance that people will recognize the name, especially people from Texas, Nevada and Tennessee [states with nuclear facilities]. The repository put Hanford on the map.

When asked what five pictures best represented Hanford from a set that I had taken, they selected the river (possibly with the bridge or water skiing), "Rattle Snake Mountain," the wheat fields and Vineyards, Columbia Park, the Sunfest, the hydro races, the golf courses and the colleges ("to get the educational angle"). The science museum was the only Hanford image to be considered. But they all agreed they would definitely not take a picture of the high school cheerleaders in front of their mushroom cloud emblem representing the Nagasaki blast.

When I showed a mixture of photos from Richland that included some nuclear imagery, they were oblivious to atomic symbols on a sculpture at the "Up Town Shopping Center" and on the Richland city billboard or the nuclear power plants in the background of striking agricultural scenes or the picture of fisherman casting off into the Columbia River. Their reaction to the Atomic torch picture on the façade of Bell Furniture was simply that it was too dated.

> That is a really tacky sign. The uptown is so '50s. Like the atomic torch. In Richland, power was real good; the atom was not offensive. It was used widely – but, these buildings, like the atom, are now dated.

Although these pictures were mixed in with residential, tourism and institutional shots, by the time we got to a picture of the "Atomic Lanes," the group had caught on to the fact that I was testing their reaction to nuclear-themed pictures.

> It's unfair to generalize from this. They make it seem like that is all there is to the Tri-Cities. This is a leftover from the old focus. But it's not the focus anymore. There are five "nuclear names" around town. They are always picked to represent Tri-Cities.

A strong response was garnered, however, by a picture of "Bombers Stadium" with the mushroom cloud logo in view. Reflecting that the image was always associated with stigma, one League member snapped, "They need to change that logo." But, it was explained, an effort by the coaches to change to uniforms without the mushroom cloud had been opposed by the children, who organized a "we're proud of the bombers" campaign. Other images were responded to because of the stigma. A cartoon in which Downwind farmers were shown selling admission to see their deformed cattle was seen as "offensive – it stereotypes the community.... People think that we'll do anything for money." An article about a nuclear truck accident was seen as unfairly exaggerated "because it was nuclear" (Cockle, 1985). In each instance, the threat was the stigma, not radioactive release.

Richland League members also defended Hanford against critics who blamed the complex for their health problems. They rejected the idea that farmers downwind had been harmed due to radiation from Hanford releases. Instead, in one League member's words, "Any problems were from farm chemicals...." Another added,

> People across the river here are not in more danger than farmers anywhere. They misuse chemicals. To blame it on Hanford! They have no leg to stand on.

Beyond the Franklin County farmers, Spokane had emerged as a center for outspoken Hanford criticism. The Spokane League's report, "Radiation in Eastern Washington: Following the Trail of Nuclear Waste" demonstrated an active risk to the Spokane region from Hanford. A Hanford scientist who belonged to the Richland League dismissed the validity of the Spokane League's work on grounds relating to the linear/threshold debate about nuclear risk.

> They believe that any percent of radiation is dangerous to your health. They are influenced by Physicians for Social Responsibility. So it is erroneous information that has people scared to death.

Given such errant information, it was understandable that Downwinders were "needlessly worried." Moreover, "once fear is created, it becomes hard to address." Furthermore, now "people are so afraid of radiation they are irrational." Another Richland League member complained,

> People in Spokane separate out Hanford from any other environmental problem. They are concerned with Plutonium. Hanford contributes 1–2% to background radiation. But they weigh Hanford radiation more than this. They see getting rid of Hanford as being as important as stopping smoking. It's a very real fear – they are terrified of radiation. Some people even refuse x-rays.

In fact, the Richland League members suggested, this fear, not nuclear risk, had itself become the problem, hurting the regional economy. And the source of the fear was not local.

> The real problem downstream is fear. After Chernobyl, there was fear about the Franklin county food. That is the real problem. Chernobyl raised the food issue. And that has led to the economic effects.

As evidence emerged in the later half of the 1980s to support Hanford opponents, these new developments were difficult for the tight-knit Hanford nuclear community to take. A one-issue, local pro-nuclear group, the Hanford Family, was formed by professionals to "get the facts out" to counter outside groups. According to a Richland League respondent,

> Since the Hanford bashing began, people have their backs up. There is a revival of Hanford support. Now people take it personally when the DOE or Hanford are criticized. The criticism is of "me" now; that "I" don't care about safety. The community didn't feel strongly about this until recently. The Hanford Family started because they were held responsible for things they didn't do. It makes people mad that others think we don't consider safety when they don't know what we do for safety. Now people stick together.

It was interesting to see this defense of Hanford at work. When the son of a Downwind informant shared information from the Spokane paper with his school class, the head of Hanford personally showed up to rebut the article and convince the children that Hanford was safe. School children and their parents were given tours of the Reservation. And children from Hanford crossed the river and "adopted a cow to show how much Hanford people like farms."

In sum, neither Mayak nor Hanford was publicly honest about risk. Secrecy prevailed. A nuclear culture became ingrained in their respective closed cities. Expertise was venerated. In Richland, the community acted to defend the reputation of the nuclear facilities where they worked and the community where they lived without the overt controls governing people's lives in Ozersk. Part of this defense was the rejection of outsiders' fears as being unrealistic and irrational, mirroring Russian criticism of radiophobia (see Filatov, Klauchek, Britanov, & Klauchek, this volume).

HANFORD'S DOWNWINDERS

Being a "Downwinder" would emerge over time as the defining characteristic for some outsiders to the nuclear community as people redefined themselves as toxic victims exposed to a hazard. Many Downwinders had witnessed clues that all was not well from accumulating mysterious events, rumors and uncertain health problems. But, given the secrecy, these clues were not enough to establish that Hanford was a hazard to their health. Tangible evidence of being "downwind" was a first step towards making the puzzle fit together.

Distant Downwinders – Spokane, Washington

Residents of the Spokane area and beyond did not traditionally see themselves as being at risk from the activities at Hanford, 150 miles to the southwest. That is, until Oregon's Mt. St. Helens erupted in 1980. Richland League members explained that ash from the eruption had fallen over a wide region encompassing the downwind areas from Hanford. Noting that the ash made Hanford's Downwind area visible, a League member quipped, "Mt. St. Helens did us in." This described pattern of ash release presaged later Downwinder identity. A Mesa farmer explained why the ash particularly connected Hanford to distant Spokane.

> The ash went from here north. Pasco got none. We got 1–2 inches. Further north [toward Spokane] it was much thicker yet.

Subsequently, the Spokane League's report on Hanford documented that Distant Downwinders bore the greatest risks. For example, the Spokane milkshed revealed considerably more Strontium-90 than found closer to Hanford because of greater rainfall in the Spokane region than found in the semi-arid Hanford area. The Spokane Downwinders lacked a concern over past events but were hyper-concerned about new dangers related to current health threats, Chernobyl and the safety of the N reactor, that arose after they understood the Hanford connection.

Major opposition to Hanford came, not from the Spokane League, but from a citizens group called HEAL (Hanford Education Action League) that emerged from a study group inspired by a Spokane clergyman's 1984 sermon. HEAL networked with key experts on nuclear risk, invited them to speak and publicized their words through its publications. It forcefully opposed the Repository, demanded that Washington ban nuclear waste shipments, and sought information on historic releases. HEAL and its allies brought about a virtual Hanford glasnost.

The Hanford Glasnost

In January 1986, HEAL joined the Environmental Policy Institute and other regional groups to file a Freedom of Information Act request.[8] As a result, within a month, the DOE released 19,000 pages of formerly classified documentation. The new information fueled intensive media coverage by reporter Karen Dorn Steele of the Spokane Spokesman-Review, bringing Hanford under national scrutiny (see Steele, 1985a, 1985b, 1986). The articles were credited by a Tri-City regional official as the milestone event causing an abrupt turn from "extreme support" for Hanford to "concern and opposition." Another Freedom of Information Act request by HEAL in April 1987 resulted in the release of another 20,000 pages of documentation (Thomas, 1990b). The information intensified anti-Hanford actions. A HEAL researcher noticed similarities between Hanford's N reactor and the Chernobyl reactors, leading to the N reactor and PUREX being shut down. A 1989 HEAL health conference formed a new organization, the Hanford Downwinder's Coalition, headed by a victim of Iodine releases. Other Hanford opposition groups formed. All opposition groups were fueled by distrust of the DOE (Lovrich & Whiteley, 1999).

Steele pursued additional Freedom of Information requests until some 70,000 pages of formerly secret information had been publicly released, documenting environmental releases and legitimating health concerns. Among the outcomes were the Hanford Environmental Dose Reconstruction Project, a retroactive study of radioactive exposures, and the Tri-Party Agreement (TPA) between DOE, EPA and the Washington Department of Ecology promising site cleanup by 2018. Public participation was deeply ingrained in these processes.

As more information became public, a crack in the closed Hanford community appeared. A significant number of insider whistleblowers passed yet more information about health and environmental problems to activists. The combination of an effective citizen "watchdog" over the DOE with access to open government process and a lifting of the veil of secrecy allowed close monitoring of DOE compliance with agreements to assess and clean up the Hanford site (Lovrich & Whiteley, 1999). Schreckhise (1999, pp. 135–136) credits citizen action with causing

> the Hanford Nuclear Reservation to be transformed from a source of great pride among those living in eastern Washington to a social and political pariah institution.

Hanford's changed status brought new scrutiny of Hanford by Congress. The GAO (Government Accountability Office, formerly General Accounting Office), Congress's investigative arm, had filed only three reports on the Hanford facility prior to 1986, none addressing the potential for undisclosed contamination or operational problems. Now, as information accrued on the historic cover-up at Hanford, a series of critical GAO reports commenced. The importance of this Congressional scrutiny lay in the role of Congress in funding Hanford and an end to its previous tendency to turn a blind eye to problems there.

Conclusion
Distant Downwinders drew on a consensus in opposition. Based around a major city, they were able to organize and mobilize to a high degree of sophistication. They had no constraints as Hanford opponents.

Proximate Downwinders: Gradual Awareness and Conflicted Reactions

Ironically, it was much harder to be a Hanford opponent in Franklin County, just across the Columbia from the Hanford Reservation.

Threatening Cues
Mt. St. Helens had impacted the Proximate Downwind area as well. At the Mesa Café interview, a local farmer was described who had "gone broke" between the ash from the 1980 Mt. St. Helens eruption and Iodine releases from Hanford. "He lost 1,000 head. They had intestinal, lung and breeding problems." But there were other concrete indicators of impact from Hanford for Franklin County farmers. First of all, the Columbia was right there, with the Hanford Reservation and various Hanford facilities clearly visible. One of these, the State of Washington's only working plant in its WPPSS complex, strongly contributed to the understanding of "downwind" for Franklin County farmers.

First, there was the warning radio issued to everyone within a 10-mile radius in 1983 by WPPSS. Residents were given training, emergency instructions and a response map. Sitting in some visible spot inside one's home, these radios were tangible symbols of a real threat. The sheep farmer's wife explained. "We never thought about Hanford until we got the radio in 1983." To her, the radio defined the region at risk "A Downwinder is anyone within the 10 mile zone issued radios." Beyond this boundary, as defined by the farmers at the Mesa Café, the Downwinder definition was expanded to include anyone who could see the visible plume of steam emitted from the WPPSS 2 nuclear power plant's exhaust stack. "You can see which way the wind blows the smoke."

Old-timer's Experience

Franklin County residents could also draw on a long history of Hanford-related incidents in defining their victimization. Mostly World War II veterans who strongly supported the original Hanford mission to build the bomb, the gathered Mesa farmers had remained passive during the Cold War period. Only in retrospect, knowing now what they did not then, could they begin to define themselves as Downwinders. In 1986, when I went to the Mesa Café, Tom Bailie was the only one using this label. But all viewed themselves as "old-timers," for them a prerequisite to being a Downwinder, as one farmer noted.

> Those arriving before 1970 fear that they have or will have cancer, heart problems, other health problems or they lost a child.

Old-timers had their share of worrisome memories. First, was the dust bowl experience recalled by the sheep farmer's mother.

> In the '40s and '50s there was no irrigation. The dirt blew. People had to wear bandanas.

Dust was unleashed from Hanford construction and from efforts to eradicate wild plants that absorbed airborne radioactivity (Gerber, 2002). But the major cause was improper "sod busting" farming techniques according to Uncle W., a "dryland" farmer when he first arrived in 1936. Proper dryland farming was protective of the land. Original Downwinders, mostly ex-servicemen, were new to farming. They plowed the earth – and it blew. Uncle W. recalled the blowing dust and witnessed whole fields of wheat blow out. Fences would silt shut. There was a five-foot accumulation in one dust storm.

> You don't know what a dust storm is! In 1948, there was such a severe wind storm that one guy couldn't find his house. We had dirt on the table.

This inspired another Mesa farmer to recall his grandmother setting the table with the plates turned upside down, "People would come in and real quick turn the plate over and eat."

Uncle W. suggested that the blowing dirt was a potential source of radioactive exposure. "This is how the radionuclides moved."[9] Irrigated farming had now changed the climate.

> Now the area is humid…It used to be 120 degrees in the day and cool at night. Dryland farmers understood nature. Every region of Washington is different – each valley. There is 7″ of rain in the winter here; up north there might be between 12″ and 25″.

"Uncle R" added that the Columbia River used to freeze over every year before Hanford, "It hasn't frozen over since. They used to drive sheep and wagons over it."

As a child, Tom Bailie found weather balloons in the fields during the "Green Run."

> My neighbors saw weather balloons drop. They probably were taking air samples. These people are now in their 50s and 60s. They had a young baby die. His lungs weren't developed. They remember having planes fly over and drop stuff on them. Her daughters had problems in their thirties. Thyroid cancer, leukemia. They're Mormons – they don't drink or anything. Yet a lot of Mormons have cancer.

Bailie believed his exposure to Iodine-131 at the time was the cause for his later stillborn children. He said that Uncle W. had received more radiation than anyone else. At age 75, the oldest Downwinder left alive, Uncle W. had farmed and erected fences in the downwind area most of his life and stole watermelons across from Hanford. He recalled seeing a pink iodine cloud come wafting over but never knew what was happening. The story inspired Uncle W. to show me his stomach, right there in the Mesa Café, crisscrossed by scars from four cancer surgeries.

Closer to Hanford, the sheep farmer's mother recalled that they never thought about Hanford when they moved into block 15 of the Columbia Basin project when it opened in 1954–1955. Yet, the family had many potential exposures, as her son explained.

> When we first came there was a cherry orchard across from Hanford. We would go pick. Then they fenced it off. It was restricted. We used to sneak in and eat the fruit.

His wife picked up the thread, explaining that her husband also played in the dirt from the time he was a year old.

> A group of guys would go back in the restricted area motorcycling. Now they're all bald in their 20s. We don't know if it is related. The generation before wasn't bald.

They had started paying attention to Hanford in the late 1950s, after a big spill at PUREX.

> The spill only became public after doctors in Spokane began to wonder why they were getting cancer and heart disease cases up there from the Tri-Cities. They pushed for release of information.

Locals had also observed testing of people, animals and the local environment. At the Mesa café, the "apple grower," an original homesteader from 1954, "the year of the ruthenium release," brought up a story that caught his eye while flipping through a national newspaper. At its mention, the entire assemblage chimed in to simultaneously tell the story. A duck waddled into a farmer's garage and set off a Geiger counter. The resulting investigation revealed that ducks landing on the Hanford reservation became contaminated. When they flew across to eat corn in the Mesa fields, they were often shot and eaten. Uncle "R" recalled going to a DOE meeting years ago where the agency asked for duck heads without explaining why.

Tom Bailie elaborated that, from 1946 onwards, the thyroids of every animal killed were tested. Asserting that Downwinders' children were irradiated, he paralleled the Muslumovo activist's discussion of secretive health monitoring (Edelstein & Tysiachniouk, this volume),

> data on them is also in the files.... We're like residents in a laboratory. They monitor our health.... Battelle is doing research on this. They control the dissemination of data.

When Bailie then described how the DOE had "killed" a proposed tumor registry proposed after the 1963 discovery of well contamination, a listening farmer looked directly at me and said, "Do you see why we don't trust them?"

The farmer/nurse also believed that deliberate radiation experiments had been carried out.

> We were used as Guinea Pigs. We know that we were exposed to radioactive releases from Hanford on an experimental basis. The DOE deliberately released radiation and checked to see what effects it had. Maybe they collected data all this time, we don't know.

Such comments underscored the belief that the cause of their health problems was not accidental, but deliberately caused.

Lifescape Change in Hanford's Victims

These impact issues illustrate a psycho-social dynamic characteristic of environmental contamination, namely a change in lifescape, or the normal assumptions that underlie everyday life. Lifescape changes associated with contamination routinely involve inversions of the sense of heath, control over one's life, security of the home, safety of the environment and trust of government and others to be protective and helpful.

Health Problems

When the Mesa Café informants began to recount health problems, their list seemed long, indeed. Like the local electric co-op official, a 23-year resident, who had lost his wife to leukemia, they all wondered if there was an association to Hanford.

> I can't help wonder if our proximity to Hanford had an influence on that. It gives you cause to wonder.

He noted that such thoughts were always just below the threshold of consciousness.

> People don't lie awake at night over this, but when they see an article or someone gets cancer, they realize that it might be related.

An old farmer then began to discuss his family's medical problems. All four boys in the family had stomach cancer and his sister's breast cancer was spreading. As the others at the Mesa Café got into a discussion about how much of their intestines were left, the farmer asked,

> Did that crap [Hanford radiation] have something to do with it? We're like the sheep in Nevada during the war [used in experiments to test radiation effects].[10]

Illustrating the kind of active shift to health concern that accompanies radioactive or toxic exposure, several informants elaborated these themes with their own stories.

1. The Waitress' Story. Tom Bailie introduced one of the Mesa Café waitresses whose husband was a Downwinder. She had lost three babies, including "One I lost nine days before it was due." Bailie told her straight out that the babies died due to "old" Hanford. He said there had been genetic disruption of her husband's sperm. Bailie added that he himself had no sperm count and described how he faced up to his sterility and adopted children. He had just returned from a conference of radiation victims in N.Y.C. where he had learned that in Japan no one will marry the daughters of bomb victims because they produce deformed children. The waitress replied,

> This makes sense. I grew up elsewhere. I had a healthy family. It's my husband's family that has problems. They were lost due to health problems. They just fell apart what with the money for all the doctor bills. He lost the family. They are scattered now.... Everything had been going well for us until the DOE came into their lives. We had land, we

> were making money, we were happy. When the DOE left our driveway with samples, I always said that they were using us for experiments.

Tom agreed sympathetically. "We are a laboratory experiment, but they should at least pay the costs." When the waitresses asked why her husband was always so tired, Bailie replied, "We have nuclear lethargy ... people from the 1950s die of heart trouble or they feel sleepy." For this reason, he explained, Downwinders often become ill and lose their farms.

> The Hibakusha in Japan are discredited as being lazy. Here, the FHA forecloses on farmers and says that they're lazy.

2. The Farmer/Nurse's Family. When the farmer/nurse's family moved to Franklin County in the late 1950s, they didn't know about Hanford. Everybody around came from a different part of the country. She described seeing airplanes dragging big boxes and not knowing why. No one made a connection between "the airplane tests" and his or her "getting a dose." Her little children would sit in the dust on the site and watch the planes fly over. She was pregnant. When her baby was born in March 1960, it had respiratory birth defects that she likened to "the type found at TMI." The baby died. Later they found out that in 1959, Iodine-131 was released daily from Hanford at levels comparable or higher to releases at Three Mile Island (TMI). She blamed the loss on the fact that she had worked on the farm all the prior year and had inhaled I-131. There were many miscarriages then.

Half a week after they buried their baby, one of their daughters developed throat nodules that had to be surgically removed. One and half years later, their second daughter had the same condition. Now she was concerned that both her daughters are ill. "Their immune systems are shot. One had a stroke at age 26!" Her daughter's doctor suspected an environmental cause for the stroke, but he didn't know what. Her older son suffered from headaches. And her husband developed sores on his head. He wore a hat with holes in it when he was working in the field. The sores continued for several years until the dust blowing subsided. The doctors didn't know what it was. But she read about similar sores observed in Nagasaki.

They had only recently learned about the Iodine releases; it explained why six of the seven surviving members of their family had to take synthetic thyroid stimulants. The doctors had not understood why their lost baby had so many birth defects. But learning how much radiation was released, they understood the baby's tragic death. At their farm, animals often died due to deformities. They had one calf born without a regular stomach. Their animals got affected by cancer. They now saw Hanford as causing all their problems. "I feel that this was unnecessary. I know that my family is contaminated."

3. The Farmer/Nurses' Patients. Nine of ten patients the farmer/nurse tended as a hospice and home care nurse suffered from cancer. She knew of high cancer levels throughout the Downwind area; a new case appeared every month. On her road alone, four of eight families had cancer. Her 36-year-old neighbor – a woman with five children – had just died of cancer. So had her daughters' best friend, a 31-year-old woman also with five children. She rattled off the names of four more neighbors who died similarly. A young boy had been born with a brain tumor. Another child had leukemia. Other neighbors lost their fathers. All the early land-levelers but Uncle W. had died of cancer. A month ago, a neighbor's 15-year-old child got cancer. Her friend with lung cancer was not a smoker. And she, herself, had skin cancer. It was overwhelming.

> I've seen so much that I have had to start pushing it out of my mind so that I could raise my family. I am frightened for the Downwinder area.

These observations were based on local illness incidence where she lived. In her work experience, she was treating seven laborers in their 20s dying of cancer. The hospice business was brusque. She said she had five job offers in the prior four days.

> I am appalled at the hospice by how much cancer there is. The hospice cannot reach enough people even with supportive care. I work for a professional nursing service and for the state. I am very busy taking care of a lot of people.

One of her patients had been a Hanford employee, son of a Hanford engineer. He died of cancer at 28. At age 18, he had been part of a work team of five sent out to a hazardous area of the Reservation. According to his widow, only one of the five was now left alive.

> Yet Hanford claims that no one has ever died of radiation there – a lie! The DOE covers up, hides, misrepresents. I take care of people who die, and the DOE says no one dies. How do you explain the discrepancy?

She noted that some people with cancer connected their illness to Hanford. One old man that she cared for "cursed Hanford to his dying breath." Her husband said that his neighbors and friends with cancer all thought Hanford was to blame. She agreed. "When the subject of Hanford comes up in light of a patient's cancer, they never deny it." She noted that Downwinders from her community understood that there was more than one potential cause for their cancers, including pesticides relied upon for farming.

> We are not narrow-minded! It is our livelihood at stake. We built our home and our farm from woods and blowing dust. We are willing to listen to the fact that we use too many pesticides that get in the air and cause cancer.

But she was not sure others were equally unbiased. "Why is it that the DOE tests don't show the rise in cancer?"

4. "The Death Map." A 1983 article got the sheep farmer's mother thinking. That night she sat down with her son over the WPPSS emergency plan zone map.

> And we marked the map with blue for all the heart problems. The heart problems were men 45–55 that began in 74–5. The heart attacks were all in a ten mile radius. The cancers showed up later. All the people had worked in the fields, and the dirt would be blowing. Before irrigation, we ate so much dirt.

She showed me the map. Red dots indicated cancers. These were further differentiated according to cancers of the lung, breast and glands and

> Also a lot of leukemia further north – in kids. A big percentage of women have had hysterectomies. Children suffer from skin and lung disorders. The neighbor's kids – one had cancer and the other had open heart surgery by age 4.

But, once the map was drawn, not knowing what to do with it, she "just put it in the drawer." Tom Bailie told Karen Dorn Steele, the Spokane reporter, about the map. Not long after Steele published it, the state's chief epidemiologist had come to see the sheep farmer's mother.

> He talked to us like we were kids. He drew a map which showed that there were no excess deaths and cancers.

A state study soon appeared that she characterized as sampling so large a population as to wash out the local effects.

> I read in the paper that they did a study of one million people, 340,000 people in this area – they must have gone pretty far. The results were negative. If he had done the study on people right here, he would have found what we did.

Some of the health problems were represented right in her own family. The sheep farmer's father and sister had thyroid problems. All the women in their families had or needed a hysterectomy. Her mother had three stillborns there, all full term deaths. These results defied the demographics, as she explained.

> Mormons have a generally lower cancer rate – but not here. There are a lot of thyroid problems. Two-thirds of the people have thyroid problems.

The sheep farmer had an ulcer when he was 17. His eight-year-old child had arthritis. And radiation was credited as the cause of widespread stress by the women. The sheep farmer's mother added,

> We have heart diseases where there is not high cholesterol. Bypasses and open heart surgery! There are no old men from out here. They all die in their 50s. If you watch Hanford obituaries, people die at 60–63 if they have lived here a long time.

And farmers "not downwind of Hanford" lived longer.

Loss of Control

The entire Hanford saga is one of lost control on the part of community members and usurped control on the part of the DOE and its agents. This loss of control involved undisclosed exposures to invisible hazards over time. Uncertainty prevailed. Information and expertise was held by government; the knowledge of regular people was of little utility. There was a disabling dependence on distrusted experts and officials for protection. Over time, people got the sense that they were living in a sacrifice zone.

A Mesa farmer explained the effect of the uncertainty. "I don't know what to think.... My opinion isn't worth a hoot – I don't know." The net effect was to render as helpless those affected. It was an alienating experience, as the sheep farmer's wife recalled,

> The government comes and makes you feel like a fool. You're a dummy. You're a little peon farmer who should go back to the dirt and not make waves. They intimidate you. They talk down to you. Their message is that you shouldn't question the government. They show charts which show how safe it is. Makes you feel like a dummy.

She remembered that the press was handled the same way.

> At one meeting a reporter questioned the officials about safety. He was told to shut up and sit down.

Inversion of Home and Environment.

There were blatant cues in the environment revealing invisible dangers. The Columbia shore had clearly posted fishing advisories. A Richland League member recalled

> They used to have warning signs on the main roads into TC 'If the lights are flashing, turn around and leave.' One time we had to do that.

Even before the release of confirmatory documents, Downwinders believed that their environment was already severely degraded by Hanford, as a farmer noted at the Mesa Café.

> Old Hanford is shut down except for the tanks filled in the 1940s. They're not safe. There are no records on them. We have every right to be worried about old Hanford. A cleanup has to come soon.

The sheep farmer's wife reflected on the constant knowledge that disaster might strike, now a property of the local environment.

> If there was a release, we are supposed to put all our animals under cover – 1,000 cows and 500 sheep! And the feed would all be contaminated anyway.... The kids would be trucked into town (Pasco) if there were a release. That wouldn't happen.

In addition, joy at home was inverted. It was no longer a place of happiness and accomplishment, as the farmer/nurse explained.

> I'm not as happy with my home. I used to take so much pride in it. We gave our lives to this way of life – our money, our time and our talents. My husband built this home from nothing.

In the wake of reported water contamination in three Franklin County wells, the future safety and value of their home now rested on the outcome of expensive water tests that they ordered in their 500 foot deep well.

> Now I'm frightened to drink the water. I'll test it, but it will cost a lot of money. If it's contaminated, what will we do? I'm sitting on pins and needles. I'm angry. I can see it threatening everything that we have worked our lives for.

Contamination not only disconnects victims from place but interrupts cross-generational relationships as well. A Mesa farmer who grew diversified crops on 1,000 acres, whose grandfather was a homesteader from Oklahoma, described himself as highly conservative: "I don't drink or smoke. I'm church going. I pay taxes. And I'm no activist." He spoke of his land in light of his three sons, five grandsons and granddaughter.

> I always felt that there was no better place than the Columbia Basin. I want to be here for generation after generation. I want to save the farm for my sons and their sons.

But he has been pried loose from the land by loss. He now speaks of second thoughts.

> The nuclear thing. The cancer. There's no doubt that it's here, affecting a lot of people. I've talked to my sons about it. I have a grandson who was born deformed and died. We don't know if this had anything to do with it, but it sticks in the mind. I talk to the family about this. If we could get out with something and move on and try for another area.

Hanford is the only reason he'd leave. But he believes he may not have a choice.

> We don't know how far we are contaminated already. Maybe it's too late to reverse the trend. This may just have to be a waste area all the way to Portland if we want to make a

> nuclear dump for the world. The Columbia is already contaminated. They may have to move us out of here anyway. It's past the point of no return.

The apple grower indicated that many were ready to leave.

> A lot of people in the area would sell out if they could. People can't sell the ground because of the publicity about the ground being contaminated. They think it's a threat.

The apple grower continued, issuing a plea for proper relocation that exhibited helplessness.

> They may find that these homes that we worked to build over the years are dangerous. There is no question that Hanford knows exactly what is going on. If this place is worthless, don't do what you did to the old-timers at Hanford, move us to a safe farming place.

It should be noted that fear of the environment and inversion of place and home combine, in many of these cases, with impacts to livelihood. Virtually the entire region was dependent upon Hanford, either for jobs or commerce or for avoiding stigma destructive to their crops. The result was that people were victimized by Hanford in virtually every aspect of their lives.

Trust

Trust is essential to the acceptance and support of a potentially hazardous facility (Dunlap, Kraft, & Rosa, 1993; Edelstein, 1991–1992, 1993). Trust for Hanford arose originally from respect for government, the effort to win the war and the resulting culture of acceptance. A Mesa farmer noted that the local population was accustomed to believing what they were told, at least until the Chernobyl disaster occurred.

> Chernobyl made people think about the N reactor. Safety was not a local issue. It was a jobs issue. People here believe what the government tells them.

Several factors helped Hanford retain trust. Personally knowing plant personnel was a key ingredient to feeling safe. The Farm Bureau representative made it clear that trust for Hanford was based on local relationships and not trust of government or their contractor, Battelle.

> Everyone has a relation or friend working at Hanford. They realize that they are not trying to hurt us. You have to have a certain amount of trust. We do not trust the government. They did a cover-up. Also Battelle did not handle information well.

Another local farm official felt reassured after getting to know the safety officer of one of the facilities.

> After you get to know people like her, you get comfortable. It is safer than the agricultural industry I work in.

The safety officer, a member of the Richland League, echoed the importance of relationships.

> We know the people who do the monitoring. We could monitor it ourselves. They live here; their kids live here. They retire here.

Other Richland League members echoed the same theme.

> You know that people are good, reasonable, intelligent and technical. And they are not afraid of it. Their kids play with your kids.

> I am not afraid of it. I trust the people who work there because they live here. They're not afraid of it or they wouldn't be living here with their families.

Even the revelation of wrongdoings was viewed as an impetus for trust.

> The DOE contractors don't cover up anymore.

Some Downwinders respected and deferred to Hanford experts, "Scientists work here. They have expertise." Essential to the trust was that scientists, "had children." A school librarian similarly trusted Hanford contractor Battelle to give her family radiation screening (Larson, 1985).

> They have their families there also. They want a safe environment too. I can't see any purpose to be served by not being honest.

And a downwind farmer was sympathetic to the experts there because, as a pesticide user, he also faced risk-based criticism. He felt deference should be paid to the experts. "I support the nuclear industry because I understand what they are up against."

Knowledge of workers could cut both ways. Franklin Downwinders were most likely to know Hanford laborers rather than scientists. Tom Bailie had mentioned use of dope and cocaine by Hanford workers, a theme also discussed by the sheep farmer's wife.

> We have friends working at Hanford. It was a disgrace. They were sleeping when they were supposed to weld and pour concrete. I know a welder who smoked dope on the job. The workers laugh at what they can get away with. They bring over "hot" tools. It scares me half to death. Last week they arrested two people for using coke. They were fired.

Of course, historic events at Hanford had already left a legacy of downwind distrust. The farmer/nurse first questioned the DOE after the macabre discovery on her farm of a dead cow with its organs removed. The papers blamed a satanic cult, but she was convinced that the DOE was doing research. Such distrust was then corroborated by the release of information about Hanford contamination.

> The DOE has lied. When they have been questioned, they give bold faced lies. They deliberately withheld facts that they should know. The documents released showed this. We should have been told that we were exposed to radioactivity. The DOE is dishonest. They turn everything around.... The DOE gets into trouble if they say too much. If we trusted the DOE, if they had given the facts, then maybe we would believe when they say it's all right. They lied before and they will lie now if it suits their purpose. They are unscrupulous.

A Mesa farmer cited as a basis for his distrust new information on Iodine releases,

> Iodine has been known for 15 years but was withheld. Hanford has not been candid.

The sheep farmer's wife stated similar conclusions but bluntly,

> We don't trust the DOE. I don't think anybody does.

This failure to disclose known exposures was a source of anger for Franklin Downwinders. It was also a source of worry. The farmer/nurse and her husband think about radiation a lot in situations that seem mundane: when the dust blows onto her grandson or when they drink their excellent tasting water from their 583′ well that cost them so much to dig. Worry begets anger. She commented,

> Don't minimize how I feel. I'm angry at the documents. They knew this. And they didn't tell until they were forced to. Look at how long we drank it after they knew! Or whoever else drank it. What right does DOE have to do this to people's lives? We pay our taxes. They should be very concerned. They've become so powerful that they got carried away. The individual families don't matter – they didn't tell us. How long have they been doing this without warning? They flew over when I was pregnant. They didn't care that my baby died. I hate the DOE. They hurt people and never said a word. Maybe people will get cancer and maybe they won't. They didn't have a choice. A lot of people feel very strongly but they won't say.

No rationalization could offset the profound disappointment and distrust of the DOE or the belief that what occurred was wrong. She continued,

> Occasionally in the national defense people may be dispensable. Maybe the Downwinders are the ones that they can do without. I don't want our children or grandchildren or neighbors to be sacrificed. I am tired of seeing people die of cancer. I care that people drank the water. Why didn't the DOE check? If they need to, they should fix it so it will be safe. Or remove the people. Why don't they care about people a little bit? The DOE never gave us a reason to trust them. The DOE – We know that we can't trust them.

And she had a specific revenge in mind for those who hid the danger.

> They ought to make the people who knew about this sit beside the people who are dying of cancer. Then they'd see how smart they are.

The negative DOE history called into question any data they provided. The farmer/nurse did not take her family for testing because she did not believe the tests would show anything.

> The DOE was just playing a game trying to make themselves look better. It won't show up. Maybe it will make some people feel better. See what they do to appease people. The harm is done. It would be better if they just said that they screwed it up. We should pay for our own tests and really see what we find.

But the tests were too expensive. However, she did decide to have the state test her water.

Folks at the sheep farm interview further deplored that DOE tended to blame problems on lifestyle and not on radiation.

> The DOE last year sent a bus to measure radioactivity in the body.... They tested 35 people. The scores came back negative. We were sent a report that said that ... the only ones with a problem were "people who lived in the fast lane."

The sheep farmer's wife refused to send her kids for testing because she heard

> that there were better tests available, and I didn't trust the government to give honest results. It was being sent to a government lab.

The repository proposal confronted a context defined in the words of a Pasco farm businessman, "We are depending on a government that we don't trust." A Mesa farmer made a similar point.

> If the DOE where more honest they would have more support. There have been lies and cover-ups. Why should we believe them?

Another Mesa farmer expressed the same thing in infinitely more picturesque language.

> I wouldn't trust them any further than I can throw my big bull by the tail. We are kept so ignorant of the facts of things. The repository will not affect us if we don't know about it.

Tom Bailie retorted, "The DOE treats us like a mushroom. They keep us in the dark and feed us bullshit." Another farmer added, "We keep our heads in the sand and we get kicked in the ass."[11]

Activism in Two Downwinder Contexts

The experience of activists highlights the difference between the Spokane and Franklin Downwinder cultures. The sheep farmer's wife lamented of Franklin County, "Toward Spokane, people talk to each other, Here, they STILL don't talk, they still don't discuss." Clearly there were differences in two cultural settings that influenced the formation of grassroots efforts.

Spokane

As noted, HEAL, the Spokane group opposed to Hanford, was an innovative community-based organization that arose organically from cultural activities in a vibrant city. It attracted scientists, producing expert reports, as well as newsletters and media releases. It networked with national organizations and the press and took primary action. HEAL enjoyed a community consensus. Spokane had little to gain from Hanford's operations. Given the operative views of risk prevailing there, it had everything to lose.

Franklin County

Activism for the Franklin Downwinders was a different story. It involved a one-man battle against Hanford waged by activist Franklin County farmer, Tom Bailie, whose wild assertions were eventually confirmed by released documents and downwind illness. It was interesting to see Bailie in action at the Mesa Café presiding over an unofficial daily Downwinders meeting. Farmers clipped newspaper articles and many agreed with him on his major points. But, whether they believed that Hanford caused health problems, as one put it, we all "pay Hanford prices."

On the morning that the Mesa Café was turned into my field research laboratory, Bailie was busy explaining a recent Spokesman Review article about Iodine-129 contamination of the area's water supply to a family farmer from Mesa over scrambled eggs and coffee. He certainly did not talk down to his audience as he launched into a lengthy explanation of the geology of the area and then drew a map on the back of a used envelop showing three aquifers at varied depths.

> The pollution is drawn under the farming area as the farmers dug deep wells creating a hydraulic pull. The DOE has known about this since 1964.

Much later, he quizzed an apple grower about the incidence of cancer and heart problems along the roads near his house, drawing a map on my tablet that had "x's" by the houses with health problems. The apple grower was reluctant because he really didn't know were there was illness, just where people talked about having illness. Later, Bailie again shared news of the Iodine contamination with the shifting gathering.

> Three weeks ago the DOE released the news that I-129 was found in Franklin County wells. It moved under the river. Our wells are contaminated. It has already happened.

The "apple grower" listened in disbelief, not having heard this before. While he had the farmer's attention, Bailie went to his truck to get more documents: a 1984 monitoring document from Battelle showing tritium contamination in the aquifer, minutes of a 1973 meeting where scientists discussed how to control damage from the pollution, a chart showing I-129 contamination in wells between 1964 and 1979 and an article on soil contamination by Alan Bensen (of HEAL) called "Blowing in the Wind." When the apple grower acknowledged that his son drank two gallons of cow's milk a day, Bailie read information regarding milk contamination. When the apple grower argued that he lived further downwind then where the measurements were taken, Bailie countered "It's even worse where you are."

The Freedom of Information Act disclosures had suddenly granted Bailie credibility.

> Seven months ago people talked about that "fucking Tom." They were afraid that they wouldn't be able to sell their food. Now they see – they ask questions. At least now it's up for discussion.

Bailie explained that he had always gotten "heat" for his efforts. But,

> I'm almost over the peak of criticism. Now the tide is turning. I saw these two older women yesterday. They were both dying of cancer. They said, "Tom, what you're doing needs to be done – you're right." Their husbands didn't agree. But the women said, "There are a lot of things that are a lot more important than land values – like our grandchildren. We know that you're a watchdog for us. We know that you are there."

Bailie agreed that he served more as a watchdog than an anti-Hanford activist. He asked a lot of questions. He was very mindful that his public questioning might hurt his farm business.

> Sometimes we have to make moral decisions. I had to make a decision that going into this is more important than my economic well being. I don't want my children to say that my dad could have done something.

Bailie had received much of his support from women.

> Women have always been suspicious about Hanford. They calve and raise children. They deal with sickness. They see the patterns. They are much more sensitive.

The farmer/nurse corroborated the importance of Bailie's educational efforts, "when Tom told us things, we found his information to be totally accurate." He was at the corner one day and asked how they were. When their thyroid problems came up, he immediately made a connection to the 1959 releases that they hadn't known about.

Bailie had not always enjoyed much support. His own father dismissed his efforts, "My dad just says that there ain't nothing there." Mesa was a tight community. The "Cattleman," citing discord over the repository, expressed his fear that community harmony would be lost. "You will see dissension. Half believe what they are told (that it is safe) and half don't." Bailie was in a perfect position to attest to the controversy.

> You can't do what I've been doing. It splits the community. I've been getting people to talk about it, to face something that no one wants to face.

Bailie had informally organized several meetings of the Downwinders. Three or four showed at the first, 10 at the next, eight or nine at the last. Bailie's challenging efforts to marshal the stubborn farmers rested on one expectation, "When the farmers finally decide to fight, the DOE is in trouble."

But not all Franklin Downwinders saw eye-to-eye with Bailie. The sheep farmer's wife believed that Hanford had become much safer. She was pragmatic in her concerns.

> Our main concern is, how this is going to affect us? What happened? What precautions are needed? What should we watch the kids for?

Just before my second visit to the sheep farm, Bailie had told the farmer's wife and mother that plutonium and uranium had been found in three local wells, including one used by the mother and the local school well. The news had created parallel anxieties in the two women. The sheep farmer's mother worried about how she would get potable water if the contamination of her well was confirmed. His wife was concerned about her children drinking the school water. Given the uncertainty, she was in a quandary over how seriously she should take the issue.[12]

Even Bailie faced uncertainty. He sold much of his crop to the Japanese. The crop had just been thoroughly tested and found to be clean. But he faced a perpetual jeopardy demanding continual testing. "Maybe it's okay, maybe I can farm, but I'll have to test." Bailie was not only a victim of contamination from Hanford, but of Environmental stigma as well.

ENVIRONMENTAL STIGMA AND HANFORD

Environmental Stigma reflects the discrediting of the victims, places and products linked to contamination (Edelstein 1991, 1991–1992, 1993, 2001, 2004; see also Flynn, Slovic, & Kunreuther, 2001). The Hanford study illustrates the power of stigma as a motivating factor in the social dynamics surrounding both hazards and potential hazards. We can distinguish between victims of contamination, per se, and victims of stigma due to contamination, categories of victimization that may or may not overlap. In our discussion, we will first discuss stigma of place and product, then social and personal stigma.

Stigma and the Post Cold War Reinvention of Hanford

Just as Mayak and other Russian weapons facilities (see Filatov et al., this volume) have been forced to rationalize their mission post Cold War, Hanford and other American weapons sites faced similar challenges. Hanford had been downsizing since the 1960s, when eight reactors shut and the workforce was cut by one third. By the end of the 1980s, the end of the Cold War and Chernobyl had brought about a shutdown of Hanford's plutonium production and the N reactor and the loss of upwards of 20,000 jobs. While, as one observer noted, "It was impossible to fail in business during the heyday," it was now clear that the heyday was over. As a local official elaborated, "Hanford has brought us PR, Notoriety. Name recognition. It put us on the map.... But now we need help."

The battle to keep Hanford facilities operating had meant an exhaustive effort to maintain defense appropriations in cooperation with the Washington State Congressional Delegation. But the repository issue pushed the state's electorate to oppose nuclear power en mass, causing the narrow defeat of a Senator supportive of Hanford. These ever-more difficult battles shook local faith in the nuclear enterprise as the basis for the economy, favoring a strategy of diversification, as a tourism official described.

> A new generation grew up and old-timers retired. They had a different philosophy. Publicly we believed in the site [Hanford]. It was everything to the community. We all said, "without it, Tri-Cities is nothing." But in secrecy, behind closed doors, there was no consensus.

The situation demanded a clean break with Hanford, as another official added.

> We are so dependent upon the DOE that we can't plan. We engage in crisis management. You get a sense of the real world politics that drives the site. We chose diversification for the future. The repository is part of our history.

As Hanford ceased to be part of their future, leaders in the Tri-City undertook to reinvent the economy. Some of their efforts built on the region's nuclear experience. A development official described the successful push to build a fast breeder reactor and to diversify into food irradiation. But a promotional effort to redefine the Tri-Cities' economy by playing down nuclear industries was already underway. The director of Tridec, the regional agency promoting tourism and industry, estimated that

> The economy has diversified from 90% Hanford industry and 10% agriculture in the 1940s ... [to] 55% ... Hanford related.... Now the economy is very diverse.

However, the Hanford Reservation was not so easy to forget. Hanford's mission was now defined, not by the manufacture of nuclear weapons but by the assessment and cleanup of its highly contaminated property. All of a sudden, its million plus cubic yards of plutonium production wastes – half of U.S. transuranic waste and two thirds of the nation's high level nuclear waste (Whiteley, 1999, p. 50) – became an asset to Hanford. Its new raison d'etre would become its own environmental cleanup. Hanford, and its parallel sites in the U.S. and Russia, had outlived their utility within a temporal context that lasted a half century, but spawned environmental degradation and hazards that would now live on for tens to hundreds of millennia.

However, the nuclear legacy of Hanford, made visible by the repository proposal, threatened to stigmatize the efforts to reinvent the Tri-Cities. The new diverse economy of industry, tourism and agriculture were all potentially susceptible to stigma.

New Industry

An economic development official denied any stigma effects, listing a string of new diversified businesses arriving in the Tri-Cities, although "We don't know who didn't call us." He said that potential businesses routinely asked about Hanford.

> All potential businesses ask about Hanford. There is name recognition but they do not know much about it. It means "big federal project." Money not risk.

Thus, as officials noted, Hanford gave name recognition that the region might lack. At the same time, contamination begets contamination. Hanford was truly an attractive site for facilities that might benefit from location in a site already viewed as unpristine due to a combination of truth and rationalization. The adverse impacts of contaminating a contaminated aquifer are obviously less than the impacts of contaminating a clean one. And, one can justify a polluting activity with less dissonance. It was no accident that one of the first new industries attracted to town was a food irradiation facility. A blending of the regions' two assets – nuclear know-how and farm products – irradiated food might be stigmatized even without association to Hanford.

A local farm official advocated renaming food irradiation as the less translatable "Pico-waving."

Tourism

One local official optimistically predicted of the Tri-Cities "There is an unlimited potential for tourism. It is a good clean industry." But stigma due to the contaminated environment threatened this hope, the official explained.

> Wine growing and tourism are the only industries viable now. People now perceive it as contaminated. We draw tourism from the west coast where there has been bad publicity.

The chief boosters currently marketing tourism in the Tri-Cities region described an "image campaign" to promote and sell the community. The glossy pamphlets they had developed promoted golf, the wide Columbia River, local wineries and hotels, but not nuclear facilities. However, they reported encountering no negativity to the Hanford complex in their surveys. Some tour groups even offered tours of the repository.

> We never lost a convention because of nuclear concerns. There are between 268 and 402 groups here a year. Hanford is not a substantial tourist detriment. The Iodine 129 scare was not viewed as credible.

A young adult involved in the convention industry interviewed on August 7 and 13, 1985 however reported that a few convention groups would not come to Tri-Cities because of Hanford.

> Ignorance! They were uneducated to the Hanford area. One lady even said that she wouldn't drive through the Tri-Cities. It's like the story about how Tri-Cities is such a good place for night fishing because if you try to catch a fish after dark, it glows. People think about the negative – but Hanford has nuclear energy, not just bombs. Why does the news always pick on Hanford? In California, I saw a nuclear plant on the San Andreas fault. That would scare me!

She was more concerned that the local economy had gone sour. In 1981, the hotel "was full 4 nights a week. Now maybe twice."

Agricultural Stigma

Stigma, to occur, requires a mark or devaluing characteristic, an identity that the threat can be attached to, an observer with the potential to perceive the stigma and respond to it, a vulnerability or harm that might occur due to the threat and publicity or visibility so that the threat becomes known to the observer. Franklin County farming had all of these characteristics. The expansive agricultural district occupying the east bank of the Columbia River and anchored by Pasco was largely irrigated with water provided by the Grand Coulee Dam project. The region was not only across the river from the Hanford Reservation but it was downwind. Moreover, much of the produce and products from Franklin County were marketed nationally or even abroad. As problems with Hanford became public and publicized, the region's livelihood became vulnerable to stigma, as noted by the Farm Bureau representative.

> The damage to us is not from the hazards at Hanford but from the rest of the world seeing it. We serve a Pacific rim market.

The energy official eating at the Mesa Café offered an important distinction over the nature of the feared risk. Unlike the Spokane Downwinders, who genuinely feared Hanford health effects, the Franklin County farmers were worried about stigma.

> People confuse historical Downwinders's thoughts with others who are concerned. My concern is that bad publicity will kill the economic growth from Hanford clear over to Northeast Montana.

First of all, the farmers were at risk from an accident or even gross violation at Hanford. A Farm bureau representative emphasized her concern that a radioactive release would stigmatize agricultural products going to Asia, particularly Japan, with its well-founded aversion to such threats and the unpleasant historical links to Hanford.

> Hanford people did not realize what they do affects farmers. This came out at a meeting of health physicists. Engineers at Hanford feel so safe that it is hard for them to believe that people perceive a risk.

Of most concern to farmers was the possibility of an accident along one of the highways leading to Hanford through the agricultural region. A farming representative remarked, "I am afraid of the highway, not Hanford." Another added, "A spill in Franklin might raise a fear of food contamination." Not only was the risk serious but it was considered likely, as a Mesa interviewee noted, "We're waiting for a truck to tip over. They drive 24 hours and pop bennies." A vintner from Franklin County saw the issue as growing worse with waste import.

> I live near highway 395. My customers would be coming to the winery following Radioactive trucks.... I am concerned about an accident and the loss of the farm.

The vintner had already experienced the effects of stigma during the public scare over Iodine-129 in Washington State wines (Seattle Post-Intelligencer, 1987) and he commented on the result.

> There is a "glow in the dark attitude." Our land has the most productivity in the world. One of our biggest fears is that people will boycott us. We had a wine conference and two speakers said, "It's nice to be here in the land of the glowing wine." I am waiting for a wine magazine to cover that. Our sales will plummet...

A farming representative perceptively noted that the threat of stigma had been magnified because the identity of specific wines was targeted, "The stigma is attached to a product, like a wine label, which people worry about." This was a higher level of stigma risk than found with pesticide use, she added, because they are used in agriculture everywhere, "pesticides are not stigmatizing even though there is concern about pesticides use and health."

The vulnerability of wine identified as from Washington came from adverse publicity not from any enhanced danger from drinking the wine. Any mark opens up the possibility that others would exploit the situation for their own benefit, as a farming official explained, "Agriculture is so competitive that it makes money because someone else has a disaster." Moreover, there was concern that Franklin farmers were a "soft spot" exploited by non-farmers seeking to gain political advantage in the battle over Hanford's future. In this view, the safety of farm products had become a weapon in the battle between Downwinders and Hanford.

> There is a conspiracy perception. We realized that people were using our industry to stop Hanford. HEAL in Spokane. Physicians for Social Responsibility even said not to eat anything raised here.

The fear of stigma by Franklin County farmers intensified after the Chernobyl disaster, when the similar N reactor at Hanford received national attention. Evidence of a categorical rejection of Hanford region crops was cited by a Mesa farmer.

> There are buildings in Spokane that say "do not buy radioactive wheat." There is a billboard of a dead cow from radioactivity in Seattle.

The sheep farmer's wife similarly noted the threat to the wheat they grew.

> The elevators have never refused to buy our crops. If you get the fanatics starting, it could happen. Farms don't want publicity. Then, they won't be able to sell their crops. Everything else is killing us, we don't need publicity.

Such vulnerability demanded extraordinary measures to assure consumers of product safety. The local farm bureau began monitoring for radioactivity to assure markets that their milk was safe. When Franklin County scored lowest of four sites tested statewide, one farmer noted, "We have the ability to tell the world we guarantee we are not radioactive." But he went on to explain the perpetual vulnerability faced by farmers to adverse publicity, even if unfounded.

> If people get the idea that there is a problem, the market will drop. Last year, there were reports that Washington potatoes were contaminated and they dropped $2 in the Chicago market. The issue to me is how to make it safe and communicate that.

While the head of the farm bureau termed this stigma effect on potatoes as "an unfounded...coffee shop rumor," she was uncertain about the underlying product safety.

> The Hanford people have not been tracking agricultural exposures. Possibly, they don't want us to know.

Social Dimensions of Stigma

Stigma is often manifested as social pressure on peers, as was evidenced both Upstream and Downstream at Hanford.

Social Pressure on the Richland League

Stigma affected Hanford supporters, initially in the context of Downwinder opposition, and, later with the repository proposal, which broadened opposition to virtually the entire state and region. Stigma was a weapon in the battle between the Richland and Spokane Leagues. The Richland League perceived that the repository issue at least opened up a space for dialogue.

> We're happy that the repository issue has been raised so that we have a chance to respond. We usually aren't asked; we are lepers.

However, the exchange was not always appreciated. The Spokane league released a study (League of Women Voters of Spokane, 1985) that the Richland League viewed as "...full of technical errors ... leading statements and bias..." that was anti-Hanford. In response, the Richland League offered what they viewed as constructive criticism. "We gave feedback to the Spokane people but could tell they were offended...."

The National League of Women Voters supported neither the Spokane nor the Richland League. However, having supported the Nuclear Waste Policy Act, it was bound to support the site characterization and selection process for Hanford (League of Women Voters Education Fund, 1985). In contrast, the Washington State League became embroiled in the Richland/Spokane conflict. A state booklet (League of Women Voters of Washington, 1987) was viewed as a week effort at balance. Members of the Richland League perceived the state organization as unsupportive and generally anti-nuclear. They felt rejected by other statewide representatives.

> On this issue, people in the state assume they know what the Tri-Cities members think and, therefore there is no point in discussing it with us. There is a lack of knowledge base among a lot of them. They are rabid on the issue [and] ... aren't willing to talk about the science, it's all emotion....

When Richland representatives entered the room at the state convention, others became tense.

> They give the Tri-Cities people the cold shoulder. We'll be sitting at a table at the convention and someone asks, "Where are you from?" Then they say, "Oh, how can you live there? Aren't you afraid?" Or else they just change the subject.

A simulation game developed by the state League to open up discussion between the factions had subjected Richland League members to comments such as "They are only in it for the money," "Why doesn't it look like they care about safety?" and "They don't care." No one asked them what they actually thought, "We were fenced in."

Social Pressure Downwind

The sheep herder's wife theorized that the Downwind farming community's fear of stigma gave Hanford a way to dissuade their vocal opposition. "Hanford put out a rumor that our crops would not be bought just to shut us up." The mechanism for insuring this silence was the strong social pressure that proximate Downwinder activists faced, not just from Hanford and its supporters, but from their own neighbors in the agricultural community. The sheep farmer's wife succinctly summarized the reason, "People are concerned with their crops." Her mother-in-law saw stigma as in itself a form of contamination. To be labeled as contaminated was to be contaminated.

> People don't talk about this. They're afraid that if they discuss it, the crops will become contaminated. They're afraid to talk – that no one will buy. If you go visit someone – no one will ever talk about it. Even my husband doesn't want to.

The farmer/nurse knew people who were afraid to eat the local food or drink the milk. But

> these same neighbors get angry at me for being vocal because the agricultural community will be hurt if contamination becomes known.

She and her husband faced employer pressure and community shunning.

> We don't push the DOE because we are tired of getting so much flack. I've tried to stay away from this. When I speak publicly, my employer takes it out on me. So I hesitate to say much. And people tell us they don't like it when we talk about it. It's not fun to have your neighbors snub you. One neighbor won't speak to me. They want to sell. It's their retirement. It hurts them bad to see things in the media.

Such pressure prevents them from being more active.

> We would go to more meetings and find out more things. We're tired of being hassled, snubbed and labeled. The neighbors don't like to hear the truth. They don't like to be frightened. When there is money involved, people lose their rationality. They think "Never mind, it will happen to the other guy, not me." People have to make a living. The Hanford base out there is the main livelihood of the Tri-Cities. When you take away that livelihood, you run up against strong feelings. People have debts and responsibilities. They don't want to hear about the deaths. I run into this all the time.

At first she had even hesitated to talk to me for fear her name might be published. And, she sounded defensive, as if she was used to people questioning her motives.

> I'm not after money or anything. I just want them to clean up their act. I want to give my grandchildren and other peoples' a chance to live. I have no ulterior motive.

Despite their silence, she was certain that many agreed with her actions.

> A lot of people feel this way but they won't talk about it.

Then there was the case of the few who did speak up. Tom Bailie had his supporters and detractors. But the farmer who said on television that he would not eat his own crops got only negative reviews at the sheep farmer's table, "He has put his foot in his mouth but he's cutting our throats."

The local squab farmer discussed the problem of stigma due to anti-Hanford activism. He was a pig farmer when "the nitrates scare occurred and I had to kill 100 pigs.... The economics of it scares me." Now he opposed Tom Bailie's actions because

> I don't like scare tactics.... Tom scares me when I read articles that quote him. I've seen this wipe out a community.

Similarly, he saw the same dynamics operating in the recent issue over Iodine-129.

> The problem that I had with the statement from Spokane about wine and Hanford was that it scared people. It hurt the wine business.

The squab farmer fears anything that generates publicity or legitimates the hazard, even the lawsuit filed on behalf of some of his neighbors against Hanford.

> It scares me. If the government gives a settlement and admits that there is a problem, it will kill the area. This is aside from whether the lawsuit is the right thing...

He claimed that these scares had already begun to take a toll.

> [Buyers] now have contracts that say that they have a right to refuse to buy [our products]. Factories have begun to move away from the downwind side...

The squab farmer would even like activists to be liable for stigma effects.

> I'd like to see people who make statements and write the articles be responsible for the damages.

This was clearly a place were it was socially hazardous to make waves.

Coping with Stigma – Humor and Anger

An important facet of environmental stigma is that, not only is the contaminated place stigmatized, but so are the people, places, animals and products of the area. It is interesting to see how people cope with being objects of stigma. Humor is often used, either jokes directed at victims by outsiders or insider jokes shared amongst those impacted. There is often a fine line between humor and anger and resentment. Stigmatized humor is not really funny.

Thus, a group of farmers at the Mesa Café good naturedly ribbed, "Uncle R," observing, "He was so irradiated that he lost his sex drive." This taunt was discussed at length before one of the farmers raised a counter argument, "Actually, radiation increases the sex drive." The conversation continued on, crisscrossing the line between humor and seriousness, only to end with a lengthy discussion of an alleged Hanford study of downwind cow turds measuring whether they were a viable habitat for flies. Tom Bailie concluded with his childhood wisdom "there is little difference between cows and people except that cows are smarter."

Then there were the "glowing" jokes. One was offered by a Mesa Café discussant.

> I see your wife's pregnant. You're going to have another one of those kids that glow at night.

A Richland Leaguer similarly commented. "We all joke about glowing in the dark. People think that we do." And the sheep farmer's wife told a story about being in Spokane with her friend, a nurse working at Hanford, and her friend asked her not to divulge where she worked. "I said, 'we should just tell people that we glowed. It's what people think anyway.'" Her story invited others in the interview to tell a string of "glowing jokes." Suddenly, the sheep farmer's wife, who had a very direct manner, asked if I thought the jokes were funny, to which I cautiously responded "not necessarily." Her response disclosed the function of the jokes for their coping, "We just don't get excited about it. There is nothing we can do anyway."

Other dimensions of stigma emerged from the Richland League interview. A League member's stepmother unknowingly committed an affront when she came to visit and brought her own water with her. Interestingly, Richland residents felt stigmatized by Downwinders' anti-Hanford response. League members understood Downwinder concerns over "dead babies." They owned, as one put it, that "The problem may be that Hanford did not originally address such concerns." And they acknowledged that many outsiders unavoidably had a minimal understanding of Hanford, viewing it as one entity. Unless one worked there, it was hard to gain an accurate understanding. One member admitted that she herself lacked an understanding of Hanford until the League undertook a project to explain Hanford to the community.

Nevertheless, bitter complaints were made regarding a *Spokesman Review* story about Tri-City high school students, showing pictures of the Atomic Lanes bowling alley and the high school's "Bombers" sports insignia (the Nagasaki mushroom cloud), that "made them out to be monsters." Another article spoke of the "sinister streets of the Tri-Cities." Such stigmatizing criticism had begun to take a toll on Hanford supporters, as a League member explained.

> There is too much defensiveness, lately especially. People are worried about Hanford bashing. There is anger and vindictiveness.

But even here, a humorous parody appeared announcing formation of a group called SEAL (the Spokane Education Action League – to spoof HEAL) intended to "inform Tri-Cities people on the hazards of spending too much time in Spokane" and study the naturally occurring radon gas issue in Spokane and tell Spokane people what they should do about it.

The Repository as a Stigmatizing Element

The proposed high-level nuclear waste repository gave Hanford's past ills a future face. The proposal generated widespread opposition that achieved a near statewide consensus. Riding on a fear of radiation, distrust of the DOE and stigma, the opposition was so strong as to suggest the impossibility of siting it using an open, Democratic process (Dunlap et al., 1993). Hanford contractors wanted new opportunities. And so did the nuclear work force decimated by plant closings. The sheep farmer's wife noted simply that, "With the repository, those who work for it are for it." Moreover, Richland area officials did not foresee the proposed high-level nuclear waste repository as a physical threat. "Risk is not an issue. It is mitigated." The nuclear facility health officer disdainfully noted that critics acted as if the wastes would be in liquid form when in fact they would be solid, and thus safe. The extreme opposition garnered was explained by two local officials as a problem of the unfortunate timing of the repository question, coming as it did in the wake of Chernobyl.

> Public perception is different now. It used to be there was no public clamor. The public is emotional. Chernobyl and TMI adversely impacted emotions.... The national media never publicizes one positive piece of information. How can you overcome this negativity?

Nevertheless, most observers were not sanguine about the prospects of safely storing more radioactive wastes at Hanford. The Downwinders I spoke with worried that the repository would become a chronic source of additional contamination. The sheep farmer's wife predicted, "No dump can hold that stuff." At Mesa, Uncle R succinctly observed, "Any time you put crap in the ground it's gonna get away from there." Several other Mesa farmers worried that the dump would be too close to the Columbia River. Other farmers cited the potential for earthquakes and mudslides. The squab farmer was concerned that burial 4,000 feet underground for thousands of years

> means death to the area, as soon as it starts to leak, and it will ... they have no control over it. It will leak out at great depth and travel.... That's why nobody wants it.

His friend, the electric coop official agreed, "God knows what will happen underground if it gets into water sources." The sheep farmer's mother also worried that the repository would contaminate the water, as did the farmer/nurse,

> The first thing I think about is water. The Columbia is already contaminated. The fish are too. I worry about the water. We can't control it. The barrels are already leaking. What will happen to people 1000 years from now?

This last point was echoed by another farmer's concern, "We are shifting the problem to our grandchildren. It will come back."

What was interesting in these negative predictions was that they were shared even by those who otherwise thought that Hanford's current operations were relatively safe. The Richland League was also not pro repository for reasons that almost echoed the Downwinders, although they might use more polite terminology. "We call it a repository; everyone else calls it a dump." League member opposition was strictly on the merits, or lack of them: "There is a sense that this is not the most technically sound place to put it." They feared a deep repository would be hard to monitor and would threaten the Columbia River. A League member put it bluntly, "The bottom line is 'Can it be built safely here?'"

Ever the pragmatists, Richland League members preferred a different approach, "We should push for Monitored Retrievable Storage and spur the cleanup of the defense waste" (see Shrader-Frechette, 1993). And they definitely did not like the emotionalism and not-in-my-backyard elements that characterized their view of Downwinder opposition to the repository.

> The real issue is not will there be radioactive waste. It has to be disposed of responsibly. Instead, it gets made into a NIMBY issue. People tangent to Star Wars and Chernobyl.

Another distinguishing characteristic was that League members also favored completing the site characterization studies, viewed as an opportunity to benefit from intense studies of the site without obligating a choice as the repository site, a view shared by local development officials who did not want to lose a billion dollar research project. The public, in contrast, had confused a referendum on allowing the characterization study to proceed with supporting the repository.

The Richland community did not share in the anathema to the federal government's presence at Hanford. However distrust was another major factor driving opposition to the repository. The past legacy at Hanford had provided ample reason for broad distrust of the DOE and its contractors. But even a local development official wanted the region to gain freedom from government control.

> We are tired of being jerked around by the Feds and tired of Washington State determining our economics.

Overall, fear of stigma was the most common reason mentioned for opposing the repository. For example, the regional economic development officer rejected the repository. "Tridec opposes the repository because of the stigma. We want other programs." A tourism official who had denied stigma due to Hanford had a very different concern about the repository.

> Our report hides the repository; it is harmful to the wines and agriculture. The repository has a special image problem because the wastes were not generated here. The waste repository is not positioned like a power plant. It is a "dump site." The existing problems are not getting publicity whereas the repository is.

As at Mayak, a stigmatizing aspect of the repository involved the import of outside waste. There was the stigma of taking what others rejected. A Mesa farmer captured the issue well,

> They want to bring everybody else's waste here and bury it. People will fight over the repository.

The Richland League discussion explored the social and economic impacts of the repository, overlapping worries about stigma found in the farming community.

> Farmers will have trouble marketing their produce. Tourism will be affected. The attempt to diversify the economy will be hampered. Some people won't want to live here if we are the "garbage dump" of the country. The economic advantages of the repository aren't of equal weight to the stigma. Jobs will be temporary. The jobs won't transfer to unemployed people.

Here was common ground between the nuclear culture and the Downwinders. The prospect of Hanford being designated as a national high-level nuclear waste repository sent chills through the agricultural community, even those who had historically supported Hanford. The repository offered them no benefits. Moreover, the publicity it generated threatened the ability to market products. A regional farming group advocated study of potential agricultural impacts.

> The local farm bureau wrote a policy. They asked for a DEIS (a Draft Environmental Impact Statement) to be written on the agricultural impacts. They contacted Deaf Smith regarding stigma.

At the Mesa Café, the "Cattleman" listed stigma as a reason for opposing the repository.

> If the dump is put there, world opinion will focus on a small area of Washington State – our lives are done here. We will have a hard time surviving in our industry (farming) if people know that the Columbia Basin is the dumping ground for the world's nuclear waste.

Farmer opposition depended, according to the Cattleman, on what crops they marketed.

> The ones that are very negative about the thought are the ones selling produce to larger cities. The first people affected are those selling fruit, where people's attitudes are really influenced. If you're selling to Spokane, New York, etc., you may not be able to sell the next week.

Tom Bailie, who raised potatoes, had been invited by his Japanese customers to discuss whether his crops are safe. He agreed that not all farmers were at equal risk for stigma, singling out "The Asparagus, orchard, potato and onion farmers..." The Cattleman contrasted these crops to "Anything that you can process and change the identity of is okay. Wheat, silage..." One of the eight farmers who now had joined the conversation added a further distinction.

> Hay farmers don't want to talk about Hanford because Iodine 129 in hay will show up in their customers' milk. In contrast, grain and cattle farmers will talk because I-129 will not show up in meat.

Examples of how stigma affects farming were rattled off. The energy official commented,

> There is a real heavy concern among farmers. The cranberry scare broke a lot of people.... With pesticides, some have no ban, others are banned. It is strictly political.

Another discussant feared characterization as well as the repository.

> The farmers are against the repository. There is nothing in it for us. I do not want the characterization because once they study it, they will do it. Scientific evidence is only as good as the people who receive it.

Another farmer had hope, however, that another site would prevail, "that's why god gave us Nevada." Humor aside, farmers were caught in a bind described by the Cattleman.

> Either we keep the dump out or we all have to get on the band wagon and say there is nothing wrong with having it.

And, as a further twist, farmer opposition to the repository, itself, raised a fear of stigma.

> Most people agree that they better not say anything. There is undercover opposition. If you keep it quiet, people will forget about the iodine. The local Downwinders are harmful. They hurt agriculture.

The repository was a source of major publicity that kept open old wounds. Prior to my 1986 visit, a local farmer announced on television that his own produce was not safe to eat and he, himself, would not eat it. Such publicity represented a "milestone" that had led, in the words of a farming official, not to a fear of Hanford, but to a

> fear of the press. We understand there is nuclear waste and that we have to do something about it. But since last year, I wonder if Agriculture can survive this.

These were also traditional reasons for silence. An outspoken farmer, leery of being seen as an opponent of Hanford, feared retaliation against his wife's business if he was too public.

> It is damn tough to oppose Hanford. You are opposing the other person's job.... There is a potential for backlash because of the jobs. The feds have big guns.

There was a further backlash that caused fear. For many officials and careful observers of Hanford, successful opposition and failure of the repository proposal was itself a major threat to the Tri-City region. It was as if the negative nuclear legacy had involuntarily wedded the region to a nuclear future. The basic issue was simple. Unless Hanford was kept open, the Reservation was a liability to the region. One local official put it this way,

> If Hanford is shut down, there will be a ... disaster. It will be a ghost town. When industries decide where to locate, why subject their employees to Hanford?

In short, Hanford minus operating industry was just a highly contaminated place – a negative for the appeal of the regional economy and future safety of the environment.

The repository concept suggested a double bind for all of the concerned parties, whether Hanford booster or opponent, whereby it was necessary to bring in more waste to assure that the existing contamination was cleaned up. It was evident from my interviews that a diverse spectrum felt that, absent the repository, there would be little impetus, in the words of a Richland League member, to "clean up what is already here." A local economic official agreed. "I believe if there is no production mission, the feds will leave it." A local farmer raised the same fear, "We still have Manhattan Wastes (i.e., from World War II bomb development). If they do not choose Hanford, these will not be dealt with." A Hanford booster agreed.

> If there is no program here, DOE will walk away from us. In their mind, the waste has been there 40 years and hasn't hurt anybody.

CONCLUSION

The double bind discussed above illustrates a form of Environmental Stigma that I call Perpetual Jeopardy. Contamination invites, in this case even demands, more contamination. And it is this reality that Hanford, and its mirror complex at Mayak confront. In a sense, as we have seen, it is a boon. The need to reign in the damage done and control the materials on site is an opportunity for self-perpetuating eco-industry. For thousands of years, there will be jobs (assuming a society to fund them) in cleanup and containment. Additionally, there will be an invitation for proposals to concentrate existing hazards in a place that is already so hazardous and to develop new hazardous activities. Such locations attract danger like magnets with iron filings.

For the downstream and downwind populations, this continuing legacy is damaging in the deepest psychological and physical ways. The closed era of secrecy at least afforded an opportunity for ignorance and denial, albeit in an unethical and unacceptable manner. But Hanford and Mayak revealed represent an important and interesting challenge to two societies – one considered a seasoned home of democracy and the other a potential newcomer. Do we tolerate in the open conditions that evolved under the most closed conditions?

NOTES

1. The October 10, 1957 accident at the British plutonium plant at Windscale released some 20,000 curie of radioactive Iodine 131, forcing the destruction of all milk coming from the 200 square mile area around the plant. The public was not informed about the event until 1987, when thirty years had passed (Dickson, 1988).

2. Issues of environmental justice are raised by the Hanford case, just as they are raised at Mayak. From its inception, Hanford placed at risk members of the nearby Umatilla, Yakama and Nez Perce Indian tribes who had treaty protected hunting and gathering rights to their reservations, fished salmon and engaged in rituals at local sacred sites (Whiteley, 1999). The risk was not only related to health, but also to the integrity of the land and of the bond between tribal people and their sacred landscape (Edelstein, 2004). Underscoring their permanent and eternal bond to the land, members of the Yakima assured a researcher, Kai Lee, who was concerned with how to warn future generations about Hanford wastes. "Don't worry... We will tell them where it is" (McDonnough & Braungart, 2002, p. 89).

3. GAO (2005) cites a considerably lower figure for the air release.

4. See Jacob (1990) for a full discussion. Also, Freudenburg and Youn, 1993.

5. Gerber (2002, p. 72) wonders if such announcements were not an effort to share "unvoiced concerns."

6. The dose reconstruction project clarified in 1997 that fish and shellfish are the main exposure pathways of concern.

7. Ironically, failure to address the contamination problem in the Environmental Impact Statement hurt the credibility of their overall case for Hanford as a suitable repository site. However, an internal DOE taskforce later found that I-129 released from Hanford had reached confined aquifers 1,500 feet below the surface, raising doubt whether a high level nuclear waste repository located in the basalt layer some 3,000 feet below Hanford could be segregated from groundwater for the required millennium (GAO, 1988).

8. A 1966 law intended to give the public access to information held by the U.S. government.

9. Confirmed by a classified 1948 AEC safety report (Gerber, 2002, p. 71).

10. Another discussant recalled how, during WWII, his company had been deliberately exposed to radiation. He had been away and thus avoided becoming an "atomic veteran."

11. Trust in the repository was impossible given this legacy (Dunlap et al., 1993; see also Mushkatel, Nigg, & Pajawka, 1993; Desvousges, Kunreuther, Slovic, & Rosa, 1993; Easterling, Kunreuther, Mushkatel, Nigg, & Pijawka, 1993).

12. The sheep farmer's wife suddenly turned to me in the midst of our discussion and asked me if I thought finding uranium in the school water was dangerous. My honest reply that it was worthy of being taken seriously led to a later rebuke by the Farm Bureau for rousing alarm.

REFERENCES

AP. (1986). *Probe Lists Radiation Experiments.* Tri-City Herald, Sat, October 25, p. a2.

Bruscas, A., & Wilson, D. (1987). *New Obstacle at Hanford: Issue of Radioactive water could doom chance of N-dump.* Seattle Post-Intelligencer. August 4, pp. 1, A6.

Cockle, D. (December 19, 1985). *N-refuse truck slips on I-84 ice.* The Orgonian, p. M2.

D'Antonio, M. (1993). *Atomic harvest: Hanford and the lethal toll of America's nuclear arsenal.* NY: Crown Publishers.

Dalton, R. J., Garb, P., Loverich, N. O., Pierce, J. C., & Whitely, J. M. (Eds). (1999). *Critical masses: Citizens, nuclear weapons production, and environmental destruction in the United States and Russia.* Cambridge, MA: The MIT Press.

Desvousges, W., Kunreuther, H., Slovic, P., & Rosa, E. (1993). Perceived risk and attitudes toward nuclear wastes: National and Nevada perspectives. In: R. E. Dunlap, M. E. Kraft & E. Rosa (Eds), *Public reactions to nuclear waste: Citizens' views of repository siting* (pp. 175–208). Durham, NC: Duke University Press.

Dickson, D. (1988). Doctored report revives debate on 1957 mishap. *Science, 239*, 556–557.

Dunlap, R. E., Kraft, M. E., & Rosa, E. (Eds). (1993). *Public reactions to nuclear waste: Citizens' views of repository siting.* Durham, NC: Duke University Press.

Easterling, D., Kunreuther, H., Mushkatel, A., Nigg, J., & Pijawka, D. (1993). The vulnerability of the convention industry to the siting of a high level nuclear waste repository. In: R. E. Dunlap, M. E. Kraft & E. Rosa (Eds), *Public reactions to nuclear waste: Citizens' views of repository siting* (pp. 209–238). Durham, NC: Duke University Press.

Edelstein, M. R. (1991). Ecological threats and spoiled identities: Radon gas and environmental stigma. In: S. J. Couch & J. S. Kroll-Smith (Eds), *Communities at risk: Community responses to technological hazards* (pp. 205–226). Worster, MA: Peter Lang.

Edelstein, M. R. (1991–1992). *Mitigating Environmental Stigma and Loss of Trust in the Siting of Hazardous Facilities.* Report prepared for the Yucca Mountain Studies Series, U.S. Department of Energy.

Edelstein, M. R. (1993). When the honeymoon is over: Environmental stigma and distrust in the siting of a hazardous waste disposal facility in Niagara Falls, New York. In: W. Freudenburg & T. Youn (Eds), *Research in social problems and public policy*, (Vol. 5, pp. 75–96). Greenwich, CT: JAI Press Inc.

Edelstein, M. R. (2001). Crying over spoiled milk: Contamination, visibility and expectation in environmental stigma. In: J. Flynn, P. Slovic & H. Kunreuther (Eds), *Risk, media, and stigma*. London: EarthScan.

Edelstein, M. R. (2004). *Contaminated communities: Coping with residential toxic exposure* (2nd ed.). Boulder, CO: Westview Press/Perseus Books.

Flynn, J., Slovic, P., & Kunreuther, H. (Eds). (2001). *Risk, media, and stigma*. London: EarthScan.

Freudenburg, W. R., & Youn, T. (Eds). (1993). *Yucca Mountain High Level Nuclear Waste Repository Studies.* Research in Social Problems and Public Policy, 5.

GAO. (November 4, 1986). *Nuclear Waste: Unresolved Issues Concerning Hanford's Waste Management Practices.* RCED-87–30. Retrieved from www.gao.gov on August 23, 2006.

GAO. (May 25, 1988). *Nuclear Waste: DOE's Handling of Hanford Reservation Iodine Information.* RCED-88–158. Retrieved from www.gao.gov on August 23, 2006.

GAO. (November 24, 1993). *Nuclear Health and Safety: Examples of Post World War II Radiation Releases at U.S. Nuclear Sites.* RCED-94–51FS. Retrieved from www.gao.gov on August 23, 2006.

GAO. (September 20, 1999). *Nuclear Waste: DOE's Hanford Spent Nuclear Fuel Storage Project–Cost, Schedule, and Management Issues.* RCED-99–267. Retrieved from www.gao.gov on August 23, 2006.

GAO. (November 4, 2005). *Department of Energy: Preliminary Information on the Potential for Columbia River Contamination from the Hanford Site.* GAO-06–77R. http://www.gao.gov/new.items/do677r.pdf. Retrieved from http://www.gao.gov on August 23, 2006.

Gerber, M. S. (2002). *On the home front: The Cold War legacy of the Hanford nuclear site* (2nd ed.). Lincoln: University of Nebraska Press.

Jacob, G. (1990). *Site unseen: The politics of siting a nuclear waste repository*. Pittsburgh: University of Pittsburgh Press.

Kutepova, N. (August 9, 2002). Paper presented to the St. Petersburg conference on contrasting Russian and U.S. responses to environmental contamination.

Larson, R. (November 21, 1985). *Turnout high for radiation tests.* The Tri-City Herald.

League of Women Voters Education Fund. (1985). *The nuclear waste primer: A handbook for citizens.* NY: Nick Lyons Books.

League of Women Voters of Spokane. (March 1985). Radiation in eastern Washington? Following the trail of nuclear waste. Washington: Spokane.

League of Women Voters of Washington. (May 1987). Nuclear waste in Washington State, League of Women Voters Education Fund. Washington: Seattle.

Loeb, P. (1986). *Nuclear culture: Living and working in the world's largest atomic complex*. New York: New Society Publishers.

Lovrich, N. P., & Whiteley, J. P. (1999). The government response at Hanford. In: R. J. Dalton, P. Garb, N. O. Loverich, J. C. Pierce & J. M. Whitely (Eds), *Critical masses: Citizens, nuclear weapons production, and environmental destruction in the United States and Russia* (pp. 287–332). Cambridge, MA: The MIT Press.

Marshall, E. (1987). Hanford's radioactive tumbleweed. *Science, 236*(26 June), 1616–1620.

Marshall, E. (1990a). Radiation exposure: Hot legacy of the Cold War. *Science, 249*(3 August), 474–475.

Marshall, E. (1990b). Hanford releases released. *Science, 249*(3 August), 474.

McDonnough, W., & Braungart, M. (2002). *Cradle to cradle: Remaking the way we make things.* New York, NY: North Point Press.

Mushkatel, A., Nigg, J., & Pijawka, D. (1993). Nevada urban residents' Attitudes toward a nuclear waste repository. In: R. E. Dunlap, M. E. Kraft & E. Rosa (Eds), *Public reactions to nuclear waste: Citizens' views of repository siting* (pp. 239–262). Durham, NC: Duke University Press.

Rhodes, R. (1986). *The making of the atomic bomb.* NY: Simon and Schuster.

Schreckhise, W. (1999). The development of interest group activism at Hanford. In: R. J. Dalton, P. Garb, N. O. Loverich, J. C. Pierce & J. M. Whitely (Eds), *Critical masses: Citizens, nuclear Weapons production, and environmental destruction in the United States and Russia* (pp. 133–164). Cambridge, MA: The MIT Press.

Seattle Post-Intelligencer. (August 4, 1987). *Hanford's iodine 129 traced to wildlife, milk and wine*, p A6.

Shrader-Frechette, K. (1993). *Burying uncertainty: Risk and the case against geological disposal of nuclear waste.* Berkeley: University of California Press.

Steele, K. D. (July 28, 1985a). *'Downwinders' – living with fear.* Spokane Spokesman-Review, pp. 1, 10.

Steele, K. D. (July 28, 1985b). *The night the 'little demons' were born.* Spokane Spokesman-Review, p. 10.

Steele, K. D. (January 28, 1986). *Hanford's emission reports will be released.* Spokane Spokesman-Review, p. A6.

Thomas, J. (Fall 1990). Did They Know? In Hanford Education Action League: Perspective: the Human Toll: Measuring the Impact of Hanford Radiation, 1,3, pp 3–4.

Thomas, J. (Fall 1990b). Radiation from Hanford: A Chronology. In Hanford Education Action League: Perspective: the Human Toll: Measuring the Impact of Hanford Radiation, 1,3, p. 7.

Whiteley, J. M. (1999). The Hanford nuclear reservation: The old realities and the new. In: R. J. Dalton, P. Garb, N. O. Loverich, J. C. Pierce & J. M. Whitely (Eds), *Critical masses: Citizens, nuclear weapons production, and environmental destruction in the United States and Russia* (pp. 29–58). Cambridge, MA: The MIT Press.

PART 3: MITIGATIONS

SECTION 3: INTRODUCTION

How are identified instances of contamination addressed, assuming they have been identified and disclosed? The U.S. has evolved an activist ethos with regard to contamination – both in terms of identifying it and, in the case of identified contamination, taking some engineering action to remediate the hazard. In Chapter 13, *A Grassroots Perspective on the Brownfields and Superfund Programs*, veteran New Jersey grassroots leader Madelyn Hoffman offers a thorough review of the Superfund program that, since 1980s, has guided U.S. cleanup of sites identified as contaminated at a threshold demanding remediation. She provides a detailed overview of the process identifying the problem and for developing and comparing alternative approaches for remediation. Public involvement is integral to this process, as is the assessment of health risks for local populations. Hoffman also describes the more recent process of Brownfields remediation added as an adjunct to Superfund to more speedily return contaminated lands to productive use. If Superfund has been the subject of political opposition for its regulations, costs and for tying up property, Brownfields offers the political remedy by streamlining cleanup in order to foster reuse. The downside of this streamlined mitigation, however, as Hoffman stresses, may be the sacrifice in the quality of restoring the contaminated land, leading to future toxic exposures. Finally, throughout her discussion, Hoffman stresses the vital role of the grassroots network of organizations in the U.S. that "watchdog" local environments. Under the leadership of Lois Marie Gibbs, this grassroots network has become a force fighting for environmental cleanup and supporting newly discovered contaminated communities as they grapple with the attendant issues. In this view, it is the combination of laws and regulations and political will with an active level of citizen oversight and participation that makes the environmental cleanup process work.

Although Russia has dealt with a number of major nuclear accidents discussed in the prior section, where issues of remediation and restoration of victims' way of life were addressed in limited ways, there appears to be no comparable process to Superfund (or Brownfields) addressing residential environmental exposures.

In early summer 2002, Volgograd city councilman Alexander Sherstyuk was kind enough to arrange for us to meet with the heads of environment for Volgograd oblast. I wanted to find out how they addressed issues of pollution, in particular in order to protect victims and address their needs. Given the proximity of polluting factories to residential areas, I was curious whether whole residential blocks were relocated, decontaminated or otherwise addressed after a hazardous release from any of the myriad of plants in the city of more than a million. In a two hour pleasant but contentious exchange, I was repeatedly told simply that Russia has excellent environmental laws, that all laws are obeyed and that, therefore, there are no toxic releases exposing the public to hazards and, accordingly, no case studies to cite as to how such problems are addressed. Like a mantra, these words were repeated and repeated.

This was a twofold exercise in denial. As Alexander Sherstyuk explained over lunch afterward, the environmental officials I spoke with were holdovers from the Soviet period who illustrated the role of the apparatchik (powerful functionary). They were bureaucrats adept at covering for themselves, making assurances that all requirements were met regardless of the actual circumstances. A second layer or level of denial was also evident. There was little experience with contamination because, despite a complex of requirements, no one looked for residential exposures to hazards; if they were identified, no one publicly admitted them (i.e., they were covered up) and, therefore, there was no experience in dealing with the attendant problems. The goal was not environmental cleanup as much as avoiding admission of mistakes or failure to meet objectives.

Russia has begun to deal with hazardous waste disposal on a large scale as a result of its responsibilities under international treaties limiting nuclear, biological and chemical weapons. In Chapter 14, *Cleaning Up from the Cold War: Medical, Ecological and Psychological Aspects of the Russian Experience with Chemical Weapons Destruction*, Boris Filatov and colleagues draw on their long experience addressing health and ecological issues associated with the production of chemical weapons to provide a detailed analysis of the issues involved in the destruction of these materials. They begin with a summary of the use of chemical warfare agents in the past century, showing the civilian issues created by military weapons' development. They then methodically catalog the issues faced in meeting international standards for waste destruction that is protective of ecology and local populations given the high likelihood of releases and hazards. What is chilling about this account is that, in its thoroughness, it exposes the extraordinary vulnerability of local populations to releases, especially considering the fact that chemical waste disposal facilities retain their active threat throughout their entire lives, until liquidation or conversion.

The U.S. Superfund program is the model used for granting Russian federal and local authorities responsibility for protecting nearby populations in the event of a toxic release. Because extremely toxic and hazardous compounds are used as feedstock, facilities *are not only potential sources of atmospheric pollution, they are real sources as well.* And the unique character of each waste stockpile forces a unique approach to its destruction, keeping the field *at the frontier stage of innovation and testing rather than maturing to known procedures.*

Filatov et al. give excellent attention to the consideration of local population impacts. This concern is addressed, in part, through creation of a series of radiating boundaries extending outward from the hazardous facility. First is the "sanitary protection zone" serving as the "ecological safety radius" or safe zone during normal facility operations. Then there is the "radius of psychological security," one or two magnitudes greater than the radius demanded for ecological and emergency security. Finally, monitoring is done for those living in the "Protective Action Zone."

Of particular interest is the analysis of chemophobia, (i.e., the irrational fear of chemicals), a topic that gained importance emerged after community opposition shut down one of the first weapons destruction facilities before it ever operated. After listing other examples of chemophobia, Filatov et al. discuss the steps needed to avoid opposition to the siting of chemical waste destruction facilities. The irrationality of such fears (discussed also by Zykova, this volume) is contested by many American researchers and activists (Edelstein, 2004; Hoffman, this volume; Martyniak et al., this volume).

Acute toxic and radioactive disasters occur with the result that someone has to deal with them. Such events create opportunities for magnificent heroism and sacrifice. What is often not reflected upon at the onset, however, is that the sacrifice may be long term and gradual. The temporal frame pits short run heroism against long-term victimization. In the U.S., this issue is most associated with the heroic firemen, policemen and other rescuers who braved the hazards of the World Trade Center on 9/11, many to perish or be injured. As that event shifted from an acute to chronic disaster, however, it has taken on a new meaning in terms of the long-term health consequences for those exposed during and after the event (see Edelstein & McVay-Hughes, this volume).

In Russia, such heroism is without a doubt associated with the "liquidators" who took on the burning and out-of-control Chernobyl reactor in April 1986 and the weeks and months thereafter. In Chapter

15, *Chernobyl: A Liquidator's Story*, Ludmila Smirnova and I relate Sergey Potopov's harrowing experiences at Chernobyl, offering an important snap shot of the world's worst nuclear disaster and what happens when individuals are given responsibility to make personal sacrifices for the common good. Potopov describes at length his training and preparation for chemical warfare. But Chernobyl was a different kind of enemy. Parallel to the emergency workers at the WTC, the story underscores how easy it is to lose site of personal protection in the face of momentous events. When pushed, the folk remedies of vodka and banya were looked to for protection from one of the most complex technological disasters imaginable. Liquidators were made supreme national heroes in the USSR, but they were also stigmatized for their exposures. Most importantly, fear of cancer and other illness clouds their futures.

Chernobyl demanded massive relocations, as had the Kyustym 57 disaster before it (see Edelstein & Tysiachniouk; Zykova; and Mironova et al., this volume). But Russian toxic victims are often forced to continue to live in the polluting circumstances that surround them. The contrast between this expectation and the American perspective became apparent from a project undertaken in the 1990s by the U.S. EPA to assist the Russian Federation in addressing some of its most serious contamination issues. One of the sites identified by EPA as a high profile environmental exposure situation was the area of northern Volgograd they termed the "toxic triangle." Caught between three heavily polluting enterprises – the Red October Steel Plant, the Tractor Factory and the Aluminum Plant – the workers and residents of the toxic triangle were being exposed to extraordinary levels of particulate matter and other pollutants. When EPA next undertook to investigate the source of these exposures, it became clear that they predominately occurred because of fugitive emissions from outdated facilities. Not only would carbon filters need to be added to the stacks of these plants, but the plants would also have to be sealed to allow air filtering to succeed. In fact, just sealing the plants to avoid the fugitive emissions would make a significant difference alone. Fortunately, the bill for the baseline remediation was extremely modest – in the few hundreds of thousands. EPA presented its analysis and then sat back and waited for the Russians to act (U.S. EPA, 1997). However, facing economic downturns and not being predisposed to spend money on environmental mitigations, the Russians did nothing. After waiting, EPA felt compelled to act. The U.S. paid for the remedial efforts (U.S. EPA, 1997).

What EPA confronted was the Russian tendency to accept conditions, no matter how dire, as the way that things are, and then to do ones best under the circumstances. This tendency has given way to a whole different way to approach contaminated communities than that in the U.S. I first encountered the approach in winter 2000 in the Russian city of Volzhsky, just across the Volga River and upstream from Volgograd. Working on a previous Trust for Mutual Understanding Project, I had spent the morning speaking with the management of a local steel plant about environmental regulation in Russia and touring a huge industrial park which discharged into a natural wetlands system buffer area and lake. The brand new steel plant, built with European investments, was failing because of the unstable topography, undermined by the huge impounded Lake Volgogradsky, formed by a dam across the Volga. One drove over this dam to reach Volzhsky, built as a satellite city of Volgograd when the dam was created.

I spent that afternoon with a remarkable physician, head of the Volzhsky Maternity Hospital. He and his top staff took hours to explain to me their prophilactic approach to maternity.

The concept of endoecology governs the final two chapters in this section. In Chapter 16, *Into Thin Air: Training Children to Live in Polluted Air*, Oleg Glazachev shares the results of his striking health studies after Chernobyl. Across Russia, he found nearly 90% of children to be in dire need of health improvements (with one-quarter to one-third of children already impaired for life). Only 10% of the children studied are fully healthy! These data form the background for Glazechev's efforts to train school children to function in polluted air using hypoxia training, or training in air containing less than a normal amount of oxygen. Glazechev describes an investigation demonstrating that hypoxia training, included within a school health course, positively influences a child's homeostatic and psychophysiological functioning and reduces their anxiety. This chapter can be read at the technical level it is presented, or the reader can step back and understand the meta-message of this work. Namely, faced with polluted conditions that are not likely to change for the better, one must improve the adaptive abilities of children to survive and thrive under adverse conditions. For those 90% of children already less than healthy, Glazechev presents such endoecological efforts as a prerequisite for their future.

A different endoecological approach is showcased by Ludmila Smirnova in Chapter 17, *Social Remediation: Contamination as an Impetus for Ecological Learning*. Smirnova relates her experience helping to design an innovative school in a hypercontaminated area of Volgograd called the Ecological Gymnasium. Presenting a safe and whole-life experience for students required rethinking the curriculum, health promotion, student activities and the educational ideology and pedagogy of the institution in order to promote physical, psychological and cognitive adaptation to the contaminated environment. On a hopeful note, the focus of the school on ecology and sustainability is designed to prepare the students to reshape the future.

Some serious differences in response to contamination emerge in this section, ranging from the U.S. approach to engineer solutions to the Russian approach to find a way to live with the problem. These approaches are further discussed in the concluding chapter by Edelstein and Smirnova.

REFERENCES

Edelstein, M. R. (2004). *Contaminated Communities: Coping with residential toxic exposure.* Boulder: Westview Press.

U.S. Environmental Protection Agency (U.S. EPA), Volgograd Health Risk Assessment Working Group. (1997). *Assessing chronic health risks from stationary source air emissions in Volgograd.* Russia: A Case Study in Quantitative Health Risk Assessment, DDP 601.

Michael R. Edelstein
Editor

A GRASSROOTS PERSPECTIVE ON THE BROWNFIELDS AND SUPERFUND PROGRAMS☆

Madelyn Hoffman

INTRODUCTION: THE PROBLEM OF CONTAMINATION

In the United States, nearly 50,000 pounds of waste per person is produced annually, for a total of approximately 6 billion tons of waste, one ton for each person living on the planet Earth. But not all wastes are created equal. U.S. businesses generate some 100 pounds of toxic and hazardous waste per day for every American resident. The impacts of municipal solid waste combine with the legacy of toxins released through pre-production, industrial production, transportation, releases and spills, direct use, byproduct wastes, and end use wastes.

As these chemicals are released to the environment along every avenue of their production, use, and disposal, they have exacted a terrible toll on people. The National Cancer Institute (2004) and the National Institute of Environmental Health Sciences estimate that some 80–90% of all cancers are environmentally caused, either from carcinogenic chemicals present in

☆This chapter draws heavily upon the archives, experiences and newsletters of the Grass Roots Environmental Organization, property of the author, its Director from 1983 – 1998.

Cultures of Contamination: Legacies of Pollution in Russia and the U.S.
Research in Social Problems and Public Policy, Volume 14, 313–331

ISSN: 0196-1152/doi:10.1016/S0196-1152(06)14013-2

our air, water, soil, and food or from smoking. In 1995, the American Cancer Society reported findings suggesting that every man or woman in the United States has a 40% chance of developing cancer during their lifetime. The risk is not equal for men and women, with one-third of women developing cancer but half of men (National Cancer Institute, 2006). These rates are even more serious among African Americans and Latinos. Statistics like this add urgency to citizens' concerns about the need for proper and complete clean up of the nation's hazardous waste sites.

The primary U.S. program for addressing these waste sites is Superfund, which just celebrated its 25th anniversary in December 2005. The Superfund law (technically called the Comprehensive Environmental Responsibility, Cleanup, and Liability Act) was signed by President Carter on December 11, 1980 in response to the growing concern about the public health problems associated with exposure to toxic chemical pollution. The premise of the law was simple. The Environmental Protection Agency (EPA) would be enabled to protect public health and the environment from actual or threatened release of hazardous substances by drawing on funds collected as a tax from the chemical and petroleum industries that are the major sources of the hazard.

Over these 25 years, 936 Superfund sites have been remediated (CHEJ, 2005). As of May 2005, the United States EPA's (2005) National Priorities List contained 1,550 Superfund sites in some stage of cleanup located throughout the country. These sites are identified as those demanding immediate clean up because of the potential environmental and health risk posed by the presence of hazardous wastes.

In this chapter, I discuss the importance of the grassroots movement created in response to the contamination of communities throughout the United States. I place this discussion in the context of evaluating two major regulatory frameworks developed to address the contamination problem, Superfund and Brownfields.

THE IMPORTANCE OF THE GRASSROOTS

Love Canal, Lois Gibbs, and the Grassroots Environmental Movement

Lois Gibbs, an ordinary housewife from Love Canal, located near Niagara Falls in upstate New York, was minding her own business, living the American Dream, when her perfect world was shattered by continuous and recurring illnesses of her two children, David and Melissa. The doctors' only answer to her continued inquiries about the cause of the illness was to attribute the problem to unfortunate coincidence.

Many others would simply have accepted the doctors' explanation. But fortunately for her children and for thousands of others in the United States and around the world, Lois was not satisfied with that answer. Undeterred by having only a high school education, she followed her instincts, learning that her home and community were located next to 20,000 tons of buried toxic chemicals. The health problems of her children and neighbors' children could be explained by exposure to those toxic chemicals.

Spurred on by her concern about her children, Lois overcame her fears of talking to her neighbors and speaking in front of government officials. She and her neighbors formed the Love Canal Homeowners Association to understand the scope of the problem they faced. They also determined strategies for bringing these problems to the attention of public officials and decision makers who could presumably take action to resolve the situation.

The Love Canal activists placed the problem of exposure to toxic chemicals squarely in the national spotlight, spawning the "new environmental movement," a movement made up of ordinary people throughout the country and the world living in communities directly affected by toxic chemical pollution.

The efforts of the Love Canal Homeowners' Association enabled 833 families at Love Canal to be moved out of their homes and relocated at government expense. The work of the Homeowners' Association inspired many. The publicity surrounding the dangers and risks of living near Love Canal helped concerned individuals and communities to realize that others faced similar health problems. But now, unlike before, residents in affected communities around the country and the world had a group to whom they could turn for advice and assistance in holding government and industry accountable for their actions.

The Ironbound

In 1979, the Ironbound Community Corporation, a nonprofit service and advocacy organization, responded to neighborhood concerns about exposure to toxic chemicals. The Ironbound section of Newark, New Jersey is a largely immigrant and densely populated neighborhood, nicknamed the Ironbound because it is bounded on three sides by railway. The Ironbound faces an uncommon number of environmental challenges derived from its multiethnic makeup, its location nearby the industrialized Port of Newark, and the consequent attraction of environmentally hazardous projects in the neighborhood. In fact, a 1980 task force established by the City of Newark identified at least 12 hazardous waste sites needing immediate clean up located in the Ironbound. The "dirty dozen" list revealed a pattern of attempts by industry to avoid taking responsibility for its contamination and a pattern of inaction by national, state, and local government.

Lois Gibbs shared her experiences with dozens of Ironbound residents early in 1980, leading to the formation of The Ironbound Committee Against Toxic Wastes. The committee developed as a strong, multiethnic grass roots organization dedicated to protecting the Ironbound from toxic chemical pollution and unafraid to pressure government officials to do the right thing. The committee took on and defeated a string of high-profile threats to the Ironbound. It stopped construction of a proposed hazardous waste incinerator within a few hundred feet of people's homes and, in coalition with groups from surrounding communities, prevented construction of what would have been the world's largest sewage sludge incinerator. Perhaps its greatest accomplishment, however, was the defeat of plan by a company called At Sea Incineration to ship dangerous chemicals to the Ironbound for processing. The chemicals would then be loaded onto an incinerator ship and transported through one of the world's busiest ports out to sea to be burned. The Committee provided technical and organizing assistance to every community where At Sea Incineration tried to sell its proposal. Eventually, the EPA, declared the proposed process to be too dangerous to permit and it eliminated incineration at sea altogether as a permissible disposal method. Today, the committee continues its work to obtain the clean up of toxic waste already in the Ironbound and to protect the neighborhood from proposals to bring in more.

Center for Health and Environmental Justice (CHEJ)

As evidenced by the Newark visit, Lois Gibbs realized from the onset that neither government agencies nor traditional environmental organizations were set up to assist communities confronting toxic chemical pollution problems at *the grassroots level.* Once the crisis at Love Canal was addressed, she vowed never to leave another community group or concerned parent without support or direction. She drew on her own experience being self-taught the science behind toxic chemical exposure. Working with scientists independent of both government and industry, she created the Citizens' Clearinghouse for Hazardous Wastes in 1981, later renamed as CHEJ.

Her organization serves as a resource for the sharing of information and organizing strategies. CHEJ provides the skeleton and nervous system serving a large often-effective decentralized grass roots environmental movement that helps to protect communities from the health problems associated with exposure to toxic chemical pollution.

GREO

Lois Gibbs' 1980 meeting with Ironbound residents also involved individuals from throughout New Jersey who later became strong community activists elsewhere. An important result was the formation of the Grass Roots Environmental Organization, Inc. of New Jersey (GREO) in 1983. For the next 15 years, while this author was its director, GREO assisted over 200 grass roots groups in New Jersey concerned about toxic chemical pollution problems. Some of those groups in New Jersey remain active today, while hundreds of groups continue their work throughout the country.

The Grassroots Alternative Program

The hundreds of grassroots organizations throughout the country that Lois Gibbs helped to organize have reshaped policies relating to the clean up of hazardous waste sites. The organization's grass roots organizing strategies have allowed only one new commercial hazardous waste landfill to be built in the U.S. since its efforts began. Instead it has sought to effect hazardous waste reduction at the source by getting industries to examine their production practices, finding ways to reduce or eliminate the amount and types of toxic chemicals used in production, eliminate unwanted discharge of toxic byproducts to the air, water, and soil, and eliminate the creation of toxic products. Grassroots activism also resulted in better policies regarding solid waste disposal, waste reduction, recycling and composting, as well as disposal of low-level radioactive wastes and sewage sludge (see Sarokin, Muir, Miller, & Sperber, 1986).

Ultimately, raising public awareness that exposure to toxic chemicals could cause health problems led to the passage of a series of laws that now serve as a rather inadequate legal framework for remediation of contamination problems. Gibbs is often called the "mother of Superfund" for her efforts to get the law passed and then monitor its success. She has also influenced the development of Brownfields policies (CHEJ, 2006).

SUPERFUND LEGISLATION

The term "Superfund" is not just used to refer narrowly to the fund created for the clean up of hazardous waste sites. It more broadly encompasses all solid waste emergency and long-term removal and remedial activities by EPA. These activities include establishing a National Priorities List of the sites most demanding remedial activities. Sites must be investigated for inclusion on the list, their priority determined, and cleanup and other remedial actions conducted or supervised. Superfund authorizes EPA to undertake emergency actions to protect life and to prevent release of hazardous substances to the environment at National Priority List sites. It establishes procedures for identifying potentially responsible parties and holding them responsible for the costs of cleanup. The trust fund, itself, is intended for use when no responsible parties can be identified. Technical assistance grants were also made available to affected communities in amounts up to $50,000 for technical assistance with evaluation of cleanup options, environmental, and health studies. These programs derive their legislative authority from two laws, CERCLA (the Comprehensive Environmental Response, Compensation, and Liability Act of 1980) and SARA (Superfund Amendments and Reauthorization Act of 1986). Superfund was authored by former New Jersey Congressman and Governor James Florio.

The Love Canal disaster was the most famous impetus for Superfund; however it was not the only one. For example, on April 21, 1980, the Chemical Control facility in Elizabeth, New Jersey went up in flames. Chemical Control was a purported hazardous waste processor, but it amassed a huge warehouse containing 24,000 barrels of chemicals, many illegally stored toxic wastes without ever processing any of them. Before any government agency acted on this hazard, an explosion ignited the chemicals, creating a toxic disaster of huge proportion. The fire burned for ten hours, destroying the warehouse and enshrouding a 15-mile area of New Jersey and Staten Island, New York in a thick plume of black smoke and ash. Burning drums of waste were launched 200 feet in the air before bursting like fire works. Amidst fears of widespread toxic contamination, schools were closed, residents were told to stay at home with doors and windows shut, and fishing was banned in a 40-mile area (EPA, 2006).

These and other incidents around the country pointed to a crying need for a shift in the way the hazardous waste disposal was monitored and regulated. Superfund legislation represented important progress in the area of hazardous waste contamination and cleanup. But, while offering hope for the residents of contaminated communities, it has also presented many challenges.

Many of these challenges are endemic to any regulation attempting to control the clean up of contaminated sites. The challenges involve several key questions: *How clean is clean? How effective are clean up technologies? Who are the responsible parties and what is their role in clean up? And how important is the on-site and off-site impact?* I will address these core challenges.

How Clean is Clean?

This question and the problems presented by it can be best answered by thinking about your bedroom when you were a child. How many times were you asked to keep it clean? How many times did you take clothes or books or toys and push them under the bed or pile them in the closet or put them anywhere, except out of your parents' range of vision? What happened here was that your room may have *looked* clean, but if you opened up a drawer or a closet or peered under your desk or bed, you would have found something there that did not belong. If each time you were asked to clean up your room you added to the piles, your room would still *look* clean to the casual observer, but, obviously, you were tolerating more and more of a mess. Perhaps, one day, you would take the hours necessary to take care of all the little hidden messes you created, but it just didn't seem like an effective use of time each time you were asked. In short, you presented the *illusion* of clean, while, in reality, you simply tolerated more and more clutter, mess, and dirt.

That, in a nutshell, is the underlying problem with discussing the question "how clean is clean?" It arises in the context of the required Remedial Investigation and Feasibility Study undertaken before any cleanup can occur at a Superfund site. Invariably, the responsible party or agency determines that the practical approach to cleanup is the one that provides the most cleanup while minimizing costs. Questions of health, safety, and restoration of the contaminated setting are secondary. Science barely factors into the debate that follows. Like many of our parents regarding our childhood rooms, government policies do not compel responsible parties to truly clean up their mess. This leaves us with a legacy of increased tolerance for greater contamination.

Cleanup Technologies: How Effective Are They?

Another major problem with Superfund legislation is that there exists no technology that can completely eliminate much of the contamination. Most scientists agree that there is no safe level of exposure to a carcinogenic (cancer causing) chemical. Many of these contaminated sites are filled with a variety of carcinogenic chemicals and chemicals that cause other health problems. No one knows the synergetic effect of exposure to a combination of these dangerous chemicals.

Despite what may be the best of intentions, the technology/method of clean up does *not* exist that can completely eliminate the risk. One example of this was the situation at the Lipari Landfill, located in Pitman, New Jersey. The Lipari Landfill was not only New Jersey's #1 Superfund site, it was also listed as the nation's worst Superfund site. Thus, it received a lot of attention during the process of determining the best method of cleanup.

Neighbors of the Lipari Landfill were very concerned about how the landfill's contamination would be remediated. They formed a grass roots organization to monitor the choice of cleanup method and to propose cleanup methods of their own. They were also appalled that the elaborate process for determining a cleanup plan failed to include any provisions for offsite remediation or assessment of health problems that may have been caused by exposure to the chemicals dumped in the landfill.

The initial cleanup plan proposed for the nation's worst Superfund site was a technique called "flushing and pumping." The idea was to pump out the contaminated groundwater, flush it free of contamination, and return it to the groundwater. However, there was one major problem with this idea. The EPA admitted that even after 15 years (and millions of dollars) of flushing and pumping, there could be no guarantee that the groundwater would be clean. The EPA decided upon this technique despite knowledge of alternative technologies brought to Pitman, New Jersey by local residents themselves!

Ultimately, due to citizens' pressure, the EPA reevaluated their plan and adopted a technology that seemed more effective and less risky. This technology involved heating and drying out contaminated soil. However, concerns surfaced about this technology and its effectiveness. While not completely satisfied with this plan either, the community felt that this stood a better chance of success than the EPA's original "flush and pump" proposal. At least those most impacted by the decision had finally been heard.

Who Are the Responsible Parties and What Is Their Role in Cleanup?

One major problem with Superfund is that it is often difficult to determine the responsible parties, particularly at landfills like Lipari. Many companies can be potential responsible parties (PRPs). These companies attempt to keep their involvement at arms' length and instead blame the disposal companies. Other companies claim bankruptcy.

When there is a toxic soup like that which exists at many disposal sites, companies try to shirk responsibility by claiming that there is no way to prove that their operations or their chemicals actually are to blame for the contamination. Each potentially responsible party will point the finger of blame at the other and will produce studies showing how their operations or their products could not have contributed to the contamination. Joint and several liabilities can be applied only in those instances where the responsible parties can be identified, are not bankrupt, and accept their liability.

Obviously, the more the potentially responsible parties duck their responsibilities, the more the surrounding community is left to suffer the consequences. While the Superfund legislation contains a provision that allows the state and federal government to conduct the cleanup and then recover three times the costs from those responsible, governments rarely utilize this provision.

Another problem with Superfund is illustrated by what happened at Burnt Fly Bog in Marlboro, New Jersey. The process of completing feasibility studies and determining remediation plans may take years. Allocating money for cleanup may take additional time beyond that. Once clean up begins, additional contamination may be found, but no money is allocated to address that problem. So while some of the contamination at an individual Superfund site might be addressed, it will take years before anyone determines what to do about the newly discovered contamination.

How Important is the On-Site and Off-Site Impact?

In general, the EPA is primarily concerned with on-site toxic chemical contamination and needs to be prodded into addressing off-site contamination problems and potential health problems in the community as a result of that off-site contamination. Here is another instance where community involvement can make a big difference.

In the case of the Lipari Landfill mentioned above, the grass roots citizens' group conducted its own health survey and hotline for those with health problems they felt could be attributed to exposure to contamination from the landfill. A registry of health problems was created and some degree of monitoring of community health problems was ultimately included in the EPA's final cleanup plan.

This was an important addition to the plan. Independent studies conducted in the neighborhood surrounding the landfill revealed excessive levels of leukemia in adults and abnormally low birth weights in children born to mothers living around the landfill. (Rachel's Hazardous Waste News, 1989). A news article published in the mid-1990s highlighted a history of reproductive problems experienced by young girls and women attending a Girl Scout camp located adjacent to the Lipari Landfill during the years of its operation. In February 2000, a Superior Court judge certified a class action suit that sought medical monitoring for Gloucester County, New Jersey, residents – including former girl scouts – who feared they were poisoned by toxic waste dumped at Lipari, one of the state and the country's worst Superfund sites (Lovell, 2000).

Unfortunately, these kinds of efforts rarely take place without the community taking the initiative and pressuring the EPA into implementing them. This was the reality in the United States in the 1980s and 1990s. Legislation on the books was only as good as communities would make them.

Many were inspired by the previously described work of Lois Gibbs who overcame many obstacles, both personal and political, to find out that hers were not the only children developing illnesses due to the construction of the local school on top of a hazardous waste dump. Lois derived strength from knowing she was not alone. In talking to her neighbors, she was able to draw a map showing a connection between the plume of contamination from the former chemical dump and the illnesses in the community.

What was initially branded as "useless housewife data" was the basis for a grassroots environmental movement that challenged the chemical industry and the government to take responsibility for toxic chemical contamination. Lois Gibbs learned what many of us have learned since, that the chemical companies will not take responsibility for the mess they create unless they are pressured to do so by government and government will not pressure the industries unless communities pressure the government.

The Superfund program is replete with examples of government officials playing politics with the program. Cleanup money is allocated to political allies or in the hopes of creating political alliances instead of based on the severity of the contamination problems existing at particular hazardous waste sites. In addition, the process of developing feasibility studies and remediation plans, as well as creating a record of decision through a series of public hearings on the proposed remediation plans can often take years, leading to frustration on the part of local residents and surrounding communities. Sometimes, the paperwork may take years to complete. Meanwhile, the contamination and its health risks remain.

BROWNSFIELDS

Reflecting the slow, incomplete, and political nature of the Superfund process, Brownfields Legislation was passed in 1995. The target of this legislation was to stimulate the expansion or redevelopment of abandoned, idled, or under-used industrial and commercial sites held back by concerns over the costs and liabilities of real or perceived environmental contamination. Initial funding came from Superfund, but the rules of this program were created to be very different. The Brownfields legislation limits the liability of those conducting the clean up of contaminated sites and utilizes lower standards of clean up than established under Superfund. It quickly emerged as a popular program for urban redevelopment.

It is hard to argue with the motivation for Brownfields legislation, the rapid conversion of abandoned, contaminated properties into land that can be used for development that would otherwise occur on Greenfields, previously undeveloped and usually uncontaminated open spaces. The program applies to far more properties than Superfund because it can be used on properties where PRPs cannot be identified. It can also be used on properties that either might take years to be placed on Superfund's National Priorities List or may not qualify to be cleaned up under Superfund. In short, the legislation gives public officials in economically depressed areas another tool for converting contaminated land to land that would be considered "safe," at least according to regulation, for certain kinds of commercial and industrial development. Today, the EPA estimates there are approximately 450,000 such sites throughout the United States.

Beyond old industrial sites, Brownfields redevelopment may also occur in commercial buildings with little or no environmental contamination. Types of property that can become Brownfields sites include former service stations, former dry cleaners, factories, warehouses, parking lots, hangers, lots where heavy machinery was stored or repaired, abandoned railroads, former railroad switching yards, air strips, bus facilities, and landfills.

Many feel that the stated goals of the program are admirable and essential. However, by its nature, the program establishes an unworkable tension between the desire of communities to collect tax revenues from all available land and the basic need for all communities to be protected from the dangers of toxic chemical contamination. There are, in fact, some notable downsides to the Brownfields approach. Predictably, the pressure to redevelop prime real estate in economically distressed cities and rural areas created a situation where public health concerns took a backseat. In addition, some supported Brownfields legislation because it reduced or eliminated the liability of those remediating the contaminated sites. Limiting liability during the clean up and after remediation may reduce costs for those involved, but it puts communities at greater risk. Most significantly, it shifts the costs from those responsible to those who simply live nearby. The legislation also reduced governmental oversight of the remediation process.

How Sites Are Selected for Cleanup

Serious questions and problems arise from the basic structure of the legislation. Unfortunately, most Brownsfields sites were not slated for clean up because of their levels of contamination, but because some industry was interested in purchasing the land or the city was impatient to obtain a certification indicating that the land was "clean" so that the land could then be sold to a developer for commercial purposes. As a result, there was little built-in tolerance for delays in the cleanup process to make sure that all contamination was properly and completely identified, to ensure that best available cleanup technologies and methods were being used, or to address questions about the health and safety of nearby residents. In other words, once the process began, there was virtually nothing that could stop it. The pressure to stamp the property(ies) as clean is stronger in many cases than the need to protect the community from excess chemical contamination in either the short or the long term.

The consequences of addressing contamination problems in this manner are potentially tragic. And grassroots opposition to a cleanup and reuse plan faces an uphill battle. Unlike Superfund, with its extensive opportunities for public information and input, Brownsfields is about business. As a result, Brownsfields redevelopment may proceed apace without regard to public health problems potentially caused by on-site and off-site contaminants.

A graphic example occurred in Bloomfield, New Jersey in the early 1990s, where Cooper Industries/ McGraw Edison operated a battery making plant. The nearly century-long manufacture of batteries on that site led to the contamination of the property with unhealthy levels of lead, cadmium, mercury, and chromium. When adjacent residents, led by a contractor named Frank Sico, learned exactly what chemicals were present on the property, they realized that illness they were suffering might be related. They demanded a pause in the cleanup process to allow state and local officials to determine the extent of the contamination and develop a more appropriate cleanup plan to both remove the contamination and protect the neighborhood from more exposure.

With the support of Lois Gibbs, Mr. Sico and local residents brought increased public attention to the issue. However, the residents were blocked at every turn. Finally, threatened with legal action if the organizing continued, the community opposition crumbled. The former Cooper Industries/McGraw Edison property was converted into a large shopping mall. By that time, Frank Sico had died of multiple myeloma, a crippling disease that normally does not occur in people so young. For all intents and purposes, the cries and concerns of the neighborhood were forgotten in favor of economic development.

Clearly, as this situation illustrates, the exigencies behind Brownfields legislation can exacerbate potential health problems caused by both the existing contamination and the cleanup process itself. In dealing with the health problems caused by exposure to toxic chemicals, it is dangerous to place greater emphasis on proclaiming a site clean than on verifying that it actually is clean and free of risk.

Cleanup Standards Used

Unfortunately, the cleanup standards used at Brownfields sites were not always as stringent as those applied at Superfund sites or the clean up of other contaminated areas. In fact, one particularly problematic outgrowth of Brownfields legislation was an effort to establish different cleanup standards depending upon the eventual end use of the remediated property.

If the remediated property were going to be sold for commercial or industrial uses, the level of toxic chemical contaminants in the soil could be higher than if the property were going to be used for residential purposes. Likewise, water running underneath the property would not have to be purified unless the groundwater had the potential to be utilized as drinking water. The result is that the environment is left at a level of degradation having potential long-term consequences, even if short-term plans for the land or water appear to tolerate it not being cleaned to a higher standard.

Community Participation in Cleanup Process

Community participation is essential to effective remediation. When dealing with toxic chemical contamination at Brownfields or other contaminated sites, it is unrealistic to expect that all the contaminated soil can be removed or that all sources of on-going contamination can be completely eliminated.

However, while those with some knowledge about toxic chemicals accept this as part of doing business, the surrounding community is often none the wiser. In fact, nearby residents are often lead to believe that after cleanup occurs, there will be no further risk to them from the adjacent property. The once contaminated property is certified as "clean" and the surrounding community breathes a collective sigh of relief.

The problem is that in most instances this is simply not true. It is wrong for residents to be given such a false sense of security. Furthermore, while it is likely that the neighborhood may prefer that contamination at a site be addressed to some degree rather than being left derelict and contaminated, the community should be included in the process of determining to what degree. This means explaining honestly, accurately, and thoroughly the types and concentrations of chemicals on the property and the dangers posed by their presence. It means providing a complete assessment of the pathways of contamination and the routes and sources of exposure to that contamination. It means providing an honest evaluation of the effectiveness of the cleanup technology to be utilized. It is only when provided with this information that a community can decide which risks are acceptable to them, not to a governmental agency and the industry it regulates. This commitment to public consent involves treating the community like equal partners in the cleanup process and assessment of its risks. It also means that the process of cleanup may take longer than the parties anticipate and that meetings could become contentious as both sides argue about what the cleanup plan will be. It may mean a more expensive clean up than what the responsible parties would prefer.

Instead of adopting such provisions, however, concern about "inconveniences" resulted in the gradual erosion and elimination of community participation provisions in Brownfields legislation. With these changes, sadly, the last remaining safeguards for communities surrounding Brownfields sites were also eliminated.

Double Standards

The cleanup standards for Brownfields properties are less strict than the standards established for Superfund sites. The rationale for this, while logical on its face, may not always be justified. The thinking is that most Superfund sites pose a greater public health risk than Brownfields sites, since the Brownfields site is located in an already industrialized area. It is assumed that in many instances the underground water supply in an industrialized area is not used as a source of drinking water. Lastly, it is believed that in a so-called industrialized area the potential for the pollution affecting people and their residences is far less than in the areas surrounding many Superfund sites.

More importantly, there is the perception that overwhelming pressure to prepare a piece of land for sale and further development renders science obsolete and irrelevant or at least forces science to take a backseat. So, while the establishment of a dual standard for cleanup may appear to make sense, we need to consider the long-term consequences of such an approach.

- Even though an underground water supply does not appear to be suitable for use as drinking water, what happens if that changes for any reason? What happens to the aquifer, in general, if the contamination continues to spread? Can drinking water some place downstream be affected? Clearly this is an exercise in the expediency of short-term thinking about issues with long-term consequences.
- Even though a particular remediated site is not going to have a residential use, many Brownfields properties are located in the middle of densely populated areas. When it rains, or when the wind blows, soil from one property can be moved to another. Establishing a standard for property to be used for commercial or industrial purposes adjacent to property used for residential purposes is short sighted. It is designed to save money for those conducting the cleanup, while short-changing those who live in the neighborhood surrounding the Brownfields site.

Unfortunately, given the nature of Brownfields sites, most residents who would be affected are often low-to-moderate income people of color. This brings us to the next point.

Disproportionate Impact on Low-to-Moderate Income Neighborhoods

Residents living in proximity to toxic waste sites are affected by soil contaminants tracked by trucks or other vehicles, blown by the wind, transported by water during a heavy rainstorm, or even picked up on the shoes or fingers of an inquisitive child. Thus, the benefits of redevelopment under Brownfields legislation benefits those who are rich (the developers and building owners), as they make and keep the profits made by the new industrial or commercial facility, while the costs of their profit-making endeavors are borne by those on the other end of the economic spectrum.

It is not surprising that many Brownfields sites located in the middle of industrial areas are adjacent to low-to-moderate income neighborhoods in economically distressed or depressed cities or towns. The pressure for renewed development in areas like this is intense. Under such intense pressure, there is increased likelihood that local governments will go along with any policy that accelerates the cleanup process and transforms otherwise vacant and idle land into a tax ratable. Unfortunately, in this process, short cuts are taken to benefit the local authorities and developers, but penalize low-to-moderate income people living in proximity to the Brownfields site.

According to the National Research Council and the Commission for Racial Justice, minority and poor communities bear a disproportionate burden of environmental pollution (March on Washington Fact Book, 2006). They are heavily over-represented in U.S. cities with the largest number of abandoned toxic waste sites, communities with the greatest number of commercial hazardous-waste facilities, and areas with documented air pollution (Congressional Black Caucus Foundation, Inc., 2006). A chilling reality of the U.S. political system is that often elected officials ignore the concerns of low-to-moderate income people because of those groups' inability to provide large contributions to political campaigns. By placing a greater priority on preparing abandoned and contaminated land for sale and development, rather than on comprehensive and thorough cleanup, economically depressed cities with large minority populations become frequent targets of Brownfields' cleanup projects.

The Costs and Benefits of Brownfields' Redevelopment

GREO of New Jersey had misgivings about Brownfields from the outset. Beyond all the shortcomings already listed, an overriding problem is the misleading of the public due to economic and political expediency. Under Brownfields, the public is misled into thinking that contaminated properties can be cleaned up, sold, and redeveloped without presenting an on-going health and safety risk. This kind of thinking allows chemical companies and polluting industries to continue business as usual, thus increasing the level of contamination in our communities.

In addition, without having full and accurate information about the health risk the "cleaned up" site continues to pose, residents are unable to make informed decisions about their lives. Unfortunately, the consequences of such lack of information and inability to participate in the decision-making process are potentially devastating.

Entire shopping areas, like the Newport Mall in Jersey City, New Jersey are constructed on top of soil once terribly contaminated with cancer-causing chemicals like chromium-6. Although local residents did extensive research into the levels of contamination that existed and offered suggestions for minimizing dispersal of contaminants throughout the community during construction, many of their recommendations were ignored. Thus, serious questions remain about the environmental and public health impact of the construction process and the current risks faced by both employees and shoppers at the mall.

Another large and now well-frequented mall, Jersey Gardens in Elizabeth, New Jersey was constructed in a heavily trafficked area on land that was previously contaminated with some of the most dangerous contaminants known, those in the dioxin family. Many questions were raised prior to and during construction about the ability and willingness of the developers to determine and document the extent and severity of the contamination problem on the construction site. Without the baseline data, it was impossible to determine whether or not the pre-construction clean up was adequate. A lawsuit brought by citizens' groups against the New Jersey Department of Environmental Protection (NJDEP) and the developers for failure to adequately address contamination concerns was dismissed and construction continued. Inaction in this matter could lead to increased worker and shopper exposure to contamination, as well as the gradual leakage of contaminants into the Arthur Kill, an adjacent waterway.

Nevertheless, the inherent tension between the need for comprehensive clean up of contaminated land and promoting economic development remains. J. Christian Bollwage, Mayor of Elizabeth at the time of completion of Jersey Gardens, quoted in a February 2000 report of the U.S. Conference of Mayors entitled "Recycling America's Lands," is clear which side of this equation dominates local decision making.

> With our considerable success with Brownfields development in the 1990s, Elizabeth has been able to capitalize on its growth potential.

CONCLUSION

The flaws and inadequacies of cleanup under either Brownfields legislation or Superfund were put into perspective for me when I traveled through the Ukraine in 1999 and witnessed many uncontrolled landfills, the likes of which used to exist in the United States more than 35 years ago. The environmental impact of such landfills is devastating. Even state of the art liners, while protecting the groundwater for a certain period of time, will inevitably leak. Constructing landfills with older, less protective liners or no liners at all is a recipe for disaster.

But I have shown that even enlightened programs like Superfund and Brownsfields run up against a fundamental problem. Once pollution has been released, it is expensive and perhaps impossible to corral it. The result is a perpetual burden for the environment, local communities, and future generations.

As I have shown, the key ingredient in making such programs work effectively has been the strong grass roots citizen movement in the U.S. As challenging as citizens' involvement was in the 1980s and 1990s, citizens' groups were able to register many victories. Lois Gibbs' CHEJ remains in existence today, 26 years after her experiences at Love Canal finally resulted in the government relocating residents deemed most at risk from exposure to chemical contamination. She and her organization have assisted thousands of grass roots groups during that period of time. Imagine where we would be had Lois Gibbs not decided to take matters into her own hands.

Beginning with the Bush administration in 2000, new fundamental challenges have emerged from limitations to citizen participation, reduced regulation, and free pursuit of profit at the expense of pollution. The Bush presidency is the first to administer Superfund that has abandoned the basic principle of making the polluter pay for the environmental damage it has caused, replacing it instead with an increased burden on the taxpayer and almost nonexistent requirement for the polluters to pay for their mess. The environmental progress of the past 26 years, accomplished in large measure because of the efforts of people at the grassroots level, is in serious danger of being completely destroyed. In its stead is a policy that will shift the entire burden of toxic chemical pollution, including the risk of cancers and other debilitating illnesses as well as costs of clean up, onto the public's shoulders, while those responsible for the problems enjoy ever-increasing profits. Lastly, by its disdain for science, this administration is unable to even acknowledge the link painstakingly and fortunately outlined, described and injected into the public policy arena by protectors of the public health like Lois Gibbs and the thousands of grass roots activists she has inspired.

No matter what politician is in power and no matter what that government's philosophy is toward big business and holding big business accountable for its actions, the role of communities will always remain the same – to speak out in defense of the community's health and safety. It is essential that all those who care about the health of their communities work with renewed energy and conviction to protect themselves and their families. This includes demanding the maintenance of adequate cleanup standards, attention to peer reviewed science highlighting the link between toxic chemical contamination and serious illnesses of all types, and corporate accountability and responsibility for the toxic messes it creates.

REFERENCES

CHEJ (Center for Health and Environmental Justice). (2005). *25th anniversary of Superfund, America's safety net in crisis*. Available at http://www.besafenet.com/NarrativeSuperfundReport.pdf

CHEJ. (2006). Available at http://www.chej.org

Congressional Black Caucus Foundation, Incorporated. (2006). Available at www.CBCF-Health.org

EPA. (2005). *Superfund's 25th anniversary*. Available at http://www.epa.gov/superfund/25anniversary

EPA (2006). www.epa.gov/Region2/superfund/npl/0200037c.pdf

Lovell, S. (2000, February 4). Girl scouts and others sue over toxic waste exposure. *New Jersey Law Journal*. Accessed May 1, 2006, from http://www.sskrplaw.com/publications/000204.html

March on Washington Fact Book. (2006). Available at http://www.nul.org/publications/documents/MarchonWashFactBook1.doc

National Cancer Institute. (2004). *Cancer and the environment*. Available at http://www.cancer.gov/images/Documents/5d17e03e-b39f-4b40-a214(-e9e9099c4220/Cancer%20and%20the%20Environment.pdf)

National Cancer Institute. (2006). SEER cancer statistics review 1975–2002. Available at: http://seer.cancer.gov/csr/1975_2002/sections.html, Accessed January 30, 2006.

Rachel's Hazardous Waste News. (1989, February 7). Landfill study finds low birth weight in babies. *Adult Cancers, 115*.

Sarokin, D. J., Muir, W. R., Miller, C. G., & Sperber, S. (1986). *Cutting chemical wastes: What 29 organic chemical plants are doing to reduce hazardous waste*. New York: INFORM.

U.S. Conference of Mayors. (2000, February). *Recycling America's Lands*. Available at http://www.usmayors.org/uscm/brownfields/full_report_rev3.pdf

CLEANING UP FROM THE COLD WAR: MEDICAL, ECOLOGICAL AND PSYCHOLOGICAL ASPECTS OF THE RUSSIAN EXPERIENCE WITH CHEMICAL WEAPONS DESTRUCTION

Boris N. Filatov, Valentina V. Klauchek,
Nikolay G. Britanov and Sergei V. Klauchek

INTRODUCTION

The world community has long striven for the liquidation of chemical weapons of mass destruction. The 1925 Geneva treaty "On the Prohibition of the Use in War of Asphyxiating, Poisonous or Other Gases, and of Bacterial Methods of Warfare" was the first international accord on chemical weapons prohibition. Signed by 125 countries, the USSR ratified the treaty in December 1927. The later development of the "Convention on the Prohibition of the Development, Production, Stockpiling and Use of Chemical Weapons and their Destruction" (henceforth "the Convention")

Cultures of Contamination: Legacies of Pollution in Russia and the U.S.
Research in Social Problems and Public Policy, Volume 14, 333–360

ISSN: 0196-1152/doi:10.1016/S0196-1152(06)14014-4

followed this early step and was undertaken with Russia's active participation. The Convention was signed by the Russian Federation in January 1993 and ratified by the State Duma in November 1997 with the decision to end chemical weapons stockpiling by 2007. As a signatory, Russia accepted international responsibilities for solving many interrelated problems, paramount among them was the protection of people and the environment (The Convention…, 1994, item 4).

Compliance with the treaty represents a significant challenge if chemical weapons are to be destroyed. The economic situation in Russia combined with the receipt of insufficient aid from other countries delayed Russia's effort to start destruction of chemical weapons stockpiles.[1] As a result of such challenges, the treaty deadline was postponed to 2012.

Some progress has been made. Russia's first chemical weapons destruction facility has been completed with German assistance, in Gorny village, Saratov region. The first destruction of chemical weapons occurred there at the end of 2002. However, with the facility budget more than halved in both 2003 and 2004, only 600 tons of weapons material was destroyed by the beginning of 2005. With aid from the US, Britain, Germany, Norway and Canada, additional facilities are under development in Schuch'e, Kurgan region; Kambarka, Udmurt Republic and in Maradikovsky village.

Safety assurance at "Chemical Weapons Destruction" (more commonly "CWD") facilities is required under the fourth of the Convention's 10 points, both under conditions of regular operation as well as during emergencies. A scientifically based protection scheme is to be developed for both the facility's personnel and the surrounding population. The goal is to avoid or maximally reduce chemical exposure to the population and environment. Each member state to the Convention employs its own national standards and safety rules.

Chemical weapons destruction is a difficult multipronged problem to solve. The complex questions bridge legal, technical, ecological, hygienic and management matters. Specifically, optimal technologies must be selected and the best sites identified, a management system for safe facility operation must be developed and all applicable laws must be complied with (Lisovoy, Pronin, & Kalyukin, 1993; Sheluchenko, Petrunin, & Demidyuk, 1998; Kalinina, 1999; Kapashin, Krotovich, & Simnansky, 2000; Averre & Khripunov, 2001). The push to meet key program milestones may cause important ecological impacts and social and psychological consequences for nearby populations. In this chapter, we address the challenge of addressing such impacts for Russian CWD.

A BRIEF HISTORY OF CHEMICAL WEAPONS DEVELOPMENT

Weapons employing toxic properties of chemical substances have been used since ancient times. However, their appreciation as potential weapons of mass destruction is comparatively recent. Factories were first developed for the creation of modern chemical weapons during World War I. This effort stemmed from experiments at several European state laboratories, begun in 1914 that attempted to convert some chemicals into weapons. The creation of special military research centers followed. By 1916, the major problems involved in chemical weapon production had been solved. This success buoyed chemical weapons advocates who saw it as a universal means of warfare applicable for almost all tactics.

While the ultimate challenge was to overcome the enemy's protective defenses, the initial goal was to deliver effective doses to the enemy territory. The only practicable way was to contaminate air and hope that some portion of the toxic agents would reach the enemy given the uncertainties of weather conditions. Furthermore, because toxic agents were dispersed by explosive charges, the size of the charge became critical. Too big an explosion spread the toxin wide but at nonlethal doses. Too little explosive failed to disperse the chemical at all. The degree of volatility of the chemical influenced dispersal as well, with volatile substances dispersing easily and nonvolatile with more difficulty.

The first application of chemical weapons occurred on April 22, 1915, when the German command released 5,730 cylinders (about 180 tons) of industrial chlorine, half its reserve, on French soldiers in the second battle of the Ypres. Although no official accounting was made, chlorine poisoning is believed to have caused at least 5,000 deaths and tens of thousands of nonlethal injuries on the battlefield. Chlorine gas was used another 200 times during World War I, eventually at more lethal doses, but never at this scale.

Further development of chemical weapons focused upon increased effectiveness through the creation of special ammunition, more toxic and dangerous chemical substances and improved methods of spraying. Given their advanced chemical industry, Germany reorganized its dye-works to produce mustard gas, as did Britain at a slower pace.[2]

Such developments were offset to some degree by improvements in personal protective equipment, principally the gas mask, standard issue after 1917. This defense was bypassed by shifting to mustard gas (bis-(2-chloroethyl)-sulfide), the major toxic blistering agent used during World

War I. Given their fast production, Germany used mustard widely, at one point firing more than 1 million shells within 10 days. By the end of the war, France and Great Britain employed mustard as well.

The military downside of mustard gas was that symptoms were delayed 12–36 hours after exposure. New toxic blistering agents were subsequently made from chlorinated arsenic derivatives. The Germans synthesized Ethyldichlorarsine. Catholic University in Washington, DC developed "Lewisite" (2-chlorovinyldichlorarsine), considered to be a promising toxic agent given its fast burning effect, its ability to cause lung injury and its potential deadliness. In all, some 125,000 tons of toxic agents consisting of at least 45 different chemical substances (4 blistering agents, 14 asphyxiators and 27 irritants) were tested on World War I battlefields. They took a heavy toll, causing 1.3 million human exposures and 100,000 deaths.

Russia was a minor chemical aggressor, employing only 4% of the total quantity of toxic agents used by all countries directly participating in World War I. However, Russian losses in killed and wounded were the highest. They made up more than 36% of total chemical weapons casualties during the war. It was, therefore, understandable that in 1917 new Russian authorities paid great attention to the need to deploy chemical weapons. They established chemical companies within the army engineer corps and, by 1920, offered advanced chemical training for central command staff. Simultaneously, the Soviet government proposed an international agreement banning the use of toxic agents.

In 1922, the Soviet Union and Germany began a program of joint chemical weapon research and development. By 1928, a chemical weapon test facility was built in the Saratov region and operated as a joint German–Soviet enterprise through 1933. During this time, a new Soviet doctrine of chemical war emerged, recognizing chemical weapons as the most powerful and effective military agents and justifying a strategy of mass exposure of soldiers, facilities and environmental objects to toxic agents.

The accumulation of huge chemical weapons reserves was aided by the fact that production of toxic agents was relatively cheap. In the period before World War II, the Soviet Union produced phosgene, diphosgene, chloropicrin, prussic acid, mustards, chloroacetophenon, adamsite and diphenilchloroarsin. By the start of the war, their arsenal included chemical air bombs filled with phosgene; cluster bombs filled with a mustard/lewisite mixture; airborne spray tanks filled with prussic acid; chemical shells; mines filled with persistent and non-persistent toxic agents; chemical fog gasses; transportable equipment for contaminating an area; tank equipment filled with toxic agents and gas-pots. After World War II, the Soviet

Union expanded production to include the toxic nerve agents sarin, soman and Vx.

GENERAL CHARACTERISTICS OF CHEMICAL WEAPONS STOCKPILES AND THEIR STORAGE CONDITIONS

Modern Russia legally inherited the former Soviet Union's chemical disarmament issues. At the time of the breakup of the USSR, there was no recorded information on the scope of this problem. Now it is known that Russia has the biggest arsenal of chemical weapons in the world (Gorbovsky, 1997; Kalyadin, 1997; Kuntsevich & Nazarkin, 1987).

The total chemical weapons store located in the Russian Federation amounts to some 40,000 tons (see Table 1). Lethal organophosphorus toxic agents (sarin, soman and agents of the Vx type) comprise the bulk of the stockpile (32,000 tons), with most of the remainder made up of toxic blistering agents (mustard gas and lewisite and their compounds). All of the USSR's chemical weapons were of the "unitary" type, mostly chemical-equipped ammunition (shells, missile warheads, air bombs and so-called "spray capacities") containing ready toxic agents (Beletskaya & Novikov, 1995, 1996). Unitary armaments are particularly dangerous to dismantle; the testing, set-up and start-up of the requisite technologies, entails a high risk of acute poisoning.

These toxic agents are currently stored at seven specially equipped and highly guarded Defense Ministry arsenals located in the Russian Federation (the town of Pochep in the Bryansk region, Leonidovka village in the Penza region, Maradikovsky village in the Kirov region, the town of Kambarka and Kizner village in the Udmurt Republic, Gorny village in the Saratov region and the town of Schuch'e in the Kurgan region; see Table 1). What is significant for our analysis, however, is that all these sites are densely populated regions of the country, sometimes in immediate proximity to cities and settlements. And five of the storage facilities are situated in the ecologically sensitive Volga basin.

As if this problem-set were not sufficiently complex, a further factor is the age of the materials. The first generation toxic agents mustard gas, lewisite and their compounds were created prior to or during World War II. Mostly they are kept in cisterns and reservoirs. These materials are no longer weapons-grade quality. Nothing is known about chemical transformations that may have altered them during their long storage. Chemical ammunition

Table 1. Russian Stockpiles. Declared Warfare Toxic Agents Stockpiles in Russian Arsenals (as of 1994) and also Index of Chemical Weapons Stockpiled for Destruction (as of the end of 2003).

Location	Portion of Total Reserves (%)	Toxic Agents List	Ammunition Type & Capacity	Quantity of Toxic Agents (Tons)
Kambarka town, Udmurt Republic	15.9	Lewisite	Cisterns	6,349.00
Gorny village, Saratov region	2.9	Mustard,	Cisterns and barrels	691.60
		Lewisite		254.60
		Mustard/Lewsite Mixtures		195.70
Kizner village, Udmurt Republic	14.2	Vx	Artillery munitions	308.60
		Sarin		3,911.70
		Soman		1,395.20
		Lewisite		129.20
Maradikovsky village, Kirov region	17.4	Vx	Artillery munitions	4,574.90
		Sarin		232.60
		Soman		1,932.90
		Mustard/Lewsite Mixtures		149.70
Pochep town, Bryansk region	18.8	Vx	Artillery munitions	4,851.50
		Sarin		252.40
		Soman		2,394.30
Leonidovka village, Penza region	17.2	Vx	Artillery munitions	5,123.70
		Sarin		266.70
		Soman		1,494.40
Schuch'e town, Kurgan region	13.6	Vx	Artillery munitions and missile warheads	695.80
		Sarin		2,920.50
		Soman		1,840.10
		Phosgene		10.00

with organophosphorus substances, produced between the 1950s and the 1987 Geneva Accord on the Prohibition of Chemical Weapons, is of sufficient quality for its destruction to be mandated, although it is quickly becoming out of date. Sarin, soman and Vx, the most hazardous toxic nerve agents, are equally distributed between five depots (Pochep town, Bryansk region; Maradikovsky village, Kirov region; Leonidovka village, Penza region; Kizner village, Udmurt Republic and Schuch'e town, Kurgan region – see Table 2).

These chemical weapons destruction facilities are at different stages of operation. Only the facility in Gorny village has reached the stage of full operations, while those in Schuch'e (Kurgan region) and Kambarka town (Udmurt Republic) are in the construction stage and the four remaining sites are merely in the design stage.

Russia lacks a systematic conceptual and scientific analysis of many important problems related to long-term storage and chemical weapons destruction. At the same time, there is considerable experience with the medical and hygienic issues related to production of these toxic agents. Russia has ample experience with potential dangers for facility personnel and the surrounding population resulting from human-caused events during the production of toxic agents (Grachev, 2003; Nagorny, 2002, 2003). The system of sanitary and hygienic controls for CWD facilities must fully take into account the state of the art for safety in working with toxic agents. Of the many uncertainties to address, the extraordinary toxicity of these agents

Table 2. Distribution of Toxic Warfare Agents Storage in Russia's Arsenals (+ = Present).

Storage Place	% of Stock	Vx	Sarin	Soman	Mustard	Lewisite	Mustard/ Lewsite Mixtures	Phosgene
Pochep town, Bryansk region	18.8	+	+	+	–	–	–	–
Maradikovsky village, Kirov region	17.4	+	+	+	–	–	–	–
Leonidovka village, Penza region	17.2	+	+	+	–	–	–	–
Kambarka town, Udmurt Republic	15.9	–	–	–	–	+	–	–
Kizner village, Udmurt Republic	14.2	+	+	+	–	–	–	–
Schuch'e town, Kurgan region	13.6	+	+	+	–	–	–	+
Gorny village, Saratov region	2.9	–	–	–	+	+	+	–

renders impossible an absolutely safe and reliable method for their destruction. At a minimum, this uncertainty must be addressed through a uniform laboratory system and information data bank about toxic risk factors for people working and living proximate to the industrial zone and for the environment (Pirson, 1998).

Russia faces unprecedented challenges, both in destroying its entire chemical weapons stockpile and in accomplishing the task within a time frame of only 10–15 years. Adding to the technical and social challenges, over the 12 years of implementation, costs have increased 10-fold. And, with only 1% of its chemical weapons store destroyed as of June 2004, it quickly lagged way behind on its schedule. By comparison, the US anticipates the destruction of only 5% of its smaller stockpile within 10 years (Drojjin & Drojjin, 1998).

Preparations for CWD began during the former USSR with the opening of an experimental-industrial facility in November 1985 near Chapaevsk town, Samara region. The capacity of this facility was 350 tons of organophosphorus toxic agents per year. The design achieved state-of-the-art engineering features capable of providing a high level of general and ecological safety. Experience with this facility was to serve as a basis for developing further proposals for the destruction of organophosphorus toxins. However, public protests blocked operation of the Chapaevsk facility, and it was eventually redesigned as a training center.

COMBINING HIGH PROBABILITY AND HIGH CONSEQUENCES: THE SAFETY OF THE CHEMICAL WEAPONS DESTRUCTION PROCESS

A new generation of efforts to address weapons destruction was unleashed with the signature by Russia in January 1993 of "the Convention." The safety of workers and residents of nearby settlements emerged as a key issue (Alekseev, 1996; Filatov, 2003a, 2003b; Kiselyov, Nagorny, & Pelischuk, 1999; Nagorny, 1998; Pirson, 1998), as noted above.

The concept of a "safety system" for chemical weapons destruction was broadly defined to include the entire complex of interconnected human-caused, sanitary, hygienic and ecological activities. To meet the criteria of ecological acceptance, the chemical weapons destruction process would have to completely maintain the health of all nearby life forms and provide good ecological conditions for people working and resting.

Destruction of chemical weapons stockpiles demands extensive preparations. There are complex questions to resolve before such a facility commences operation. These include:

- plant siting,
- choice of technology and equipment,
- planning of the production process,
- design of the sanitary technologies for containing, trapping and decontaminating escaping toxins,
- devising methods for air control,
- organizing workers to promote safety and avoid skin contamination,
- addressing dumping of contaminants down the drain or into solid waste, and
- selecting and designing waste disposal sites (Blok & Stolov, 1998; Stolov, 2000).

Complicating these decisions is a fundamental issue. There are no absolutely safe technologies. Accidents might be caused by fluctuations in technical processes, spreading harmful substances. Even demanding the strictest ecological and hygienic requirements of the technologies employed does not preclude the potential for minute releases of toxic agents and their breakdown products and a consequent health impact.

In making such novel decisions there is one major body of experience available to draw upon. This was the design and operations model for the industrial facilities that made the toxic agents originally. Similar issues of worker safety and danger of release existed for these industrial plants (Aslanyan, 1997, 1999). The applicability of this experience is determined by the similarity of toxic agents produced to the composition of substances to be destroyed. The key resource has been the data collected through the observations of sanitarians, toxicologists and clinicians who oversaw medical and hygienic safety at the former chemical waste production facilities. The success of their studies is now the most solid guide available to anticipating occupational and environmental hygiene questions at the CWD facilities.

A comprehensive list of the medical and ecological problems to be solved in designing a chemical waste disposal facility is listed in Table 3.

Because elimination of risk from CWD facilities is not possible, particular attention must be given to emergency procedures and damage control (Filatov & Klauchek, 2000; Karlson, 1995; Kolodkin, 1995; Kodolov, Makrushin, Lipanov, & Frizorger, 1998). The prospect of an emergency is ever present during the operation of the CWD facility. CWD risk analysis identifies several points of vulnerability to emergency. During plant

Table 3. Medical and Ecological Problem Sets to be Solved in Chemical Waste Destruction FacilityDesign.

Requirement Set	Details
Ecotoxicological	Maximum permissible levels must be developed for each specific toxic agent in all possible environmental contexts, as well as guidelines for toxic agent breakdown products and wastes from the destruction process.
Worker Safety	Required for all technical decisions, the environmental impact assessment of all projects, for manuals governing worker and residence safety, and development of equipment for personal protection and exposure detection and measurement.
Worker Health	Medical care preparation for plant emergencies, creation and support of networks between specialized medical institutes, and medical worker training.
International Co-operation and Safety Control	Standardization between Convention members with regard to Chemical Waste destruction medical and ecological safety.
Legal, Socio-economic and Social-psychological	Providing safety to workers and nearby residents.

operation, there is a risk from internal factors that include equipment failure, personnel error and accident and outside factors such as natural disaster, airplane crash and terrorism. Similar threats exist outside the facility regarding the handling and transport of the chemical weapons (Udal'tsova, 1993, 1997).

American experts calculated some 3,000 types of emergency that might occur at CWD facilities, each potentially happening at escalating levels according to the quantity, type and condition of the chemical and the duration and environmental conditions of the release. Their conclusion was that even a small-scale emergency, such as an accident during transport, could lead to significant consequences. Given the high likelihood and consequences of a release, the effectiveness of emergency response is essential. As a result, careful attention is being given to the creation of optimal emergency response systems and measures (Filatov, 1993, 2003a, 2003b; Gabrichidze, 1998; Grigor'ev, Dneprovsky, & Djuma, 1995; Tarasevich, 1993).

The establishment of effective emergency response capabilities requires extensive planning. Trained forces must be in place ready to act. A decision-making support system must be in place ready to disseminate accurate information. It must be able to assess what emergency actions are most likely to succeed under the conditions presented and support day-to-day management of the emergency liquidation, diagnosing emergent issues that nullify previous decisions and analyzing alternatives to select the best. The goal of this emergency response system is to abate damage to human health and quality of life.

The US Superfund Authorization and Reauthorization Act of 1986 has served as a model for granting Russian federal and local authorities responsibility for protecting nearby populations in the event of a toxic release. Such responses are integrated with a general state system for emergency response and management. In the event of a chemical waste accident, specific responses are determined by the level of the released agent's toxicity, the speed of exposure, the need for fast notification, the agent's ability to penetrate buildings and shelters, the stability of local buildings, whether there is soil and water contamination, the provision of modern anti-chemical protection to all segments of the population (e.g., children, old people, ill people), availability of local medical personnel and the preparedness of emergency workers to work under highly contaminated conditions.

Before CWD facilities can be built, a careful siting study must be conducted to examine the baseline ecological and health conditions of an area (Grozdova, 1998; Malmigin, 1998; Pshegroda, 2003; Filatov, 2003a, 2003b). Such studies routinely include examination of ecosystem health, a risk assessment focused upon sensitive populations and exposure pathways potentially connecting these populations to hazards and an epidemiological study seeking cause and effect relations between discovered hazards and population health changes.

Experience at production sites has provided evidence of how lewisite exposure affects vulnerable flora and fauna, as well as successful restoration approaches. These results provide an early basis for impact prediction and biological monitoring in the region around a CWD facility. Regarding human epidemiology, no health effects were found in studies around lewisite stockpile sites in the town of Kambarka in the Udmurt Republic, as well as in Gorny village, Shchuch'e town and Kizner village (Alexandrov, Filatov, Petrushova, Bulanova, & Frolov, 1996, 2002; Kir'yanov, 1998; Malmigin, 1998).

Crucial to any "safety system" is the development of hygienic safety standards representing permissible toxic substances concentrations in different media. The Russian Federation list of obligatory standards for working areas and for environmental control includes maximum permissible concentrations (MPC) for air in working areas in indoor air, atmospheric air, water basins and soil, and maximum permissible levels (MPL) for technological equipment exposure and for skin contact.

Also essential to the environmental safety system is knowledge of what reaction byproducts and production wastes are created from the CWD process, as well as the related toxicity and hazard. Such information is used to inform the choice of optimal decontamination technology; maintain hygienic conditions during technological process organization, equipment setup and operation; isolate workers from toxic substances; control the completeness of waste destruction and prevent toxic substances release.

From the predesign phase, the highest priority is directed to avoiding ecological and human exposure (Demidyuk, 1998). A 1995 contest held in the Russian Federation resulted in the selection of a destruction technology for organophosphorus toxic substances that passed thorough joint Russian/American bilateral laboratory tests using both Russian and American toxic substances. The technology was found to comply with the 1992 Russian/American Agreement on safe, reliable and ecologically clean CWD.

THE SYSTEM OF COMPLEX MONITORING

Perhaps the most essential ingredient necessary to assuring worker and population safety is the creation of an integrated monitoring system to simultaneously track technology, ecology and social hygiene. Monitoring is the most expedient way to organize the collection of medical-ecological information, allowing for its assessment to identify problems and their cause. The monitoring system is thus the keystone of worker and population health protection. It is the only way to offset the high consequence/high probability nature of CWD.

Given its importance, a closer look at the basic research components of social/hygienic monitoring is in order (see Table 4). These begin with a baseline assessment of the potential health hazards of the toxic materials given their condition after long-term storage. Second, careful anticipation must be made of both natural and human-caused changes to the CWD facility's impact zone and their potential for harm to health and ecology. These changed conditions might influence the effects or needed responses to a future release. Third, a training program is required to inform officials of the impact region about the approaches necessary to deal with releases of such chemicals. Fourth, methods must be developed for active monitoring of this region for both chemical and hygienic indications that a release has occurred. Finally, health screening of workers and the regional population is required, both at the onset and in a continuing fashion.

Table 4. Components of a Social/Hygenic Monitoring Program for Chemical Waste Destruction Facilities.

Chemical and Toxicological Assay of the toxic materials to understand their condition after long term storage and the associated hazards.
Assessment of anticipated natural and human-caused changes in the region that may affect future hazards and their best response.
Training emergency responders from the impact region to understand the nature of the challenge in addressing a release.
Active Monitoring of this region for both chemical and hygienic indications that a release has occurred.
Health Screening of workers and the regional population, baseline and continuing.

Table 5. Components of a Populated Areas Monitoring System.

Stationary stations for atmospheric air pollutants control
Mobile stations for sampling and chemical control
A stationary ecological laboratory for sample analysis of the sanitary protection zone's air, water, soil and flora
An information and analysis laboratory coupled to regional information control systems
Equipment for monitoring data transfer into the data processing center

This social/hygienic monitoring regime must be combined with a technological monitoring system during the CWD facility start-up. The monitoring system must include three subsystems needed to monitor these contexts:

(1) The working area and production buildings. This task requires the use of stationary automatic continuous gas analyzers in all working areas with a sensitivity of 0.5 – 1 MPC and a quick operating automatic alarm with a sensitivity set to the threshold concentration levels for the main pollutants.

(2) The complete facility zone including the "sanitary protection zone." Documentation must be kept of ambient air quality on the facility's property and in the surrounding sanitary protection zone for analysis, possibly utilizing a system for automatic data entry into a computer.

(3) Any nearby populated areas. The populated areas monitoring must include a number of components for sampling, control, communication and data summarized in Table 5. Anticipating accidents, emergency response and notification equipment would be contained in a control center located in the residential area.

Health status monitoring for worker and community populations must focus on the identification of exposure symptoms indicating release of toxic substances. An epidemiological approach to analyze resulting data involves such steps as comparison of the "risk contingent" data with control group data, comparison to baseline data collected before CWD began and consideration of indices of background health status.

MEDICAL PREPARATIONS

The emphasis on worker and community health protection requires medical preparation for prevention, treatment, diagnosis and rehabilitation. Among the contributing factors identified from years of experience with chemical waste production, three deserve particular mention: proper use of appropriate protective equipment rated for the hazard, regulation of temperature in the facility and the inherent stress associated with working directly with toxic poisons.

MAIN FEATURES ASSOCIATED WITH THE STAGES OF CHEMICAL WASTE DESTRUCTION

There are five stages in the life of any CWD facility, as shown in Table 6. All must be adequately planned for. Each stage is discussed below in more detail.

Table 6. Five Stages of Chemical Waste Destruction Facility Life.

Design
Construction
Start Up and Initiation
Regular Operations
Liquidation

Stage 1. Designing the Chemical Waste Destruction Facility

A network of specialized CWD facilities is anticipated to be built in Russia with operational lives of about 10 years. These facilities will be in full compliance with "the Convention" (Beletskaya & Novikov, 1996; Drojjin & Drojjin, 1998; Gorbovsky, 1997; Kuntsevich & Nazarkin, 1987; Petrov, 1990).

The main CWD facility design requirement is the protection of personnel, population and environment. This concern commences with the initial step of siting the CWD facility in a location optimal both for implementation of its liquidation tasks and for minimizing any ecological and socioeconomic consequences of both normal operation and emergency conditions.

Siting for CWD facilities varies greatly from the conventional approach used for industrial plants, where cost savings for engineering and building infrastructure as well as available local labor are among the key desired locational attributes. The unique nature of CWD facilities, namely their extreme potential threat to environment and population, creates quite a different set of site requirements. The priorities are resource conservation, waste avoidance and utilization, emission and release avoidance and harm minimization should releases occur. In the latter instance, physical contact with the atmosphere, soil and water is to be kept below the hygienic standards through fast response to harmful releases, waste trapping and decontamination.

CWD facilities are unique in other ways, as well. There are no analogues to them as "production facilities." The technologies involved are untried and lack careful study. The use of extremely toxic and hazardous compounds as feedstock means that the facilities are not only potential sources of atmospheric pollution, they are real sources as well. As if this were not a formidable enough list of challenges, each waste stockpile has a unique structure requiring its own destruction technology and facility design. As a result, the field will always remain at the frontier stage of innovation and testing rather than maturing to known procedures.

Given these premises, it is necessary to plan for a sanitary protection zone during the siting process. The buffer zone dimensions are determined by the "ecological safety radius," the distance from the facility outside of which no negative impacts will affect the population or environment during normal facility operations (Demidyuk & Petrunin, 2003; Pinigin, Nekrasova, Kurova, & Fedotova, 2003). The sanitary protection zone is key to meeting the goals of protection for the environment and community.

Some design components for the CWD facility must be established prior to the design process. These include hygienic standards and approaches for waste disposal, worker and community safety and for environmental protection. A CWD technology must be selected that is capable of effective and irreversible neutralization of toxic substances. Nothing less will guarantee the safety of workers, community members and the environment. Hygienic safety standards must be developed experimentally for CWD facilities, addressing air in working areas, the surfaces of technological equipment, the key structural units and personal protective equipment. Environmental standards are required for ambient air, water basins and soil and snow cover. The regulatory framework must address both the chemicals to be destroyed and also their breakdown products. It is particularly complicated to set maximal permissible concentrations on an experimental basis for chemical waste agents that are extremely hazardous due to their physical characteristics and extreme toxicity. This extreme toxicity dictates a further problem, the need for chemical analysis methods with very high sensitivity levels for the working area and environment.

During the design phase, studies are needed of the site characteristics and health status of the local population. Continued improvements will be needed to personal protective equipment for the workers' and to automatic gas analyzers for continuous control of the toxic substances levels in the working area. The challenges of designing a safety system for CWD raise this work to the status of urgent scientific research.

Stage 2. Construction of the Chemical Waste Destruction Facility

The construction of a CWD facility makes special demands. Facility personnel must undergo in-depth medical examinations. Training and operations manuals must be developed that thoroughly explain the organization and implementation of health monitoring, exposure and chemical control, safety, emergency or contingency response and medical surveillance. It must be further assumed that local health officials will have no expertise in addressing toxic chemical releases and local medical response personnel will know little about addressing acute toxic exposure. These information gaps must be filled before the CWD facility becomes operational.

Stage 3. Start Up of the Chemical Waste Destruction Facility

As with any chemical production facility, the most dangerous time in the entire CWD process is the set-up and start-up of the facility. Experience shows that, at this stage, the potential for emergency is extreme. A very high level of emergency readiness is necessary at these points, ideally backed by computer-aided decision-making support systems. Medical and toxicological teams must be at a higher state of readiness. This is a vital point at which to do sanitary surveillance, ecological monitoring and hygienic assessment research of adverse environmental impacts and negative production factors as well as a risk assessment for workers and the community. Research outcomes can help to establish sanitary rules including hygienic requirements for chemical waste storage and destruction facilities, including siting, design, construction and operation.

Stage 4. Regular Operation of Chemical Waste Destruction Facilities

In order to prevent the real potential for occupational disease, control is needed over both the industrial environment and over worker health status. Medical monitoring requires both the routine collection of information in the workplace and also the maintenance of data for a control community. It is equally important to maintain monitoring of those living in the Protective Action Zone. Risk assessment is needed for all sources of chemical pollution. To assess population health status, a careful comparison is made of the various representative risk groups. Such research is used to make recommendations for addressing identified negative impacts.

Stage 5. Liquidation of Chemical Weapons Destruction Facility

This focus on health must continue into the liquidation phase. For, in contrast to other chemical operations, CWD facilities must be degassed, dismantled and liquidated after their stockpiled feedstock is gone. There is a line of argument that opposes such liquidation. It is instead argued that CWD facilities can be profitably converted to other economic tasks, such as the destruction of agricultural chemicals with expired shelf life and processing of industrial wastes. Whether engaging in conversion or liquidation, there is a significant risk of environmental pollution from toxic substances absorbed by building materials and equipment. CWD facility conversion requires that the whole complex of decisions will have to be revisited relating to health protection. CWD facilities developed for other aims will have to be redeveloped, necessitating reconstruction and disposal or recycling of building debris and old equipment disposal, addressing of new waste streams and the reorganization of testing, monitoring and medical surveillance for the plant and its surrounding impact area.

Past experience with converting former chemical waste production facilities to new uses provides some guidance for the liquidation of CWD facilities. While the US ended production of chemical warfare agents in 1969, production continued through 1987 in the Soviet Union. Under the Convention, Russia disclosed its chemical waste production and stockpile sites in January 1998. Included were sites harboring chemicals regulated by the convention but not produced as chemical weapons. Since the beginning of 1998, more than two dozens chemical waste production and stockpiling facilities have undergone inspection under the Convention, as have other industrial chemical plants.

Russia has identified 24 chemical weapons production facilities under the Convention, situated in five Russian regions. Rather than stand alone enterprises, these facilities tend to be constituent parts of six first-rate chemical industry enterprises. Five are located in the city of Novocheboksarsk, Chuvash Republic; eight in the City of Volgograd; three in the town of Chapaevsk town, Samara region; seven in the city of Dzerzhinsk, Nijny Novgorod region and one facility in the town of Berezniki, Perm region (see Table 7).

The decisions about conversion or liquidation for these former chemical waste production facilities raised questions and forced a careful interpretation of the Convention that is relevant to the future choices about liquidation or conversion of the CWD facilities. Table 8 summarizes factors that have supported conversion.

The Convention creates strict rules for conversion that seek to guarantee that the facility will not be returned to weapons production. All specialized equipment and features specific to weapons production must be destroyed according to these requirements. For example, at the Joint Stock Company "Chimprom" in the city of Volgograd, liquidation activities removed such features as specialized equipment and connected communications technology; armature and control systems; intensive ventilation including underground ventilation canals, makeup air and venting systems and hermetically sealed production booths made out of glass blocks and metal.

At the Joint Stock Company "Chimprom" in the city of Novocheboksarsk, specialized technological and control equipment, technology booths, intensive ventilation, tambours with "black douches" and special laboratories were liquidated at a total cost of some $99 million. The significant part of this sum was spent on demilitarization of the facility for V-gases production and related ammunition production equipment. These components were so contaminated as to represent a major threat to regional

Table 7. Characteristics of Former Chemical Weapons Production Facilities in Russia.

City and Joint Stock Company (plant)	Specialization	Year of Suspension of Production/Year of Liquidation (*indicates not Liquidated)
"Korund," Dzerzhinsk	Ammunition equipment with	
	Phosgene	1945/1949
	Prussic aid	1946/1949
"Orgsteklo," Dzerzhinsk	Ammunition equipment with	
	Prussic aid	1949/1991
"Soda," Berezniki	Mustard gas	1943/1959
"Kaprolaktam," Dzerzhinsk	Mustard gas	1957/1994
	Lewisite	1946*
	Ammunition equipment	1946/1992
"Sredne-Voljsky zavod chhimikatov," Chapaevsk	Mustard gas	1943/1957
	Lewisite	1944/1993
	Ammunition equipment	1946/1990
"Chimprom," Volgograd	Mustard gas/lewisite	1957/1957
	Ammunition equipped with	
	Mustard gas/lewisite	1951/1957
	Experimental organophosphourus toxic agents	1975/1975
	Soman	1987/1993
	Ammunition equipped with	
	organophosphorous agents	1987*
	DF	1987/1991
"Chimprom," Novocheboksarsk	Vx	1978*
	Ammunition equipped with	
	Vx	1987*
	Vx and soman	1987*

ecology. Such demilitarization includes destruction of all technological equipment as well as the buildings that housed it.

The other main constituent of the conversion process is the creation of new commercial chemical operations in areas released after the facilities' demilitarization. Russia is ready to offer partnership to western investors in new ventures located in former chemical waste production facilities. Creation of such joint enterprises will allow western companies access to the Russian market that has a demand for a wide range of chemicals products. "Clean"

Table 8. Factors Favoring Conversion of Chemical Waste Destruction Facilities.

Economic Expediency. It is not expedient to liquidate facilities, at a cost of more than $300 million dollars each that were already producing commercial chemical goods.
Integrated Industry. Most of the former Chemical Waste weapons facilities were already fully integrated parts of large chemical enterprises and, necessary infrastructure, supporting systems and communications would be destroyed affecting the rest of the enterprise.
Industrial Dependency. For crucial industries, losses amounting to as much as $1 billion have been estimated for liquidation of former Chemical Waste plants now producing products necessary to the economy. For example, one former Chemical Waste plant produces 10,000 tons per year of a stabilizing agent, acetoneanil, used in making rubber tires. Loss of this material would cripple Russian tire and vehicle production.
Job Protection. Conversion maintains employment for old workers while opening up thousands of new jobs for highly trained workers.
Inviting Foreign Investment. The destruction of weapons equipment within a plant creates capacity that can be filled with foreign investment.

modern buildings equipped with necessary communications and support equipment are required to invite this investment. Chemical warfare agents can never have been produced in these buildings, although they may have been used in the production of less toxic precursors regulated by the Convention.

Conversion offers the world community a guarantee of full irreversibility for former chemical waste production facilities connected to profitable co-ownership of new manufacturing efforts. Given this history of successful chemical waste production facility conversion, it is possible that some degree of profitable conversion can be built into the CWD process as well.

FEARS OF RESIDENTS IN LOCATIONS NEAR POTENTIALLY HAZARDOUS CHEMICAL FACILITIES

The complicated ecological situation in the areas where potentially hazardous industrial projects are sited drives communities to neuropsychic strain termed by various authors "ecological stress" (Simonov, 1993), "hidden panic" (Molyako, 1992) or "radio-and chemophobia" (Boddington, 1989; Kay, 1989; Shkodich, Klauchek, Filatov, & Klauchek, 1995c).[3] Public stress reflects the fact that the public estimation of danger often exceeds expert opinion and the scientific estimate of existing or potential ecological risk (Antoine, Antoine–Paille, & Valade, 1994; Predtechensky & Gazetov, 1989)[4]

Such stress has been evident in public opposition to chemical weapons destruction. A primary instance occurred in implementing the Russian federal government's ruling on the organization of chemical weapons destruction at Chapayevsk, Samara Oblast. Here, experts faced for the first time the real consequences of underestimating expected psychological impact on the residents, which resulted in mass protests, later termed "the Chapayevsky Syndrome" (Yevstafyev, Yatsenko, Fokin, & Braun, 1990; Shkodich & Klauchek, 1995a, 1995b) Similar public reactions were observed in response to the activity of production facilities, which manufacture household chemicals and agricultural pesticides (see Tsepilova and Kulyasova & Kulyasov, this volume).

Similar community response has also been observed with regard to functioning chemical installations. The conflict surrounding the future bazudin insecticide production facility in the Southern industrial district of the City of Volgograd (see Smirniova, this volume) can serve as a graphic example of chemophobia. Production of the chemical agent bazudin attracted mass media attention and public opposition even though the plant never operated beyond a minimal level. Representatives of the green movement charged that the bazudin production adversely affects public health. Though the production facility was never permitted to fully operate, the local media continues to publicize the view that bazudin production would have adversely affected the health condition of the residents. It is interesting that a detailed study of international experience of bazudin production by Western companies (Ciba–Geigy, Switzerland and Lurgi, Germany) clearly demonstrated that there was no negative reaction by the local residents to similar production facilities in those countries.

Other examples of psychological response have been noted. In 1990 Tbilisi, Georgia, a harmless tear gas agent was used by the riot police. Some 1,000 people were hospitalized suffering from "psychology disorders." Efforts to explain the symptoms cited the inability of people to escape to a safe place and the fear of chemical exposures experienced by the victims.

An even clearer picture of psychogenic induced conditions was observed around this time in Armenia during a period of ethnic conflicts between Armenians and Azerbaijanis. The setting was a garment factory staffed mostly by Armenian women. When an Azerbaijanian workwoman was dismissed, she left behind her work robe in a plastic bag on a windowsill. Acetate acid had been spilled on the robe while the woman had been working. Some two hours after the fired worker departed, a pregnant worker thought to have had hyperosmia caused by her pregnancy, detected a strange smell and became nauseated. Another worker announced out loud that

Azerbaijanians have been known to use toxic gas as a weapon. Before long, 49 of the 51 women present experienced psychogenic induced conditions.

As these examples illustrate, concern for human's safety from both technological and psychological points of view is of increasing importance as an issue in the siting of facilities perceived as hazardous (Petrov, 1993; Sutokskaya & Avkhimenko, 1993). Depending upon the degree to which possible accidents are forecast and the extent to which the public is informed, perception of the hazard from chemical projects determines the resulting chemophobia (or toxicophobia) level among communities (Boddington, 1989; Reshetnikov, Baranov, & Mukhin, 1990). Just a mention of offensive gases gives rise to concern about their extreme hazard. This concern for human health consequences overrides competing ecological considerations resulting from the use of alternative chemicals and chemical processes that may have much greater long-term environmental consequences.

Psychological perceptions of chemical hazard occur mostly in populations that are superficially informed. The degree of suggestibility evidenced by a person further contributes to their development of induced states, such as chemophobia (Alexandrovsky, Lobastov, Spivak, & Shchukin, 1991; Bekhterev, 1921). Suggestibility causes various sensations, emotional responses and vegetative manifestations.

It is of vital importance to gain voluntary acceptance for the siting of extremely dangerous chemical installations where the local population both recognizes the risk and the necessity of running it. There is no better example than the siting of chemical weapons destruction facilities (Evstaf'ev, Kholstov, & Grigoriev, 1993; Shkodich & Klauchek, 1995a, 1995b). In an effort to offset the chemophobia compelled by a project, a "radius of psychological security" is required that is one or two magnitudes greater than the radius demanded for ecological and emergency security (Alexandrovsky, Rumyantseva, & Shchukin, 1990; Marshal, 1989). This psychological radius is overestimated even by experts who possess full and objective information about distances necessary to afford a high degree of safety for communities in the event of an emergency (Filatov, 1993; Petrov, 1993).

During accidental situations such as a catastrophe involving a direct threat to the lives of large populations, both dramatic ecological and psychological (psychotraumatizing) consequences occur due to the realities of the situation. In contrast, chemophobia is based upon the expectation (presumed likelihood) of the threats being materialized. Chemophobia requires the ability to anticipate possible results of an event before it happens based upon the model of some prior experience with such an event (Klauchek, Delaryu, & Ivanchuk, 1993; Psychic Disorders, 1989; Smirnov & Bobrov,

1990). A lack of information assists in feeding rumors and fostering threatening fantasies (Burkle, 1983; Gabriel, 1988; Mandel, 1992; Smirnov & Bobrov, 1990), elucidating the genesis of negative psychogenic responses.

This analysis confirms the potential for chemophobia in communities settled close to potentially hazardous chemical installations in emergency situations with possible chemical exposure. Such populations are as a group considered at risk for the development of "boundary neuropsychic disorder." As a result, there is a need to implement psychohygienic and psychoprophylactic measures to address this novel problem. In additional to developing novel psychohygienic approaches, research is needed on the application of individual methods for correction of the psychophysiological sphere in an effort to effect mass psychoprophylaxis.

Following from this analysis, it is possible to state objectives and approaches for addressing chemophobia.

The first goal is the lowering of psychological strain and uncertainty for populations who are victimized by chemical hazards.

Second, we must seek to reduce the psychological strain and uncertainty in human populations due to the siting and operations of hazardous chemical facilities and due to negative personal experiences caused by exposures in chemical emergencies.

Third, we must establish conditions for the appropriate assessment of individual risks in possible emergency situations.

Fourth, we must eradicate the observable attributes of chemophobia.

And finally, we must employ purposeful psychoprophylactic measures with populations taking advantage of new information technologies and means of mass communication.

CONCLUSION

Russia has thoroughly studied the issues of medical and health protection relating to CWD facilities. It is ready to address the unique demands these facilities create for health and medical monitoring and response necessary to protect workers, members of the community and the surrounding environment. The fact that chemical weapons agents already exist requires that such protective practices begin in the current frame and extend through the difficult processes of set up and start up, and then through operations and eventual liquidation or conversion. The development of the best technologies and procedures, however, does not make safe the effort to be rid of these weapons. They were created for the purpose of human destruction. Given the proximity of human populations to the stockpiles that must be destroyed, even in the act of liquidation, they retain this threat.

NOTES

1. At this time, the total amount of foreign aid to Russia comprised about 200 million dollars, i.e., 1% of the money required for the Chemical Waste destruction program over 10 years.

2. On its part, Russia produced no toxic blistering agents, employing the irritant acetonyl chloride and the asphyxiators perchloromethyl mercaptan, prussic acid, and chloropicrin, the first agent used by Russia during hostilities (Prentiss, 1937).

3. In the west, Edelstein has written about "anticipatory fears" (1988), and there is extensive discussion under the rubric of risk perception. The American tendency has been to describe community opposition to hazardous facilities with the term "NIMBY" (Not-In-My-Backyard) Syndrome (for a full discussion and critique of this term, see Edelstein, 1988, 2004).

4. On the other hand, the public views experts as frequently underprotective, and comes to distrust them for this reason (Edelstein, 1988; Edelstein & Makofske, 1998).

REFERENCES

Alekseev, V. A. (1996). Ecological safety: The main criteria of chemical weapons storage and destruction. *Chemical disarmament – 96: Ecology and Technology* (pp. 6–8). Ijhevsk: CHEMDET.

Alexandrov, Y. V., Filatov, B. N., Petrushova, N. A., Bulanova, E. V., & Frolov, A. S. (2002). The approach to population health status monitoring in connection with CW destruction. *Proceedings of the conference: Medical and hygienic aspects of facilities connected with extremely hazardous substances*, St. Petersburg (pp. 354–365).

Alexandrovsky, Yu. A., Lobastov, O. S., Spivak, L. I., & Shchukin, B. P. (1991). *Psychological geniuses under emergency conditions*. Moscow: Meditsina.

Alexandrovsky, Yu. A., Rumyantseva, G. M., & Shchukin, B. P. (1990). Medical and psychological aid during and after natural disasters and catastrophes. *Military Medical Journal, 8*, 73–76.

Antoine, J., Antoine-Paille, M. T., & Valade, H. (1994). L'opinion publique et les risques sante–environment. *Securite–Revne de Preventique, 13*, 57–60.

Aslanyan, L. V. (1997). Organization of chemical weapons destruction facilities personnel providing medical and sanitarian services. *The third public hearings on the chemical weapons destruction problem*, Kurgan (pp. 166–167).

Aslanyan, L. V. (1999). Some questions of medical and sanitary provision of safety in chemical weapons destruction. *Federal and regional chemical weapons destruction problems* (Vol. 1, pp. 91–95). Moscow: VINITI.

Averre, D., & Khripunov, I. (2001). Chemical Weapons Disposal: Russia Tries Again. *Bulletin of the Atomic Scientists, 5*, 57–63.

Beletskaya, I. P., & Novikov, S. S. (1995). Chemical weapons in Russia. *Herald of the Russian Academy of Sciences, 65*(2), 99–111.

Beletskaya, I. P., & Novikov, S. S. (1996). Chemical Weapons in Russia: Prospect of its Storage and Destruction. *Chemical Weapons and its Destruction Problems, 2*, 15–17.

Bekhterev, V. M. (1921). *Mass reflexology*. Petrograd: Kolos.

Blok, V. G., & Stolov, A. S. (1998). The main decisions on chemical weapons destruction facility creation. In: T. G. Gabrichidze, A. S. Russkih, N. V. Kostikin & P. M. Fomin (Eds), *The fourth public hearings on the CW destruction problem* (pp. 62–68). Ijhevsk: Kizhner village.

Boddington, M. J. (1989). Risk management considerations. *Chemosphere, 18*(1–6), 57–67.

Burkle, F. (1983). Coping with stress under conditions of disaster and refugee care. *Military Medicine, 148*(10), 800–803.

Convention... (1994). Convention on the prohibition of the development, production, stockpiling and use of chemical weapons and their destruction. Revised August 8. PTS PC OPCW, p. 191.

Demidyuk, V. V. (1998). Ecological and hygienical characteristic of the Russian two-stage technology for sarin, soman and V-gases chemical detoxification (preliminary report). In: V. V. Demidyuk, I. V. Shalganova & A. Y. Shirokov (Eds), *Medical and ecological safety in the regions of CW storing and destructing* (Vol. 8, p. 63). Moscow: Green Cross Press.

Demidyuk, V. V., & Petrunin V. A. (2003). Well-founded safety is the base for Russian CW destruction facilities sanitary protection zone sizes establishment. In: *Proceedings of international conference "Medical and biological aspects of chemical weapons destruction"*, Volograd, Russia (pp. 186–187).

Drojjin, Y. G., & Drojjin, S. Y. (1998). Chemical disarmament: heavy incidence for Russian economy. *Chemical Weapons. Ecological Problems of Destruction* (Vol. 2, pp. 104–109). Moscow: VINITI.

Edelstein, M. R. (2004). *Contaminated communities: Coping with residential toxic exposure*. Boulder: Westview Press.

Edelstein, M. R., & Makofske, W. A. (1998). *Radon's deadly daughters: Science, environmental policy, and the politics of risk*. Latham, MD: Rowman Littlefield.

Evstaf'ev, I. B., Kholstov, V. I., & Grigoriev, S. G. (1993). Methodological basis of emergency hazard assessment of CW storage and destruction facilities. *Russian Chemical Journal, 3*, 50–59.

Filatov, B. N. (1993). Psychological and psychologically-hygienic aspects of chemical accidents. *Emergency Medicine, 1*(3), 38–42.

Filatov, B. N., & Klauchek, S. V. (2000). Organizing the system of decision making support for medical and sanitary care during chemical emergencies at destruction facilities for toxic substances with blistering effect. *Chemical Disarmament – Ecology and technology. All-Russia conference with international participation* (pp. 82–85). Ijhevsk: CHEMDET.

Filatov, B. N. (2003a). Organization of monitoring of population health status in the territory adjacent to the CW destruction facility in the Krasnopartizansky district, Saratov region. In: B. N. Filatov, Y. V. Alexandrov & N. A. Petrushova (Eds), *Medical and biological aspects of chemical weapons stockpile demilitarization* (pp. 189–191). Volgograd: Research Institute of Hygiene, Toxicology and Occupational Pathology.

Filatov, B. N. (2003b). Medical aspects of CW destruction: safety provision in Russia. In: B. N. Filatov, G. N. Galkin, M. F. Kiselyov & M. B. Murin (Eds), *Medical and biological aspects of chemical weapons stockpile demilitarization* (pp. 195–197). Volgograd: Research Institute of Hygiene, Toxicology and Occupational Pathology.

Gabrichidze, T. G. (1998). Decision making optimization in emergencies at CW storage facilities. In: T. G. Gabrichidze, A. S. Russkih, N. V. Kostikin & P. M. Fomin (Eds), *The fourth public hearings on the CW destruction problem* (pp. 140–143). Ijhevsk: Kizhner village.

Gabriel, R. (1988). *The painful field in the Psychiatric Dimension of Modern War*. New York: Greenwood Press.

Gorbovsky, A. D. (1997). Chemical weapons: Under prohibition on the boundary of the 21st century. *Chemical weapons: Ecological problems of destruction* (Vol. 1, pp. 50–55). Moscow: VINITI.

Grachev, V. F. (2003). Hygienic aspects of former chemical weapons productions conversion. In: V. F. Grachev, N. G. Britanov, M. F. Britanov & I. E. Pil'dus (Eds), *Medical and biological aspects of chemical weapons stockpile demilitarization* (pp. 208–210). Volgograd: Research Institute of Hygiene, Toxicology and Occupational Pathology.

Grigor'ev, S. G., Dneprovsky, S. I., & Djuma, I. O. (1995). Approaches to the problem of emergency responding system optimization at the CW storage and destruction facilities. *Russian Chemical Journal, 39*(4), 73–76.

Grozdova, T. Y. (1998). Organization of the health status medical monitoring system of population living nearby the places of CW storage and destruction and forecasting of chronic noninfectious pathologies development. In: T. G. Gabrichidze, A. S. Russkih, N. V. Kostikin & P. M. Fomin (Eds), *The fourth public hearings on the CW destruction problem* (pp. 106–114). Ijhevsk: Kizhner village.

Kalinina, N. I. (1999). Regulatory provisions for the chemical weapons destruction process. *Federal and regional chemical weapons destruction problems* (pp. 15–23). Moscow: VINITI.

Kalyadin, A. (1997). Russian politics dilemmas in the sphere of chemical disarmament. *Chemical Weapons and Its Destruction Problems, 4*, 12–14.

Kapashin, V. P., Krotovich, I. N., & Simnansky, A. V. (2000). *Scientific and technical aspects of safety provision during chemical weapons storage and destruction* (Vol. 2, pp. 85–104). Moscow: VINITI.

Karlson, E. I. (1995). Assessment of the consequences of possible emergencies at the lewisite storage facility in the Kambarka region. *Russian Chemical Journal, 39*(4), 79–88.

Kay, M. (1989). Overcoming chemophobia: A key factor in solving the Missouri dioxin problem. *Chemosphere, 18*(1–6), 1177–1180.

Kir'yanov, N. A. (1998). Assessment of health status of the population living in the region of organophosphorous CWA storage: The case of Kizner. In: N. A. Kir'yanov, A. A. Malmigin & V. M. Chuchkov (Eds), *Proceedings of the fourth public hearings on the chemical weapons destruction problem* (pp. 102–106). Ijhevsk: Kizhner village.

Kiselyov, M. V., Nagorny, S. V., & Pelischuk, V. K. (1999). The problem of complex sanitary and ecological (hygienic) safety: research on population and chemical weapons destruction facilities personnel. *Emergency Medicine, 3*, 19–28.

Klauchek, S. V., Delaryu, V. V., & Ivanchuk, L. V. (1993). SMOL: Critical assessment. *Psychological Journal, 14*(4), 126–131.

Kodolov, V. I., Makrushin, B. S., Lipanov, A. M., & Frizorger, G. G. (1998). Possible situations at fire condition on chemical warfare stockpile. In: *Fourth public hearing on the CW destruction problem* (pp. 129–139). Ijhevsk: Kizhner Village.

Kolodkin, V. M. (1995). *The forecast of emergency consequences at a CWA storage facility in Kambarka region, Udmurt Republic*. Ijevsk: Udmurt State University Press.

Kuntsevich, A. D., & Nazarkin, Y. K. (1987). *Chemical weapons: Under absolute prohibition*. Moscow: Nauka.

Lisovoy, V. A., Pronin, M. A., & Kalyukin, S. L. (1993). Personnel and population health monitoring in the region of chemical weapons destruction facility location. *Russian Chemical Journal, 37*(3), 91–96.

Malmigin, A. A. (1998). The region's population health status assessment and medical and sanitary base possibilities. In: N. A. Kir'yanov, A. A. Malmigin & V. M. Chuchkov (Eds), *Proceedings of the fourth public hearings on the chemical weapons destruction problem* (pp. 42–46). Ijevsk: Kizner village.

Mandel, V. (1992). Comment developer une consience ecologique. *Recherche, 23*(243), 664–666.

Marshal, V. (1989). *Main hazards of chemical productions*. Moscow: Mir.

Molyako, V. A. (1992). *Specificities of panic presentations under ecological emergency in Psychological Journal, 13*(2), 66–74.

Nagorny, S. V. (1998). Specificity of sanitary and hygienic control of working condition and personnel health status at planning chemical weapons destruction facilities. In: S. V. Nagorny, O. I. Miroshnikova, V. F. Silant'ev & N. I. Pshenichnova (Eds), *The scientific and applied conference devoted to the 50-year-old Federal Directorate of Medical, Biological and Extreme Problems at Russian Ministry of Health and Social Development* (p. 175). Moscow: Russian Medical Academy.

Nagorny, S. V. (2002). Complex hygienic assessment of 'former' chemical weapons productions during development and realization of the scientific medical programs at chemical weapons storage and destruction facilities. In: S. V. Nagorny, O. I. Miroshnikova, E. A. Tsibul'skaya, V. F. Silant'ev & V. P. Tidgen (Eds), *Medical and hygienic aspects of facilities for extremely hazardous substances. Proceedings of the applied conference devoted to 40th anniversary of NIIGPECh* (pp. 303–309). St.Petersburg, Russia: NIIGPECh.

Nagorny, S. V. (2003). The analysis of medical and hygienic problems of 'former' chemical weapons productions for developing approached to intoxication reduction during organophosphorous CWA destruction. In: S. V. Nagorny, O. I. Miroshnikova, V. F. Silant'ev & E. A. Tsibul'skaya (Eds), *Medical and biological aspects of chemical weapons stockpile demilitarization* (pp. 248–249). Volgograd: Research Institute of Hygiene, Toxicology and Occupational Pathology.

Petrov, S. V. (1990). Principles of the approach to the state program of chemical weapons destruction in the USSR. *Russian Chemical Journal, 35*(4), 483–485.

Petrov, S. V. (1993). Main issues of CWD in the Russian Federation. *Russian Chemical Journal, 37*.

Pinigin, M. A., Nekrasova, G. I., Kurova, I. Y., & Fedotova, L. A. (2003). Optimization of the size of industrial plants sanitary protection zone. In: Y. A. Rahmanin (Ed.), *Theoretical bases and practical decisions of atmospheric air sanitary protection problems* (p. 255). Moscow: Russian Medical Academy.

Pirson, G. (1998). Chemical weapons destruction in Great Britain. In: M. Arsky Yu (Ed.), *Chemical Weapons. Ecological Problems of Destruction* (Vol. 2, pp. 44–57). Moscow: VINITI.

Predtechensky, A. N., & Gazetov, B. N. (1989). The methodology for expert evaluation of organization and medical-ecological system of industrial personnel protection in the case

of emergency at chemical facilities in Emergency medicine – medical and sanitary protection of population against natural disasters and techno-genic catastrophes. In *Sklifosofsky research* (pp. 35–46). Moscow: Institute of First Aid.
Prentiss, A. M. (1937). *Chemicals in war*. New York: Stockholm International Peace Research Institute.
Pshegroda, A. E. (2003). Social and hygienic monitoring of a population's ecological health. In: Y. A. Rahmanin (Ed.), *Theoretical bases and practical decisions of atmospheric air sanitary protection problems* (pp. 159–162). Moscow: Russian Medical Academy.
Reshetnikov, M. M., Baranov, Yu. A., & Mukhin, A. P. (1990). The catastrophe in Ufa: Specificities of state, behavior, and activity of people. *Psychological Journal, 1*, 40–44.
Sheluchenko, V. V., Petrunin, V. A., & Demidyuk, V. V. (1998). Safe, reliable and ecologically clean modern Russian two-stage technology of chemical weapons destruction. In: N. A. Kir'yanov, A. A. Malmigin & V. M. Chuchkov (Eds), *Proceedings of the fourth public hearings on the chemical weapons destruction problem* (pp. 55–62). Ijevsk: Kizner village.
Shkodich, P. E., & Klauchek, S. V. (1995a). The problems of psychological hygiene and mass psychological prophylaxis of population living near toxic chemical substances storage and production facilities in ecological – hygienic problems of the Ural region. *Scientific-practical conference*, Ufa (pp. 145–150).
Shkodich, P. E., & Klauchek, S. V. (1995b). Main issues of psychological hygiene due to CWD problems. *Hygiene and Sanitation, 2*, 39–42.
Shkodich, P. E., Klauchek, V. V., Filatov, B. N., & Klauchek, S. V. (1995c). Psychohygienic aspects of chemophobia prophylactic of populations at ecologically problem areas. In: *Proceedings of international conference "Man at Risk"*, Saint-Petersburg, Russia (pp. 31–34).
Simonov, P. V. (1993). A man under condition of ecological and social stress. *The Bulletin of Russian Academy of Science, 63*(1), 27–32.
Smirnov, S. N., & Bobrov, A. S. (1990). Border-line neuropsychic disorders of chemical facility personnel. In: *Proceedings of All-soviet union conference "Medical and biological aspects of mental health care promoting"*, Tomsk, Russia (p. 56).
Stolov, A. S. (2000). Environment and human protection systems foreseen in chemical weapons destruction facilities design. In: Chemical disarmament – 2000: *Ecology and technology* (pp. 66–74). Ijevsk: CHEMDET.
Sutokskaya, I. V., & Avkhimenko, M. M. (1993). Considerations about risk related to adverse effect of environmental factors and how the population perceives it: The experience of foreign countries. *Hygiene and Sanitation, 4*, 60–62.
Tarasevich, Y. V. (1993). Methodical approaches to development of emergency response at the CW Storage and destruction facilities. *Russian Chemical Journal, 37*(3), 72–76.
Udal'tsova, G. Y. (1993). The ways of decision of the hazardous substances destruction safety providing problem abroad. *Russian Chemical Journal, 37*(3), 43–49.
Udal'tsova, G. Y. (1997). The USA program of chemical weapons destruction. In: G. Y. Udal'tsova, N. A. Tankovich & L. P. Lyangasov (Eds), *Chemical weapons in ecological problems of destruction*, (Vol. 2, pp. 87–98). Moscow: VINTI.
Yevstafyev, I. B., Yatsenko, A. I., Fokin, V. N., & Braun, D. L. (1990). Choosing methodology of possible sites for location of hazardous objects on country territory (by the examples of CWD facilities). *Russian Chemical Journal, 4*, 486–492.

CHERNOBYL: A LIQUIDATOR'S STORY☆

Lyudmila V. Smirnova and Michael R. Edelstein

EDITORS' INTRODUCTION

In the cold war era, perhaps there were no greater heroes of the Soviet Union than the as many as 700,000 firefighters, workers, and military personnel who fought the blazing fires at the Chernobyl nuclear reactor that burned out of control for months until smothered by a makeshift concrete structure, called the "sarcophagus" built to contain radioactive releases. Whether this sarcophagus will last as long as the Egyptian monuments, its name evokes has been a matter of grave continuing concern. And it is doubtful that its symbolism as a lasting evidence of 20th century life will be equally appreciated through the ages. But in April 26, 1986, northern Europe had been placed in dire peril by the catastrophic accident in the number 4 reactor at Chernobyl. Smoke pouring from the fires lofted high in the sky to carry radioactive contaminants eventually across the Northern Hemisphere. Some 100 million curies of radiation were released within 10 days of the initial explosion, comprising the word's worst civilian radiation release (Lawrence Livermore, 1999). And, the reactor threatened to unleash a nuclear explosion that would have dwarfed the effects of what already was

☆The text is related by Sergei Vassilyovich Potopov (Sergei is the brother and brother-in-law, respectively, of the first and second authors).

Cultures of Contamination: Legacies of Pollution in Russia and the U.S.
Research in Social Problems and Public Policy, Volume 14, 361–372

ISSN: 0196-1152/doi:10.1016/S0196-1152(06)14015-6

the world's worst nuclear accident. The situation demanded extreme sacrifice (see also Zykova, this volume).

The first shifts of liquidators, as the ranks who fought Chernobyl came to be known, were sent in to quell the fire. Plant workers and firefighters bore the worst radioactive exposures. Some 30 died quickly of radiation sickness, with others succumbing over time. Because of the extreme hazard, successive shifts of liquidators had to be mobilized to limit the accumulation of personal doses for all the workers (EGE, 2006).[1]

In the United States, we can perhaps best compare this selfless sacrifice to the emergency workers who rushed into the World Trade Center on 9/11, many to never return. Like the liquidators, these police, fire, medical, and other responders often worked without adequate protection from the environmental hazards at the site. Often, their decision to do so was influenced by the actions of their superiors and, in the case of 9/11, by the misinformation about air safety given by the EPA. The survivors and the successor workers at the World Trade Center site now bear an environmental health legacy that raises parallel issues to the cloud of health worry that continues to shroud the liquidators (see Edelstein and McVay Hughes, this volume).

Rather than a distanced study of the "liquidators," we have chosen to include an autobiographical account by Major (retired) Sergei Vassilyevich Potopov. The chapter is based upon interviews conducted in August 2002 and December 2005 in Russia and in follow-up phone calls and email communications.

THE LIQUIDATOR'S TALE

I am Sergei Vassilyvich Potopov, 50 years old, and born in Stalingrad, now Volgograd. I am a graduate of Kostroma, the Military College for Radiation, Chemical, and Biological Protection. I attended Kostroma from 1972 to 1976. I did not always wish to be a military man. I really always wanted to become a clown. This aspiration began from the first grade. We had to write a composition about what we wanted to be. When I wrote that I wanted to be a clown and work in the circus, everyone laughed at me. But I was serious. I planned to enter the arts college and work in the circus. I had held this dream for 10 years. But as I went to apply to the circus college, the rules changed. While previously, students were allowed to enter from the 8th grade, now they required completing the 10th grade. After graduation from high school, I rethought my future. Now I elected to become a military officer like my mother's cousin Gregory, a military officer assigned to rocket weapons. I determined to go to the weapons Commissariat to enlist. I decided I wanted to go to the tank military college. But I was too tall. What to do?

By chance, I met my friend Leonid Fireman. He was in the process of writing an application to Kostroma, where his uncle, who was chief of the military department of the medical university in Yaroslavl, had a chemical commission. Leonid invited me to join him. This is how I first found out that there was such a military college. Throughout my school years, all of my advisers were chemical teachers. So, I had come to love chemistry. Kostroma seemed to be a great opportunity. So I applied.

I might explain that the field of chemical weapons protection dates from World War I. The Russian scientist Zelensky designed an anti-gas mask using charcoal as a defense against the German's use of chlorine gas. Before World War II, there were military chemical services and chemical troops. The goal of the chemical troops was to protect forces from chemical and (later) nuclear weapons. They also used napalm. They provided individual troops with protection. And, before the special biological branch was created, they were involved with deactivation and disinfection of biological agents. They would also wash soldiers for lice before that task passed to the medical corps. But their main focus became tactical design and deployment of biological, chemical, and nuclear weapons, as well as dealing with the consequences of their use. There was a lot of meteorological information that had to be used, taking into account the wind, cloud dispersal, and weather forecast. These military units were considered to be very intellectual. Besides the technical education, they engaged in military studies, learned personal protection and studied the behavior of soldiers organized to use nuclear weapons.

Both Leonid and I successfully passed the pre-entrance exams, and I got the news that I was accepted into Kostroma pending passing the entrance exam. Being from Stalingrad (now Volgograd) in the far south of Russia, I now had a chance to go to the Moscow region for the first time. I thought, even if I fail, at least I will get the chance to see Moscow. Before this, all I knew about the city of Kostroma, where the University was located, was that a famous brand of cheese was made there.

I traveled with my friend, Leonid. He was personally responsible for my enrollment. My friend had passed with highest scores. Yet he was caught breaking discipline, and he was kicked out. He was not allowed to take the entrance exam. I knew the real reason for this. He was rejected because he was a Jew. I was furious. I decided to quit in solidarity.

I went to the Major. But my protest had an unexpected result. It provoked him to recognize that I had a natural quality of real leadership. He advised me not to quit. But he also did not rest on advice alone. He refused to issue orders allowing me to transfer to another institution until after it was too late to take the entrance examination. And if I failed to enter one of the special training institutions for officers, I would have been drafted into the regular army. So, I never left Kostroma despite the injustice for Leonid. But I never forgot it.

My younger brother Valery joined me three years later. He entered as a student in chemistry and became a "coursant" (a term for a military college student). After he graduated as a Lieutenant, he was immediately advanced to a Major position. Meanwhile, I had to work my way up through the ranks for 8 years. I married on January 31, 1976, in my last year in military college. After I graduated, I was sent to serve in the Mongolia People's Republic. I was made commander of a platoon of troops equipped with chemical weapons. I supervised 30 soldiers and their sergeants. Our mission was to protect the Mongolo-Chinese border during the Chinese-Vietnam war of February 1979. Meanwhile, Valery was assigned to protect nuclear weapons on the border with the West in Poland and Romania (Ivano-Frankovsk).

Because I had accounting and analytical skills, I was transferred in 1984 after four years to Borisov (the village of Krupki), Belarus, where I became a Major and company commander. After special training in Saratov, I was given responsibility for calculating the consequences of using weapons of destruction. In 1983 I was appointed chief of a chemical analytical station in Borosov. I had a captain, soldiers, and two seargents under me. If there was a war, we would have had 20 in the unit and been attached to an army of 50,000 people. It was my responsibility as major to forecast the consequences of nuclear, biological, and chemical weapons in case of war and to address contamination of water.

Allow me to explain that every army had chemical and nuclear weapons. These weapons had become a doctrine of the army and were to be used only in extreme situations of defense. My job was to forecast the consequences. In this, we did not differentiate between chemical weapons used by the enemy and by our own forces. Both sides had the power to deploy these weapons. My job was to collect data allowing me to project the consequences of use. These consequences depended particularly upon wind direction and the power of the weapon.

As a major, I was no longer commanding the troops who would use such weapons. I would only do the forecasting. We used early computers. They were primitive aides. They had formulas installed. They were developed from experience with chemical and biological weapons. In order to advise how to protect the troops, I would calculate how many people would die in what zones depending upon how the weapons were directed. I would look at the map and see where the Russian troops were. I would advise whether weapons had to be deployed in a different zone because of the wind direction. The wind direction was different at different altitudes.

I also personally participated in tests of chemical weapons. The lab conducted experiments using mice and guinea pigs exposed to chemical agents through inhalation or injection into the skin. In Saratov, I participated in studies conducted by the military academy that used dogs tied in an area where chemical weapons were released to test their effectiveness. I also knew the officers that participated in nuclear tests in Totsk in Sarotov, where they had first tested nuclear weapons in 1947. We developed the formulas and tables used to estimate outcomes of exposure to these weapons. We were certain that what we did was bad for the environment. We did our tests in special locations far from residential populations in territory fenced off by signs saying "stop." I also participated in the liquidation of older chemical substances by burning or explosion. The burning was done in the town of Shechany where there was a specially equipped institute for chemical protection.

After my next promotion, I qualified to enter the Marshall Timoshenko senior military academy of chemical protection in Moscow. It was a highly competitive academy. Forecasting was included in our course of study, particularly employing meteorological information and equipment. Military and civilians did the weather forecasting. We also had monitors near the ground and barometers. If war occurred, all the information would be sent to me by the field investigators. They would collect information on the territory occupied by the military troops. They would report to headquarters on special radio stations. At headquarters, I would keep a map of real data to compute the expected damage to people and equipment. I also would calculate where there were clean spots. We would have to remove our troops to such places while we decontaminated damaged areas, equipment, and people. If there was radioactive dust released, we would wash clothes with a mixture of special powders. We then would test the clothing in a field radiation lab. The idea was that these services enabled people to accept and understand the real nature of the hazard and to change their clothes. If chemical weapons were used, special decontamination gear was required according to the weapons employed.

Of course there would be accidents. Our professor at college told us stories of people who failed to zip up their pants while exposed to agents and who became impotent as a result. Or their hands would become infected. We also learned of accidents from other sources. The brother of Valerie's friend died because of a leak at a breeder reactor. And my friend worked in chemical weapons testing in Sukhana. They were testing new clothes and masks. Some soldiers died and my friend had health damage. And then there were the people who opposed the tests and received chemical poisoning when they encroached into areas fenced and labeled "keep out."

By the time I entered the academy in 1984, I was grown up as a military officer And I was mature. Because my background was in the defensive use of chemical weapons, I was interested in using chemicals for protection only, as in creating a mist or fog with a smoke machine that would cover the landscape and mislead the enemy. Hiding troops was my topic of research and the subject of my final paper at the academy. The project was recognized with second place in the Mendelyev Prize. I graduated with a diploma of excellence. Valery had entered the academy at the same time as I in 1984. He graduated 16th in a class of 100. In 1987, I was sent to Vyaznicki (Vladimirskaya oblast) as battalion commander (where I currently reside). In 1999 I was sent to Kursk to become Deputy Commander at the headquarters of the 27th Brigade of Chemical Weapons Protection. But I have gotten ahead of myself.

CHERNOBYL INTERCEDES

My studies were redirected in September 1986, when I was sent to Chernobyl to participate in the liquidation of the exploded fourth reactor. I first read about Chernobyl in the newspaper. It was 26 April. But I did not think it involved me. My experience in the academy had only to do with chemicals. My group was scheduled to go to Kaliningrad for two months of tests and then be sent to Germany for two months of field training. But as we got ready to go to Kaliningrad, we were told that all of the officers and all the graduates of the academy would be sent to Chernobyl for a real-life experience for the second part of our training. Before this we had only studied theory. The last day of training in Kaliningrad, we all drank to staying alive.

We were sent on 1 September and stayed until 24 September. By the time we got there, the reactor fire was out, but the fourth block was still burning and they could not stop the nuclear reaction. To this day, the reaction continues.

Our task when we came was to start building the Sarcophagus. There were radioactive releases at least twice a day. The radiation had its first influence on the breathing organs. Losing the voice was the only noticeable physical effect. Almost immediately, all the officers lost their voices. Many of us were coughing. We had to carry on. It was amazing how your body adapted to the radiation while you were there. Within 10 days you would get your voice back, and you would stop coughing. Your body got used to everything.

Because there is a difference between the deliberate use of nuclear weapons and an extreme unexpected explosion, we received some additional training about Chernobyl. The consequences of these events are the same. The damage at Chernobyl was like the use of a neutron bomb. The damage is local but the consequences depend on the form of the radiation released. In this case, the radioactive isotopes of iodine, strontium, and cesium were created, and the contamination, especially iodine was spreading.

In the very beginning, firemen were sent in to liquidate the fire. Then soldiers were sent to establish control. Investigators were sent to do tests. They pinpointed the hot spots and left signs of where there were high levels. But their equipment was not the right equipment. It had been designed for use with nuclear weapons. It was not able to identify all the isotopes found at Chernobyl. So, we did not really know what levels of radiation we had. Likewise, our equipment was also designed for nuclear weapons. And we had no exposure badges. It was not until later that they started designing special detectors for defining the danger. We were guinea pigs.

There were altogether some 150,000 workers at the site. My group and I worked at a variety of tasks. Some were cleaning core debris from the roof of the fourth reactor. The facility was disassembled piece by piece and thrown down into the reactor. Workers wore special aprons and gloves and worked quickly taking things down and throwing the debris in. You would walk right to the edge of the hole and throw pieces down into the reactor.

We tried using Japanese robots to look into the reactor. I saw this myself. I was responsible for watching the monitor from the roof. The robots did not survive. They could not function with the high levels of radiation. They would go out of order and fall into the reactor. It was suicide. So, the Russians created their own robots that could withstand the conditions. You had people and robots working in parallel. Then the robots were taken by helicopter to be cleaned.

The second type of work we were doing was on the ground around the fourth block. Chunks of the reactor were falling from the roof. These pieces had to be picked up and placed in soil inside metal containers. The containers were then collected in special vehicles and taken outside the reactor territory and buried in holes in the ground and covered with soil. We all knew there would probably be consequences later. Metal will rust. Radioactive products will be washed out to the Pripyat River to the Dnepr River and on to the Black Sea. But, under our extreme circumstances, nobody cared. We were focused on the problem immediately before us.

The third type of work involved testing the reactor containment. If levels of radiation got too high, we would bring in concrete and dump tons of it on top. We knew that this was a fruitless exercise. It was for nothing. It was a stupid job. Within 10 days, the new layers of concrete would themselves become too radioactive because of exposure to gasses and dust.

Other troops were sent to Pripyat and the other villages to wash off radioactive dust and remove contaminated puddles. They used special cars and washing aides, sprays, and brushes.

We worked an eight-hour day without rest. We had early breakfast and ate again only at 6 p.m., after eight hours of work. The second shift started at 2 p.m. And went until 11 p.m. We were issued special masks and respirators. And daily when we worked at Chernobyl station, our clothing was changed. Every day I went to a banya[2] where the dirty particles were washed off my body. At the banya, we took our clothes off. When we came out, we were given new clean clothes.

We combined two functions. We were there for training. But we comprised a military battalion also, capable of undertaking real services. There were two commanders. There was a real battalion commander. I was the second in command as a trainee. The training continued besides the work. There were also exercises. I brought the soldiers to the banya and back.

We were housed in Oranoya, a village 40 kilometers from Chernobyl station and 30 kilometers from the dead zone around Chernobyl. We lived in tents there. Buses came to pick us up. Elderly people had remained behind in this village. They did not want to leave. Chickens were running around with red butts and no feathers. And dogs were running around with bald spots.

In all, we had 17 trips to Chernobyl. For every trip, you were allowed 1 Roentgen of exposure by special order.[3] If your collector got 25 Roentgen, you were sent away with a gift of special compensation. Several active officers working with the pieces that fell from the roof collected 25 Roentgen, and the state gave them money. I do not know what happened to them.

To avoid having to pay, it became the practice that you were sent away to a civil job when you collected 24 Roentgen. I and others were only allowed to collect 15 Roentgens on our collectors so they did not have to pay. Of course we all got more than the acceptable Roentgen level, but no one knows how much. If the commander liked you, he would list you as having received 25 Roentgen.

DENIAL AT WORK

You ask if I was afraid. Fear is for when you fight with a known enemy. When it is real and you can see guns and explosions and reality. I had known fear. But when the enemy is unseen and not able to be observed, it is different. As a chemical officer, I knew the consequences. It was dangerous. But because of my high level of responsibility, I also knew I had to fulfill my obligations. Along with our responsibility for the consequences for life from the disaster, we had special training to be there. It was our mission. We all felt this way. We all had a light attitude. It was childishness. We had no fear. It is the Russian way to not take things seriously. There is the Russian term for this: "'Avos' – It will pass me." My colleagues who later got bone cancer did not know anything was wrong at the time. We were not tense at the site. We did not take it seriously. And besides we were paid well.

The childishness spread throughout the whole population of soldiers and officers. They would pick mushrooms from the dam area. They would catch and eat fish. They would work without respirators, including me. Even from Pripyat, where the soil was radioactive, they looted to improve their lives. They brought carpets and tape recorders. It was not quite stealing. I recall an occasion when the host of a dacha met an officer's car and asked for a ride to his own dacha. As payment he offered a bottle of cognac. Upon arrival, he said *whatever you like here you can take*. So the officer got carpets and a tape recorder to improve the quality of his life.

Why did not we take it seriously? Well, from the start there was the effect of our role models. At our first class at the academy and defense ministry upon arrival at the Chernobyl station, the captains who were the experts on nuclear subjects showed us an example. They picked apples from a tree and ate them without washing the fruit first. They were showing off and defying the danger. They told us that they had encountered hazardous substances and nuclear zones and had survived. Their bravado made them overlook the danger and not take it seriously. Of course no one knew at the time that they both would die within the year.

Not everyone displayed bravado. Some people were so afraid that they could not take trips to the reactor. Some pretended they were sick. They would disappear from the hospital.

Such escapes were particularly evident at another site on the Baltic where there was radioactive exposure. But they did occur at Chernobyl. Retired military officers who came to fight the fire were allowed to work a half year with no punishment for escape.

HEALTH PROBLEMS

Your body adapted while you were at Chernobyl. But later you felt it. The next day after leaving the nuclear station, when I came to the city "White Church," my right leg became sore just above my Achilles heel. Then my left leg became the same way. It was difficult to walk. A week later, I was not able to walk at all. Each step was painful to tears. I was sent to the Podolsk Military hospital in a Moscow suburb for two months. The diagnosis was an infection of the Achilles heel. I had lost my shoes at Chernobyl. Radiation made my Achilles worse. But the doctor refused to blame the radiation. There was an unwritten order from the ministry of defense to the effect that not a single diagnosis was to be connected to radiation except radiation sickness.

On 6 November, before the national holiday (October 1917 revolution day), I was signed out. When I left the hospital, I was in worse condition because the doctors did not know how to cure me. My legs were swollen close to my foot at my Achilles heel. And then my knees got swollen too. I went to Moscow. I visited Berta Borisovna Bukatova, who is now in Israel. She was the ticket cashier for the Bolshoi. She had a huge group of friends. She was resourceful and could solve any problem. She said she had a good friend who was a physician at a sports clinic. He cured famous athletes. I was supposed to do compresses on my knees and legs and use a special sticky cream to stop the infection. The next three days I was to rest and then go to the doctor. I met with the famous medical professor Belankeroff, who cured me with three deep injections of an American drug and said I could go. When I left and reached the road, I took a taxi and felt relief. The medicine worked to decrease the swelling. American medicine saved my life. Belankeroff was not afraid to say that he believed the cause of my problem was radiation. He openly violated the secret order that you were not supposed to connect disease to Chernobyl.

WERE YOU SOCIALLY ACCEPTED AS A LIQUIDATOR?

I saw no stigma. People had interest and curiosity. "How was it?" Afterwards, I lived among officers with military families. No one was afraid. But I did know someone who lived in Volgograd. They sent their uniform home. Their mother got rid of it. She was afraid of it.

At the same time there were reasons to fear radioactive products. One contribution to this threat came from the fact that so many apartments had been abandoned. People left only with their documents leaving all else behind. Some people were not honest and good. Like vandals or grave robbers, they would go to places left by fleeing people and steal jewelry or things to sell. People who worked in the nuclear station became rich. They would sell these things to people who did not know they were from Chernobyl.

Rather than stigma, the society recognized the important role of the liquidators. There were benefits in the beginning. I used to go to the front of the line for the theater. Later, the benefits ceased, but we got money. I put it in the bank. I bought several things. I bought a coat and shirt with the money. We bought a lot of books. Later the government distributed a special liquidator certificate card offering some significant benefits. These included payments of half of apartment fees; half price for gas, electric, heat, and water; land taxes were forgiven; free notary papers; free public transportation; inflation protection; an extra two-week vacation paid for with a free voucher to a resort; free car inspection; and forgiveness for the road tax. My wife also received discounts and free transit.[4]

CONCLUSION

The health concerns remain with me. To this day, I have something in my throat. Some times during the night I have difficulty breathing. I have difficulty swallowing. I left the army because of my health problems connected to radiation. I have arteriosclerosis. I feel it is due to this. But doctors are not allowed to use radiation as the basis for a diagnosis.[5]

My brother Valery never came to Chernobyl, but he served in another place protecting radiation equipment and nuclear weapons. Yet, he must have had some exposure. He died a young and horrible cancer death. No one knows what actually happened.

What guidance was I given to protect myself subsequently? I was told not to go to the south in the summer because there is so much soil contamination there and to avoid the heat. I would love to go to the south. I am not supposed to go.

And I am supposed to have my whole body check up every year. I do not do this because I am too busy. Besides, I still practice the Russian phenomenon, "Avos – It won't touch me."

NOTES

1. The IAEA reports that some 350,000 people involved in the initial clean-up of the plant in 1986–87 received average total body radiation doses of the order of

100 millisieverts (mSv), roughly equivalent to 1,000 chest x-rays, a dose five times that permitted for nuclear workers and 40 times background radiation (IAEA, 2005). Lawrence Livermore Laboratories of the U.S. Department of Energy reported that the several hundred thousand workers in the first year of the cleanup received doses as high as 70 centigrays (the international unit for measuring absorbed ionizing radiation, the gray is equivalent to 100 rads, or 1 joule per kilogram). The half million additional workers who followed in later years received much lower doses (Lawrence Livermore, 1999).

2. Banya is the Russian sauna. Combining the virtues of a dry and wet sauna, the banya is dry until steam is generated by pouring beer or water onto the superheated rocks.

3. A Roentgen is a standard unit of radiation exposure. It technically refers to the dose of ionizing radiation needed to produce 1 electrostatic unit of electricity in 1 cubic centimeter of dry air.

4. Both Russian and Ukranian liquidators have had to mobilize to keep their benefits from being reduced over time (IAEA, 2006).

5. Although the IAEA reports no definitive association of the liquidator's exposures and physical health problems, they do cite evidence of posttraumatic stress disorder (PTSD), in particular, a study that found a relationship between likelihood of PTSD and duration inside the 10 K exclusion zone around the plant and time working on the roof of the damaged reactor and forest clean-up (IAEA, 2006).

REFERENCES

EGE. (2006). Environmental Consequences of the Chernobyl Accident and their Remediation: Twenty Years of Experience. Report of the UN Chernobyl Forum Expert Group "Environment" (EGE). Last retrieved on April 11, 2006 from http://www-pub.iaea.org/MTCD/publications/pubdetails.asp?Pubid = 7382

International Atomic Energy Agency. (2005). Chernobyl's 700,000 "Liquidators" struggle with psychological and social consequences. Last retrieved on March 29, 2006 from http://www.iaea.org/newscenter/Features/Chernobyl-15/liquidators.shtml

Lawrence Livermore National Laboratory. (1999). S&TR, September. Last retrieved on March 29, 2006 from http://www.iaea.org/newscenter/Features/Chernobyl-15/liquidators.shtml

INTO THIN AIR: TRAINING CHILDREN TO LIVE IN POLLUTED ENVIRONMENT

Oleg S. Glazachev

INTRODUCTION

The deteriorating relationship between humans and the environment is a cause for our concern. On one hand, the human influence on nature has resulted in global climate change and a decline in the health of the world's oceans. On the other hand, it is evident that humans cannot adapt to new ecological conditions, as evidenced by new diseases. Is there any way out of the crisis?

Our choices are illuminated by this metaphor (Levi, 1991). Picture humans' ecological system as an old bridge – "the road of life." The bridge is full of defects and holes with people walking and periodically falling through the holes into the water. Some of them drown. The rest are saved by divers and lifeguards (i.e., modern medicine). After resuscitation, they are returned to the same bridge, eventually to fall again. Undoubtedly, for saving human beings we need rescuers and divers. But other alternatives also deserve extraordinary resources. These are: (1) to repair the bridge, (2) to provide a safety system and warning signs, (3) to inform people about the dangers of deep water if they cannot swim and (4) to train people to swim.

Modern society has focused on rescue and resuscitation to save and revive as many as possible. But a systematic approach to creating new

Cultures of Contamination: Legacies of Pollution in Russia and the U.S.
Research in Social Problems and Public Policy, Volume 14, 373–389

ISSN: 0196-1152/doi:10.1016/S0196-1152(06)14016-8

interdisciplinary knowledge is necessary to solve the problem. As it has emerged, this new thinking has focused almost exclusively on the first three alternative responses to the problem. As a result, complex measures are taken to promote environmental safety, create zero-discharge technologies, teach ecological safety and use education to promote an ecological culture and imperative that respects the coevolution of humans and nature. These approaches are problem oriented and most effective. But they are extraordinarily expensive, protracted overtime and demand coordinated efforts of the whole society that may not be easy to achieve. While these long-term strategies are pursued, another approach is needed.

This chapter explores the fourth alternative, to teach people to swim. Specifically, I refer to the need to increase human adaptive ability and resistance to unfavorable environmental factors through new approaches and technologies for health promotion. The resulting prophylactic method will be aimed at promoting the health of people who are still relatively healthy (i.e., those still on the bridge). Such methods are inherently complex, having pedagogical, social and medical aspects.

The full understanding of this approach demands an example. Here I take the case of children and teenagers living in an ecologically unfavorable environment. Instead of waiting for them to become ill and treating them, I employ a non-medical method designed to restore their physiological functions and increase their adaptation level and psychophysiological resistance. The method is called "intermittent normobaric hypoxic training." It involves adapting children to breathing air that has less than normal levels of oxygen.

THE STATE OF CHILDREN'S HEALTH IN CONTAMINATED REGIONS

For many years, my colleagues and I have examined children and adolescents living in regions of Russia contaminated by the Chernobyl disaster (Glazachev & Sharapov, 1992; see also the work of Stark & Metter, 1991). The results demonstrate that the state of health of school children is near critical. We found only 10% (between 9 and 16%) of the children we studied to be completely healthy. In contrast, between a quarter and nearly a third (between 26 and 31%) already suffer from chronic health conditions. The adaptive capabilities of these children are already permanently impaired.

The remaining half to two-thirds (53–65%) of the children are only "relatively healthy." These children are less adaptive than healthy children due to non-specific functional abnormalities. Such deviations include various autonomic-humoral disregulations, neurotic disorders, immune function disruptions and impairments to bearing and/or to bronchial potency. These children are at risk of having their functional deficits become chronic conditions from daily exposure to repeated emotional stresses, unsuitable school infrastructure, unfavorable environmental factors and the cumulative hazards of chemical pollutants.

What may be surprising, however, is that this pattern of impaired health is not limited to regions affected by Chernobyl. Rather, it is typical of many regions of the country with unfavorable ecology. According to official Ministry of Public Health statistics, only one-third of schoolchildren in Russia enjoy a satisfactory state of health. By the end of secondary schooling, half the graduates have health problems that make them unfit for some future activities, certain professions and military service (Korobeynikov, 2005). Nearly 90% of the children are in dire need of health treatments, physical therapy and rehabilitation courses.

Conventional medical treatment options make little sense when one considers the non-specificity of the abnormalities suffered by the majority of children and the fact that many will take permanent residence under ecologically unfavorable conditions in the future. The conclusion is that there is an urgent need to develop methods to enhance children's adaptive potential and physical resistance without relying on medication.

HYPOXIA TRAINING

One such technique is training children to breath air containing less than normal amounts of oxygen. Our approach is formally called "intermittent hypoxic training" (IHT or hypoxia training). A typical hypoxia training session occurs under conditions of normal barometric pressure. A subject is cycled through a regimen of 5 minute periods of breathing air with normal oxygen levels interspersed with 3–5 minute intervals of oxygen deficient air (a gaseous mixture with 11–14% volume of oxygen).

Hypoxia training is derived from a basic method used in adaptive medicine. Subjects are preconditioned to different natural stimuli (e.g., physical load, emotional stress, cold and high altitude). Such conditioning improves performance under the specific conditions involved in the training. As a bonus, it also tends to produce other general benefits (i.e., nonspecific protective cross-effects). Therefore, training the body for short-term hypoxia not only protects subjects exposed to air pollution (i.e., oxygen-poor environments), but it more generally enhances their resistance and adaptive capacities, preparing them to carry heavier-than-normal physical loads, tolerate exposure to chemical pollutants and endure emotional stressors (Meerson, 1993; Meerson, Pozharov, & Minyailenko, 1994).

Hypoxia training is targeted on the earliest stress manifestations that people experience after exposure to hazardous environmental conditions. Such exposures cause an imbalance in the interactions between the different homeostatic functional systems of the body (Sudakov & Glazachev, 2001). This imbalance can be understood according to the theory of functional systems (Anokhin, 1974) and one of its core principles, the principle of multiparametric regulation.

The central and peripheral parts of our body connect to form dynamic self-regulating functional systems (e.g., the function of breathing unites the brain with all the coordinated parts of the lungs, the blood, the heart, the kidneys the other cells, etc.). Not only the respiratory system, but also the whole organism, depends upon the cumulative activity of these individual elements to achieve an adaptive result (e.g., not losing one's breath when climbing stairs under adverse conditions). The activity of one functional system inevitably influences the ultimate adaptive results in other connected functional systems. Basic functional systems also serve as foundations for and influence the organization of more complex adaptive human functional systems, ranging from the physiological to the social and including human relationships to the environment.

The principle of multiparametric regulation posits that activity of a definite functional system requires the coordinated activities of executive "autonomic" and peripheral "humoral" mechanisms (e.g., the brain's control over breathing influences the rate at which the blood is pumped, acid-base balance, haemopoyetic processes, etc.), as well as the cooperative activity of other functional systems. At the metabolic and homeostatic levels, functional systems engage in complex hierarchic, multiparametric and successive interactions in order to achieve significant results (Sudakov, Lazetic, & Grujic, 1998).

Hypoxia training externally adds an additional executive mechanism for controlling homeostatic function. After hypoxia training, a person can breathe normally even though there is less oxygen present than they would normally require. For this outcome to be achieved, the body's different homeostatic functional systems must cooperate and harmonize in order to keep physiological parameters in the normal range despite unfavorable environmental stressors.

Hypoxia training has been experimentally proven and clinically used by scientists and physicians in Russia and the Commonwealth of Independent States (CIS) countries for many years (Kolchinskaya, 1993; Meerson, 1993;

Serebrovskaya, Swanson, & Kolesnikova, 2003). Controlled inhalation of oxygen-depleted air has been shown to effectively train many physiological functions and to increase the capacity of patients and sportsmen to do physical work (Berezovsky & Deyneka, 1990). More recent western research using placebo-controlled studies explains the increased tolerance for exercise after passive hypoxia training as resulting from enhanced stress resistance and improved oxygen delivery (Burtcher et al., 2003) and identifies tiny neuro-humoral, biochemical and biomolecular intracellular mechanisms underlying hypoxia training (Bailey, Davies, & Baker, 2000; Serebrovskaya, 2002; Davydov, Starkova, Koroleva, Ehrenburg, & Tkatchouk, 2002). Both passive and active methods for hypoxia training have been developed.[1]

At the International Institute of Social Physiology, our focus has been to test intermittent hypoxia training as a tool for helping children who live in ecologically distressed zones. We have studied hypoxia training as a tool for rehabilitation that enhances the reserves of homeostatic functional systems. And we have assessed hypoxia training as a method for increasing adaptability through improved somatic and psychophysiological function. Our aim, figuratively, is to prepare schoolchildren to swim in polluted waters. Below, I describe our research on the benefits of hypoxia training.

METHODS

My center undertook a study at the "health-promotion center" of the extended-day school N12 in Novomoskovsk, Tula region, 220 km from Moscow. Novomoskovsk has been designated as an ecological distress zone.[2] The sample included 108 relatively healthy schoolchildren aged 9–12 (58 boys and 50 girls) taking a course in Dynamics of Health Promotion. Hypoxia training was given to 64 pupils while 44 children were used as controls. All the children stayed at school for 30 days under similar conditions. They swam in the swimming-pool four times a week, had three substantial meals a day, took multivitamins (Unicap M) and, if indicated, received physical exercise therapy.

Details of the Procedure

Each child assigned for treatment was given 18–20 sessions of intermittent hypoxia training. Each session consisted of 3–6 hypoxic (11–12% O_2) periods lasting 3–5 minutes sandwiched between 5-minute intervals of

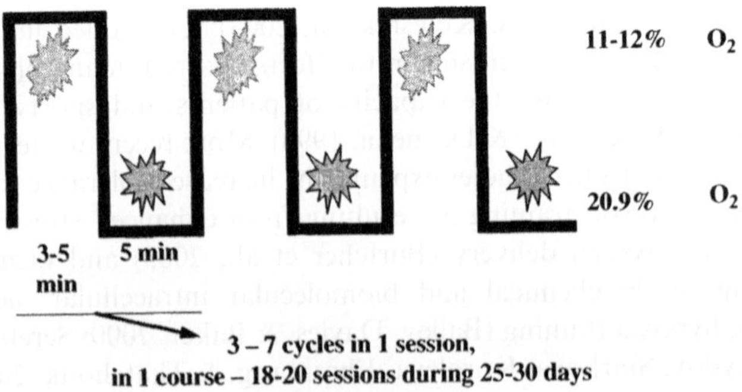

Fig. 1. Example Showing the Organization of an Intermittent Hypoxia Training Session.

breathing normal air. Students sat in a chair and inhaled the gas mixture through a face mask. Gas mixtures were made with a "hypoxicator." The breathing program was controlled by a physician. Subjects wore a fingertip sensor that allowed for continuous monitoring of O_2 saturation (SaO_2) and heart rate by a pulseoximeter. A schematic illustration of an IHT session is represented in Fig. 1.

To control for confounding conditions, a comprehensive physiological and psychological examination was made of all of the schoolchildren (treatment and control) one or two days before the beginning of the hypoxia training program and 1 and 30 days after its completion.

A complex of psychological and autominomic measurements were recorded for each child using a portable diagnostic microprocessor or "MIR." These parameters were measured:

- A pneumogram
- Respiratory rate as measured by pulsogram
- Breathing rhythm irregularity (degree of emotional tension, heart rate and heart rate variability)
- Autonomic regulation (range of variation and stress index)
- Respiratory function (vital capacity, forced vital capacity and forced expiratory volume at 1 second)
- How long a child can voluntarily hold their breath (Stange's test)
- Blood pressure (Systotic and diastolic arterial blood pressure) and dynamometry.

The analysis of psychoemotional status combined a number of measurements, including a complex sensory motor reaction test to determine the child's reaction time and the number of errors; a static tremorometric test with calculation of the time of tremor, the number of contacts, and the static tremor coefficient; the tapping test with analysis of the number of dashes made per minute, the critical frequency of light flash merging and anxiety level as assessed by the State-Trait Anxiety Scale for Children (Spielberger, Gorsuch, & Lushene, 1983, adapted to Russia).

Individual sensitivity to acute mild hypoxic stimulation was assessed immediately before and after the hypoxic program. A hypoxic mixture containing 11% of O_2 was inhaled for 10 minutes through a special mask. Hypoxic resistance was reflected in the maintenance of a minimal level of SaO_2 and maximal hypoxic reserve.

A further measure determined the effect of hypoxia training on cerebral vascular tone and blood supply. The reactivity of cerebral vessels to hypoxia was used as an independent parameter, reflecting the capacity of blood supply in the brain to adapt to different pollution exposures. Ten children from each group (treatment and control) were given a 10 minute rest and then asked to sit in a chair wearing bifrontal leads. A rheoencephalogramm (or "REG") was then recorded using a rheopletismograph RPG2-03. These "REG" tests were repeated after five and 10 minutes of hypoxia training.

REG data measured the following indices: (1) the systolic blood inflow to the brain (on the rheographic index, Om), (2) time of the ascending portion of the rheographic wave α (in seconds), (3) duration of the pulse wave T (in seconds), (4) the state of vascular tone for supplying arteries as measured by the relative duration of the anacrotic phase α/T (by percent), (5) contraction in the small arteries and arterioles measured by the dicrotic index (by percent) and (6) the tone of the veins as measured by diastolic index (by percent).[3]

RESULTS AND DISCUSSION

Psychophysiological testing revealed no differences at the onset between children assigned to the treatment and the control groups. However, numerous functional changes occurred by the end of the program.

Hand Tremor

Two-thirds (64%) of the children scored high on a test of hand tremor (the static tremorometric test) conducted at the onset. The finding was equally prevalent in the treatment group (static tremor coefficient of 1.45+0.25) and control group (static tremor coefficient of 1.47+0.23). Suggested is that the whole sample suffered from initial high levels of psycho-emotional tension and disorganized psychomotor regulation.

Hypoxia training was followed by a substantial reduction in tremors (decreases in both the number of contacts and the related static tremor coefficient) for children in the treatment group. Moreover, treatment group children had significantly fewer tremors than did control group children. The findings suggest that hypoxia training reduced emotional stress and also significantly improved a child's accuracy in differentiating stimuli (see Table 1).

Table 1. Changes of Psychophysiological and Physiological Indices in SchoolChildren after Hypoxia Training ($M \pm m$).

N	Index	Group	Stage I (before IHT)	Stage II (after IHT)
1	Time of response, ms	A	351+18	313+13
		B	321+13	310+16
2	Number of errors	A	1.90+0.24*	1.47+0.27**
		B	2.12+0.23	1.82+0.25
3	Number of contacts	A	18.7+3.4	11.1+2.4**
		B	17.7+3.3	23.6+3.0
4	Static tremor coefficient	A	1.45+0.25	1.00+0.2**
		B	1.47+0.23	1.54+0.2
5	Critical frequency of light flash, Hz	A	31.7+2.0	34.3+1.7
		B	32.0+1.6	30.7+1.5
6	Tapping test (Tapt)	A	205.7+13.1*	245.3+8.3
		B	195.7+8.4*	235.2+10.2
7	Vital Capacity, ml	A	2,319+79*	2,528+79**
		B	2,078+92	2,162+78
8	Forced Vital Capacity, ml	A	1,985+105*	2,307+85**
		B	1,592+83**	1,583+68
9	Forced Expiratory Volume, ml	A	1,880+101*	2,152+70**
		B	1,507+83**	1,569+60
10	Stange's test, s	A	36.4+2.3*	44.8+2.0**
		B	38.7+2.5	39.7+2.0
11	Hand muscular strength, kg	A	12.6+0.8*	18.4+0.9**
		B	6.4+0.8	16.3+0.3

Note: here and subsequently: A – denotes children being given hypoxia training, B – control group. *– significance of the group index value differences ($p \leq 0.05$) before and after Intermittent Hypoxic Training, ** – significance of the index value differences ($p \leq 0.05$) between the groups at one stage.

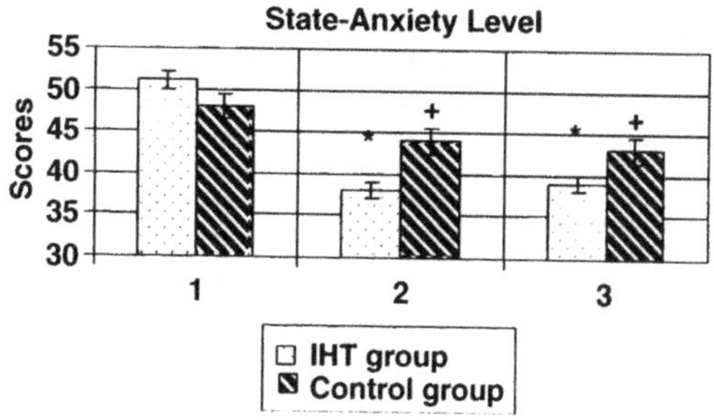

Fig. 2. Changes in Emotional State among Schoolchildren. *Note:* Anxiety level was tested using the State-Trait Anxiety Scale for Children (Spielberger et al., 1983). 1 – before IHT, 2 – after IHT, 3 – 30 days post-IHT. $^{*}p \leq 0.05$ in one group to initial data, $^{+}p \leq 0.05$ between two groups in one period of examination.

Reduction in State Anxiety

At the onset, the entire sample of children evidenced high levels of trait anxiety, a measure of subjectively perceived emotional stress. Children receiving hypoxia training were significantly less anxious than they had been before the training and significantly less anxious than the controls, a result that persisted when testing was repeated a month later (see Fig. 2). Hypoxia training was well tolerated by the participating children; no side effects were identified.

More Regular Breathing

Another indication that hypoxia training reduced emotional tension was an index of breathing irregularity. Both groups showed improvements, but not at the same frequency. While only half (11 of 21) the control group improved, most (16 of 19) children in the hypoxia group evidenced less breathing irregularity. The strength of the findings demonstrate an absence of emotional tension in the improved students. Not only was tension absent, but its psychophysiological manifestations decreased. At the same time, the hypoxia trainees showed improvement in stimulus differentiation and nervous process mobility.

Increased Oxygen Saturation by Blood

Rehabilitative effects were most evident with regard to respiratory function (autonornic homeostasis) (see Table 1 and 2). After treatment, children uniformly evidenced increased dynamometric (MS, kg) values. And external

Table 2. Changes of Autonomic Indices in Schoolchildren after Hypoxia Training (M + m).

N	Parameter	Group	Stage I (before IHT)	Stage II (after IHT)
	Systolic Blood Pressure, rnm Hg	A	103.7 + 1.7*	97.7 + 1.5
		B	99.2 + 2.1	100.6 + 1.7
2	Diastolic Blood Pressure, mm Hg	A	70.5 + 1.8*	63.3 + 1.5
		B	64.8 + 1.7	62.3 + 1.9
3	Mean Blood Pressure, mm Hg	A	81.6 + 1.7*	74.7 + 1.2
		B	75.2 + 1.7	75.1 + 1.7
4	Heart Rate, bpm.	A	83.4 + 2.7	81.6 + 1.0**
		B	87.3 + 2.0	86.6 + 2.0
5	Double Product, (Robinson's index), cond.units	A	58.8 + 2.2*	51.7 + 1.7
		B	55.3 + 2.0	54.4 + 2.2
6	Range of Variation, s	A	0.249 + 0.017	0.275 + 0.015
		B	0.231 + 0.015	0.232 + 0.012
7	SI, cond.units	A	153.5 + 26.3*	108.3 + 15.0**
		B	178.5 + 29.4	142.3 + 14.0

Note: Designations are the same as in Table 1.

respiratory function increased significantly after hypoxia training on all of the recorded volumetric parameters, including vital capacity, forced vital capacity and forced expiratory volume. Children receiving hypoxia training also exceeded the corresponding values found in control group children. These results indicate that hypoxia training increases the lung's alveolar surface and the potency of large- and medium-calibre bronchi, improving gas exchange and presumably hemoglobin saturation by oxygen.

Ability to Hold One's Breath

Hypoxia trained children also showed a significant increase in their ability to voluntarily hold their breath (from 36.4 + 2.3 to 44.8 + 2.0 seconds) Reflected are two benefits of hypoxia training: enhanced resistance of the nerve centers to hypoxia and (indirectly) increased capacity of the cardiorespiratory system. In contrast, control group children showed insignificant improvement in breath holding over repeated examination.

Blood Circulation

There were a variety of indicators that blood circulation and dynamics improved after hypoxia training. By the end of treatment, children showed significant decreases in Systolic, Diastolic and Mean Blood Pressure and a slight heart rate deceleration. The hemodynamic indices also changed after hypoxia training as indicated by a significant decrease in Robinson's index or double product. On a whole, these hemodynamic shifts influenced by hypoxia training reflect some degree of optimization of central and regional blood circulation.

Hypoxia training improved regulation of heart rhythm by the parasympathetic division of the autonomic nervous system. This rehabilitation effect was evidenced by slight increases in the "mean group Range of Variation values" for children completing hypoxia training and by significantly higher "Range of Variation values" for the treatment group by the end of training than found for the control group. The stress index of autonomic regulation in the group of children given hypoxia training decreased significantly as well, representing both vagal activation and sympathetic decrease.[4]

Hypoxia training clearly improved upon individual baseline levels of resistance to acute mild hypoxia (see Fig. 3). Considering both parameters of hypoxic sensitivity, it resulted in significant increases in minimum SaO_2 levels and decreases in maximum heart rate against both baseline measurements for the treatment group and comparing the treatment and control groups.

In studying the dynamics of cerebral reactivity (i.e., vascular responsiveness to the hypoxic test), a parallel decrease in rheographic index (systolic blood flow to the brain) and the state of vascular tone for supplying arteries (as measured by the relative duration of the anacrotic phase α/T) was observed. According to the literature (Moskalenko, Beketov, & Orlov, 1988),

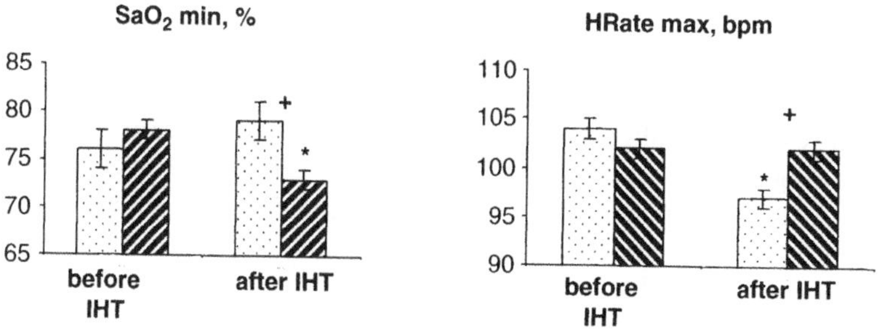

Fig. 3. Effects of Hypoxia Training (IHT) on the Minimal Value of SaO_2 and Maximal Heart Rate on a Hypoxic Test Provided before and right after the Course of Hypoxia Training. *Note:* For Designations see Fig. 2.

hypoxic exposures of such intensity do not result in substantial changes of cerebral blood flow since the threshold of cerebral receptor sensitivity to hypoxia is rather high. Nevertheless, even this load has resulted in the redistribution of pre- and post-capillary vessel tone (i.e., a significant decrease in dicrotic and diastolic was observed in both control and experimental groups at the 5th minute of hypoxic test). These shifts may be considered as indicating an improved oxygen supply in the brain with virtually unchanged intensity of cerebral blood flow as a response to hypoxia training, provided before the rehabilitation course. By 10 minutes, similar shifts were observed in the group receiving hypoxia training, while in the control group, on the contrary, the tone of pre-and post-capillary vessels was enhanced by the end of hypoxic exposure (which could be related with autoregulatory responses preventing the excessive perfusion). It is necessary to note that virtually all children displayed severe tachycardia in preliminary tests taken before the health-promoting course. Therefore, before the rehabilitation program, all the pupils displayed functionally disturbed cerebral blood flow (as evidenced by symptoms of cerebral hypotension and a pronounced increase of hypoxic sensitivity to 10 minutes of hypoxia simulation associated with a dilation of the cerebral arterioles and venostasis).

After the course of the IHT, subjects of both groups displayed an increase in systolic blood flow to the brain (or rheographic index) with a parallel decrease in the state of vascular tone for supplying arteries as measured by the relative duration of the anacrotic phase α/T (by percent), which may be interpreted as some optimization of cerebral blood flow. Attention is drawn to considerable intergroup differences in the dynamics of the dicrotic index and the diastolic index values before and after the IHT (see Fig. 4).[5] Individuals in the hypoxia trained group showed significantly increased results while the control group remained at the initial level. This fact may reflect the increased tone of pre- and post-capillary vessels with some general enhancement of cerebral blood flow, i.e., a decrease in initially revealed signs of cerebral hypotension.

In the experimental group, following the rehabilitating course with hypoxia training implementation, the profile of the cerebral vessel's response to the hypoxic test substantially changed. At the 5th and 10th minute, the changes in parameters indirectly characterizing the intensity of cerebral blood flow (rheographic index and duration of the pulse wave T in seconds) were similar to those observed before the course of hypoxia training. Note the non-significant increase in the vascular tone for supplying arteries (as measured by the relative duration of the anacrotic phase α/T), which remained unchanged in the course of hypoxic testing. We interpret this finding

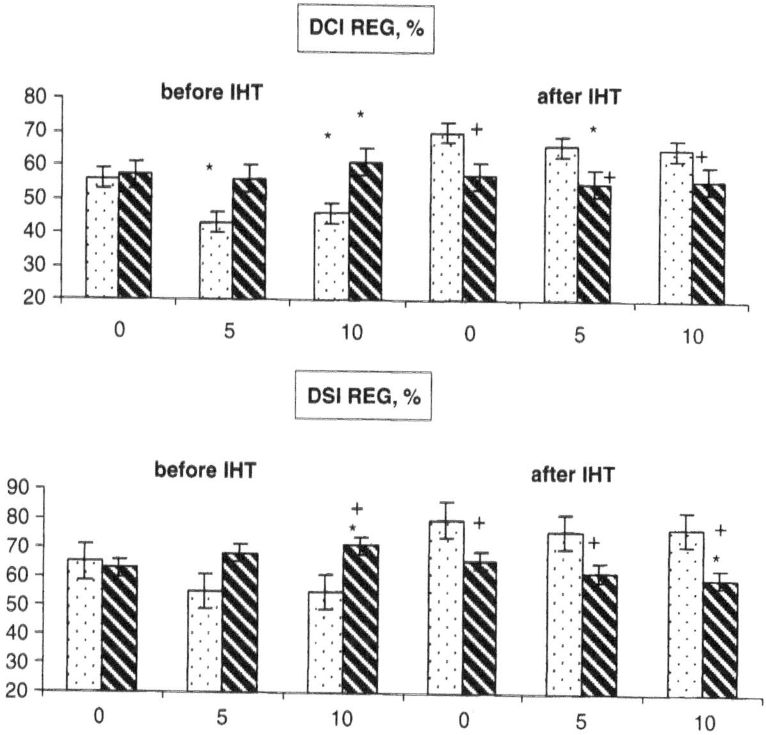

Fig. 4. REG parameters: dicrotic index (% DCI) and diastolic index (% DSI) in children subjected to the course of Hypoxia Training or IHT (light bars) and children of control groups (hatched bars) in the hypoxic test (1. baseline, 2. 5th minute, and 3. 10th minute) before and after the course of rehabilitation of hypoxia training. *Note:* *Significance of differences at $p<0.05$ within the same group relative to the initial value. +Significance of differences at $p<0.05$ between groups at the same stage of investigation.

to indicate an increased resistance of cerebral blood vessels to short-term hypoxic exposure in IHT, which is potentially beneficial to keep cerebral circulation stable in stress conditions.[6]

Thus, the investigation supports the value of incorporating intermittent hypoxia training into a course promoting health. Hypoxia training has been shown to positively affect various homeostatic and psychophysiological functions of the child's organism. The completion of IHT is marked by activation of neurodynamic processes and the clear decrease of psychomotoric and autonomic correlates of emotional tension and anxiety. These

Table 3. The Dynamics of Rheographic Parameters in Junior Schoolchildren before and after the Health Promotion Course (M + m).

	Indices	Group	Period of Hypoxic Test		
			Initial	5 minute	10 minute
1.	RI, Om	A before after	0.04 + 0.002 0.06 + 0.004	0.04 + 0.001 0.05 + 0.002*	0.04 + 0.002 0.05 + 0.002*
		B before after	0.05 + 0.003 0.07 + 0.002**	0.04 + 0.001* 0.05 + 0.001*	0.04 + 0.002 0.06 + 0.002* **
2.	α, sec	A before after	0.08 + 0.002 0.09 + 0.002	0.07 + 0.001* 0.09 + 0.001	0.07 + 0.001* 0.08 + 0.002*
		B before after	0.10 + 0.002 0.09 + 0.002	0.09 + 0.004 0.09 + 0.003	0.10 + 0.006** 0.08 + 0.004*
3.	T, sec	A before after	0.69 + 0.011 0.71 + 0.012	0.63 + 0.012* 0.65 + 0.010*	0.63 + 0.013* 0.64 + 0.013*
		B before after	0.68 + 0.013 0.69 + 0.005	0.62 + 0.006* 0.61 + 0.005* **	0.62 + 0.005* 0.64 + 0.005*
4.	a/T, %	A before after	12 + 0.5 12 + 0.3	11 + 0.3* 14 + 0.3*	12 + 0.3 13 + 0.3*
		B before after	15 + 0.7** 13 + 0.3**	14 + 0.6 15 + 0.5*	16 + 0.8 15 + 0.9*

A – group of schoolchildren subjected to Hypoxia Training (the IHT program), B – control group, before and after the Hypoxia Training course.
*– significance of differences at $p<0.05$ within the same group relative to the initial value,
**– significance of differences at $p<0.05$ between groups at the same stage of investigation.

changes are more markedly pronounced than in children rehabilitated through a conventional program.

Table 3 shows the mean values of the rheoencephalographic parameters obtained during baseline measurements and responses to the hypoxic exposure at the 5th and 10th minutes of the test, before and after the rehabilitating course. The data revealed one important baseline difference between the treatment and control groups. Children receiving hypoxia training showed a significant decrease in the α/T index not evident in control group children, suggesting that children in the treatment group had lesser elasticity of the vascular wall in large and middle-sized blood vessels. Both groups scored within age-related normal ranges. However, the dicrotic and diastolic indices and rheographic index values were at the lower limit of the age-related norm. In the case of rheographic index, this may reflect the increased lability of cerebral blood flow regulation in junior schoolchildren.

Thus, the described investigation demonstrates that IHT, incorporated into a health promotion course, positively influences a child's homeostatic and psychophysiological functioning. The result is a reduction in anxiety through the activation of neurodynamic processes that quiet the psychomotor and autonomic correlates of emotional tension. These changes are markedly pronounced over the benefits of conventional rehabilitation.

Hypoxia training has the most marked effects on the autonomic homeostasis found with sudden physiological strain common in children living under psychoemotionally stressful conditions in ecologically unfavorable regions. Training increases the resistance of nervous centers to hypoxia, improves bronchial conductivity and pulmonary ventilation, optimizes blood dynamics, decreases sympathicotonic activity and the cardiac strain due to autonomic regulation and activates vagal influences.

For virtually all children residing in the conditions of ecological trouble, the combination of the heath promotion course and hypoxia training appears to have resulted in improved pulse and thus blood flow to the brain, as well as increased arteriolar and venular cerebral resistance to stressors including acute hypoxia.

Volunteers undergoing hypoxia training completed the four week breathing program without any complaints, objective complications or side effects despite the decrease in adaptive potency suffered by the majority and the stressed or disturbed homeostatic functional systems evident in their baseline conditions.

CONCLUSION

Our studies demonstrate clear benefits of including intermittent hypoxia training within a program of rehabilitation to address the psychoemotional, cognitive and autonomic status of adolescents. These benefits include reductions in emotional tension, anxiety, grief and restraint, enhanced positive emotions, improved psychophysiological and neurological functioning, autonomic tone and neuro-humoral cardiac regulation and normalization of homeostatic cardiovascular and respiratory mechanisms.

As a result of such research, we know that it is feasible to use a pattern of "impulses" mixing normal and oxygen-deficient air to improve resistance to environmental and social impacts. The increased resistance occurs because a program of IHT helps to train neuronal centers, balance communication between central structures of the autonomic nervous system and peripheral autonomic organs and tissues, increase the balance between homeostatic functional systems and, accordingly, promote cross-adaptation and increased homeostatic and behavioral capacity.

We have learned that the success of IHT for adolescents depends upon their individual characteristics. Hypoxia training works best for individuals who are highly stressed and have lowered resistance to hypoxia and increased Autonomic Nervous System sympathetic tone. There is likely to be an optimal dosage of hypoxic stimuli for any individual. This hypoxia resistance level must be controlled precisely before commencing a course of Hypoxia Training.

With young students, IHT has been found to be accompanied by a general improvement in health and well-being. Suggested is the possible use of IHT as a rather efficient "healing" method applicable for applied occupational, pediatric projects in preventive medicine. Of particular promise is the finding that IHT serves to stimulate the organism's general, nonspecific resistance without negative side effects.

To return to the image of walking a treacherous old bridge as a metaphor for the road of life, IHT technology allows young representatives of the future generation to cross the bridge more safely and comfortably, to avoid its defects and holes, to be better prepared for swimming should they fall and to participate actively in repairing the bridge. In the end, it is this generation who is called upon to build new and more adequate rules and regulations for coevolutionary development between nature and the human ecosocial system. We must do all that we can to help them succeed.

NOTES

1. As examples, in the USA, "Hypoxico LTD" (http://www.hypoxico.com) and "Colorado Altitude training systems" http://altitudetraining.com) and, in Russia, Hypoxia Medical Academy (http://www.hypoxia.ru).
2. In Ecological Distress zones, it is not uncommon to extend the school day to provide all meals and to keep students inside away from harm.
3. The results obtained were treated by statistical methods: results are given as $M \pm SEM$, significance of difference was evaluated by Students T-test. Data analyses were conducted with the use of "Statistica for Windows 6.0" software package.
4. However, the mean group values at the end of the health-building course did not differ. Less marked differences in the autonomic regulation parameters between the groups seemed to be linked to their considerable initial variability, individual and typologicall heterogeneity.
5. The dicrotic index is a measure of contraction in the small arteries and arterioles (by percent) and the diastolic index is a measure of vein tone (by percent).
6. It is necessary to note that some children showed individual peculiarities of the REG dynamics. For instance, 4 of 10 children had pathological responses of vascular spasm. Peculiarities of the individual reactivity of cerebral vessels need further investigation.

REFERENCES

Anokhin, P. K. (1974). *Biology and neurophysiology of the conditioned reflex and its role in adaptive behavior*. NY: Pergamon Press.

Bailey, D. M., Davies, B., & Baker, J. (2000). Training in hypoxia: Modulation of metabolic and cardiovascular risk factors in men. *Medical Science Sports Exercise*, *32*(6), 1058–1066.

Berezovsky, V., & Deyneka, V. (1990). *Physiological mechanisms of mountain climate sanogenic effects*. Kiev: Naukova Dumka.

Burtcher, M., Pachinger, O., Ehrenbourg, I., Mitterbauer, G., Faulhaber, M., Puhringer, R., & Tkachouk, E. (2003). Intermittent hypoxia increases tolerance in elderly men with and without coronary artery disease. *International Journal of Cardiology*, *96*, 247–254.

Davydov, A. L., Starkova, N. T., Koroleva, A. V., Ehrenburg, I. V., & Tkatchouk, E. N. (2002). Effect of interval hypoxia on growth hormone secretion. *Hypoxia Medical Journal*, *10*(1–2), 18–21.

Glazachev, O. S., & Sharapov, A. N. (1992). *Doctor*, *11*, 24–26.

Kolchinskaya, A. Z. (1993). Hypoxic hypoxia and load hypoxia: Destructive and constructive effects. *Hypoxia Medical Journal*, *1*, 8–13.

Korobeynikov, A. (2005). Education for balanced development in school: Report of the Parliamentary Assembly, Council of Europe, Committee on Culture, Science and Education. Available at http://assembly.coe.int

Levi, L. (1991). Intervening in social systems to promote health. In: M. G. Ory, R. P. Abeles & P. D. Lipman (Eds), *Aging, Health and Behavior* (pp. 276–295). Newbury Park, London: Sage Publications.

Meerson, F. Z. (1993). Adaptation medicine: Protective cross-effects of the adaptation, Moscow, Hypoxia Medical Ltd, p. 421.

Meerson, F. Z., Pozharov, V., & Minyailenko, T. (1994). Superresistance against hypoxia after preliminary adaptation to repeated stress. *Journal of Applied Physiology*, *76*, 1856–1861.

Moskalenko, Yu. E., Beketov, A. I., & Orlov, R. S. (1988). *Cerebral circulation: Physico-chemical techniques of investigation*. Leningrad: Nauka.

Serebrovskaya, T. V. (2002). Intermittent hypoxia research in the former Soviet Union and the Commonwealth of Independent States (CIS): History and the review of the concept and selected applications. *High Altitude Medical Biology*, *3*, 205–221.

Serebrovskaya, T. V., Swanson, R. J., & Kolesnikova, E. E. (2003). Intermittent hypoxia: Mechanisms of action and some applications to bronchial asthma treatment. *Journal of Physiological Pharmacology*, *54*(Suppl. 1), 35–41.

Spielberger, C. D., Gorsuch, R. L., & Lushene, R. E. (1983). *State-Trait Anxiety Inventory*. Palo Alto, California: Consulting Psychologists Press, Inc.

Stark, S. N., & Metter, F. A. (1991). *Pediatric Research*, *29*(4), 124.

Sudakov, K. V., & Glazachev, O. S. (2001). Multiple physiological assessments of long-term stress at work and in daily life: A system approach. *Advances in Psychosomatic Medicine*, *22*, 61–79.

Sudakov, K. V., Lazetic, B., & Grujic, N. (1998). Bases of theory of functional systems: Perspectives. In: B. Lazetic & K. Sudakov (Eds), *Basic and clinical aspects of the theory of functional systems* (pp. 7–16). Yugoslavia: Novi Sad.

SOCIAL REMEDIATION: CONTAMINATION AS AN IMPETUS FOR ECOLOGICAL LEARNING

Lyudmila V. Smirnova

INTRODUCTION

Volgograd, Russia, my home for most of my life, was entirely rebuilt after World War II. Under its prior name, Stalingrad, the city was the epicenter for what many believe was the most crucial battle in the entire war. That battle came at the cost of many millions dead and wounded and the destruction of all but one shell of a building. From this oblivion, a new city arose under the direct order of Stalin, who mobilized captured German engineers for the task. Following his concept of planning, the city would be long and narrow, hugging the banks of the mighty Volga River for some 80 km. The width would involve only two main thoroughfares with side streets. The narrow profile would allow for farms and dachas to be close by on one side and the river on the other, providing bounteous and accessible food. Residences were organized in neighborhoods formed around key enterprises lining the river to meet their needs for water. The neighborhood designs allowed workers to easily walk to work. It was a truly utopian scheme.

But there was a dystopian side to the picture. Stalingrad, renamed Volgograd after Stalin's death, has since earned an unfortunate distinction as one of Russia's top ten polluted cities. For example, my neighborhood in

Cultures of Contamination: Legacies of Pollution in Russia and the U.S.
Research in Social Problems and Public Policy, Volume 14, 391–402

ISSN: 0196-1152/doi:10.1016/S0196-1152(06)14017-X

the north features major industrial facilities producing tractors, aluminum, and steel. This triangle of intensive pollution even drew the attention of the U.S. Environmental Protection Agency (1997) for its extraordinarily high levels of particulate emissions. In contrast, the most southern section of Volgograd is polluted by the petrochemical factory stacks that crowd the Krasnoarmeysk (literally Red Army) region's skyline.

I served for many years as a faculty member in the education program at Volgograd State Pedagogical University (VSPU). My focus was innovative education. The intensity of our city's pollution raised an interesting question for educational innovation. How does one help children adapt to life in a hypercontaminated environment while setting forth a program of learning that points to solutions and alternatives?

VSPU's most tangible effort to address this question was our Perestroika era assistance to community activists in creating an experimental school plunked right in the center of Krasnoarmeysk, Volgograd Ecological Gymnasium (or Public School no. 137). The project combined the objectives for school children of maximally protecting them from and adapting them to this polluted setting. At the same time, falling, as it did, within a period of broad social experimentation with democracy, openness, empowerment, and ecological awareness, the school project became a way to redefine the relationships of people to the environment through the creation of an ecological curriculum. We would put the next generation in charge of redefining this relationship through action learning, which made their very life context the subject of study.

Within Russian educational theory, the gymnasium is a model school for the future, demonstrating pioneering Russian work in ecological education, which calls for the creation of an ecological culture within the school featuring a functional civil society. In this view, ecological education is the new content of the entire educational process, defining the needed developmental path of both the individual and the society toward a state of maturity or ecological harmony. Among its features, this education has a prognostic or anticipatory character (Glazachev, 1997; Arefyeva, Glazachev, Dmitrichenko, & Zvereva, 1997).

Because this school can serve as a prototype for "the ecologization" of education and culture all over Russia and other regions of the former Soviet Union and inform the current sustainable education (or "greening") movement in the U.S., I will describe the experience of the Volgograd Ecological Gymnasium in detail to show the influence of this experiment on the development of the ecological educational movement in other parts of Russia and Byelorussia during 1990s, beginning 2000. I will also detail how the school approaches the tasks of protecting and adapting students to their contaminated surround.

CASE STUDY: VOLGOGRAD ECOLOGICAL GYMNASIUM

The Ecological Gymnasium was established in 1989 in Volgograd[1] by the initiative of several teachers: V. Arefyeva (now a principal of the school), A. Streltsov, and others under the supervision and help of the head of the district administration V. Galushkin. A conventional secondary school was reorganized into an institution whose mission was to create the conditions for formulating an ecological culture for its students. In the Ecological Gymnasium, children study from first grade through the 11th (in Russia, the final grade of high school). There are 1,700 students currently enrolled and some eighty staff.

The institution was not created in a vacuum. Rather, important partners nurtured and guided it. From the very beginning, the idea was supported and supervised by researchers from the VSPU and also the members of the ecological education laboratory of the Research Institute SiMO. The group of scholars and school teachers worked out an experimental program for cultivating ecological literacy in schoolchildren so as to create and realize a model of continuing ecological education from kindergarten through high school graduation. Their interest was not merely curricular, but also on the nature of the setting and the institutional system of support for student development. The possibility for the school to play a positive role in the polluted community was not lost on the planners either.

Key Ingredients of Ecological Literacy

There has been a careful elaboration of the normative basis of the gymnasium. These include regulations and a chart specifying the relationships of the component elements of the gymnasium's approach. Evaluation of the first decade of the Ecological Gymnasium's operations reinforced the importance of five pedagogical characteristics of ecological literacy built into the core of the program. There is an impressive mesh of these principles with the latest Western education theory (Bruner, 1996), if not always practice.

1. *Person-oriented.* A well-developed pedagogical concept calls for a carefully differentiated person-oriented ecological education. Rather than a

blanket approach for all, ecological literacy is cultivated through a developmental approach based upon assessment of students' personalities, traits, abilities and talents, and individual development.

2. *Systemic organization.* A systematic approach underlies the entire educational process. Ecological literacy is made a systemic part of the educational process, helping to unite basic and optional components building in a continuous process. In this manner, ecological literacy develops through both the accumulation of knowledge and its qualitative gradation from grade to grade. Each component not only adds more information, but it reorganizes the learner's understanding of the world.
3. *Interdisciplinary approach.* All subjects are approached in an interdisciplinary fashion. Moreover, thanks to the integrity of ecological knowledge, environmental issues are studied in all school subjects programs, ranging from conventional subjects to special integrative ecological disciplines.
4. *The integrity of global and local.* Ecological problems are studied at the global level but applied to the local environment, providing depth to ecological knowledge while making knowledge more personally relevant, and thus more stimulating.
5. *Unity of emotional and rational.* Serious ecological issues are studied with an accent on the spiritual and moral consequences of ecological catastrophe; art, music, and poetry are used to stimulate the emotional perception of nature, understanding its unique value and linking nature and creativity.

Organizational Structure

The organizational structure of the Ecological Gymnasium provides for a fluid structure capable of continual change in order to promote ecological education. At the core of the structure are two elements: the "Scientific-Methodological Council" and the "Ecological Education Laboratory." The council defines the main strategies to be used in the gymnasium during its bimonthly meetings. The laboratory monitors implementation and outcomes of these strategies, analyzes the results, and presents them before quarterly school conferences of staff and teachers, students and parents. These two processes assure the organization of subject areas to promote ecological literacy and of the Valeological and Psychological Services in order to promote a stress reducing and adaptive school environment and state of personal functioning within that environment.

Finally, there are the issues of ecological literacy (see Orr, 1994). Edelstein (2004) has defined ecological literacy as gaining the knowledge and wisdom to contribute to a sustainable world. He suggests that ecological literacy rests on greening each of what he calls the four "c's" of education: curriculum, culture, campus, and community. Considering the central role of Valeological and Psychological services, a fifth "c" must also be considered here, climate. Each of these elements must be weighed in understanding the Ecological Gymnasium.

Curriculum

As a public school, the curriculum at the Ecological Gymnasium offers conventional sounding subjects such as mathematics, physics, science, Russian Literature and Language, foreign language, history and subjects of the "aesthetic cycle." However, in this institution, the chair for each subject area is responsible for promoting the ecological development of the schoolchildren studying that subject. This is done in a typically holistic manner.

The case of the science cycle is illustrative of this approach. Each chair is responsible for bringing ecology into their subjects. The effort moves past the science of ecology to consider ecology as an ethical/value system. In this way, ecology is used to foster the development of a student's understanding of the spiritual, aesthetic, cognitive, and economic values of ecological bonds and relationships. But the goal is not merely to shape the student's abstract understanding of the world. Rather, as a human being, the study of ecology is used to promote the development of ecological relationships between person and environment. This is done at two different but parallel levels of consideration. First, the Ecological Gymnasium recognizes the importance of the relationship of a person to the external environment. And, at the same time, it focuses upon the "endoecology" or internal ecology of each person (i.e., personal health). Among the integrative ecological courses taken by students of the Ecological Gymnasium are "Foundations of Ecology," "Country Study and Ecology of the Native Land," "Health of a Human Being and the Environment," "The Ecology of a Park," "A Human Being and a Forest," and "Ecology of Russia."

The chair of science supervises the work of the society of young researchers called "Ecolog." "Ecolog" establishes mastery of empirical methods, including observation and description of the biotic and abiotic components of natural systems and determines their state of health. Students also learn to observe their personal development and human relationships, and they develop skills in creative thinking, experimentation, problem formulation, hypothesis setting, testing, and interpretation.

Using readiness- and age-appropriate materials, physics students become acquainted with contemporary methods of investigation and management of nature, developing an understanding of nature's importance in human activity, the physical basis for the modern ecological crisis, and ways to use physics in developing solutions. Physical factors relating to earth, water, and atmosphere are treated as elements of a united biosphere. Ecological information on the physical factors of natural environment and its contamination is used to concrete and deepen the knowledge of physics.

Similarly, views of cognitive science and development are given an ecological perspective. This is done as an extension of Piaget's theory in the belief that cognitive functions are realized through a child's knowledge of the laws governing the existence, organization and development of natural systems, and how such systems promote life. Thus, humans come to be seen by the child as not mere individuals, but part of a larger whole. This collective consciousness applies to the individual's part in the human learning group, as well. Thus, communicative function manifests itself in the development of the student's ability to exchange information with others.

Sample pedagogical techniques employed in this approach include:

- Discussion among students of different ecological situations and relationships. Students then learn to cooperate in their efforts to solve practical ecological problems peculiar to their local environment.
- Study of natural systems in the field in order to cultivate the skills and habits of ecologically appropriate behavior.
- Student involvement in problem solving for actual ecological issues.

In a parallel vein, mathematical skills are taught in ways that illustrate and utilize data about ecological problems. In turn, mastery of mathematical knowledge and skill is not only useful for everyday life but for application to ecological education. Solving problems with ecological content and role-playing are two techniques employed to enhance both mathematical and ecological understanding. Conferences and seminars are used to educate teachers of math and physics about ecological issues and how they might be addressed.

The Russian Language and Literature curriculum addresses the interrelation of language and the society. It includes a specially created program on ecological linguistics. Ecological education through literature is viewed as an integral part of the social, spiritual, aesthetic, and emotional upbringing of the students. A linguistic research group includes students from grades 5–10 who learn to investigate the research literature and carry out individual research work projects. The best reports are presented at the city research conference "Science. Nature. Man."

Ecological education in foreign language is realized through learning appropriate vocabulary, learning about international ecological movements, reading literature and articles on ecological issues, and participating in an integrative course called the "Nature Around Us." In history, students study environmental laws and the historical development of the interrelationship of nature and civilization. Teachers of the Department of Aesthetics use art, music, and graphics to provide students an idea of a healthy way of life, the beauty of the human body, and the ecological requirements of the home environment.

Culture

Given the immersion of students in the school environment, the Ecological Gymnasium also provides an active cultural program of training and participation. Teachers developed a center of leisure activity to develop children's ability to perceive beauty in the arts, human relations, and nature. The leisure center unites different forms of cultural activity and, integrated with the curricular program, helps to promote the goal of ecological education.

An ecological club is operated at the school, and the school serves as the base for a broader children's ecological movement, which extends to outside schools and groups. Outside partners also run programs that the students of the Ecological Gymnasium participate in, notably the programs of the local chapter of Greenpeace.

Included within the Leisure Center are the activities of the school's puppet theater "Pinocchio," the choreographic group "Childhood," the dance group "Russian Patterns," the musical literature salon "Parnas," the family club "Fathers and Children," the creative organization "Morning Star," the folk group "Ivolga," and the art studio "A Rainbow." A Center for Vocational Education is also operated to promote development of talents and abilities.

Campus

The Ecological Gymnasium has a supportive physical plant and is outfitted with equipment for ecological investigations, physical therapy, and has plans for a computer lab. There are 28 classrooms, 3 methodological offices, musical and literature studios, the laboratory of ecological education, 3 sport halls, 3 handicrafts workshops, a physiotherapy room, a relaxation room, the salt cave room, a library, a canteen, and an assembly hall.

At the same time, as there is reasonable functional support for the key tasks undertaken by the Ecological Gymnasium, the age of the project, its location in an old school building, and the limited funding that is available for state-of-the-art improvements underscores a serious limitation of the project. As the principle environment occupied by children from the Krasne Armaesk neighborhood, it would be ideal if the school building itself were equipped with the latest green and environmental technologies, both to demonstrate and achieve positive values of efficiency and renewable energy, but also to establish maximum filtration of external air, isolation from external hazards, and minimize generation of pollutants within the internal environment.

Community

The process-oriented organization of the institution creates a fluid set of relationships as part of a system of management that is itself ecological in nature. The result is an innovative means for dealing with both the external and internal community.

Among the external projects, the Ecological Gymnasium is engaged in long-term research collaborations with universities and organizations in Volgograd and Moscow. These collaborations become relevant to the activities of the gymnasium staff and often relate to the educational projects undertaken by students.

For example, what is called the "science" project is a program of research by teachers of the gymnasium that examines the process and outcomes of ecological education. Leading teachers publish the results of their work, develop workbooks for their courses, write textbooks, and participate in research seminars and conferences. The project also coordinates research activity of the schoolchildren oriented toward investigation of the local environment; the state of health of the local population; development of case studies; description and evaluation of the state of various ecosystems; studies of air, water, and soil and inventories of protected natural objects. Students present their findings in their senior courses, at research conferences held at the school, and at city conferences for young researchers.

The "education" project focuses on a variety of concerns, including the development of ecological courses and modules to fit the state curriculum, summer field internships, opportunities for ecological education through extracurricular activities and involvement in city affairs, and studios and workshops on ecological orientation held for the public.

The "business" project is run by the Children's Greenpeace Movement and guided by the gymnasium's Eco-parliament. Project activities include protecting, restoring, and improving natural landscapes, establishing "eco-routes" (interpretive trails) and protecting local species of flora and fauna. The Eco-parliament educates the public, providing ecological information through posters, photo exhibitions, ecological newsletters, lecturing, meetings, performances of the eco-theater and eco-excursions in the region.

In addition to these programs, the Ecological Gymnasium has become an important base for training of teachers in ecological education, hosting teaching internships for students from the VSPU.

Internally-focused community projects also occur. The "staff" project addresses the fact that effective work on ecological education requires a well-trained staff. The project prepares teachers to develop an integrated educational process that can stimulate ecological literacy development. Among the activities of this project are efforts to define children's readiness for ecological education, creation of academic consultation centers for upgrading teacher preparation, arranging conferences and seminars for experience exchange, dissemination of the experience in ecological secondary education, arranging visits to other ecological centers, and analysis of the levels of teacher readiness to implement the concept of ecological education.

The "school management" project focuses on how to maintain a functioning ecological school. Among the concerns are staff hiring, contract systems, special teacher training programs and continuing education, incentives and support for teacher creativity, and management of sabbatical and leave programs.

Climate

Creating a healthy and adaptive environment through valeology and psychology is a core goal of the Ecological Gymnasium, central to the efforts to create an adaptive setting that strengthens students' ability to cope with a hypercontaminated environment. The main goal is to create a stress-free and stress-reducing setting that helps to offset external environmental stresses and helps strengthen the body and mind of students in order to cope with this contaminated surround.

The first aspect of physical and psychological health attended to is the aesthetic qualities of the school environment. This attribute signifies a sophisticated understanding of the school setting as framing the overall mindset of the occupants that is rarely found in the east or west. The school environment attempts to create a pleasant and healthy ambience that contrasts and offsets the visibly-polluted industrial section that surrounds the conventional-looking school building. Upon entering the building, however, one is struck by the extraordinary steps taken by the school to immerse students in a dramatically different setting—bright, airy, and dominated by a natural aesthetics. Because children are in school for a very long day, incorporating all three meal times, and stay nearly year-round, this school interior setting is by far their dominant environment. The walls are hung with large murals of nature and with wall hangings and sculptures made of natural materials. Drapes covering large windows allow light in but keep the outdoors sealed from the classroom.

The valeological service or health regime in the Ecological Gymnasium has pushed way past conventional practice in the Russian system of education. There is an extraordinary focus on health found at the Ecological Gymnasium. This focus is demanded by the hypercontaminated environment in which children and staff both live and work. The consequences of this environment are apparent in a frightening manner. As an indicator of the issue, 100% of the children attending the Ecological Gymnasium have asthma or some other serious respiratory problem.

To address such issues of the environment that might normally fall outside of a school's responsibility, the Ecological Gymnasium has operated a Valeological Service since 1993. The main goal of this service is to address the health of the children and teachers: fostering the healing process, testing for diseases, conducting treatment and prophylactic work, and disseminating information about a healthy way of life. A completely holistic approach to health is evident.

Diagnostic work by the Valeological Service employs clinical, physiological, psychological, laboratory, and hygienic methods of investigation. Specialists at the school compile children's health maps that track development and health over time. These diagnostics include testing children's morphofunctional maturity, physical and mental development, and their ability to work. Tests are carried out right in the Ecological Gymnasium. Deviations from norms are identified and treated. Still other diagnostic procedures are used to identify sources of tiredness, stress, and irritation and to correct the daily routine and schedule in accordance with the individual peculiarities of the children. Treatment procedures include physio-therapeutic procedures, massage, and breathing exercises.

Herb therapy is actively used for children with breathing problems. There is also a special room outfitted for those needing to gain control of their breathing. Inside this room, a limestone cave has been created, emulating the physical conditions of the cave spas that have been used in Europe for centuries to treat breathing problems. A child having asthma can go and sit in the cave until their normal breathing is regained.

There is a major focus on countering immune disorders and strengthening the body to confront the environmental pollution. The concept is straightforward. Since the polluted environment demands that adaptive individuals have a strong immune system, yet such conditions challenge immune system and therefore health in many ways, a prophylactic approach must build the strength of this system. All staff and students consume herbal tea drinks intended to stimulate the immune system. Also available is an experimental program developed in Moscow that employs injections of immune system stimulants directly into the child's lymphatic system.

An active program of aroma-phyto-prophylactic work includes use of herb teas, polyvitamins, relaxation techniques, and opportunities for solitude and decompression. There is a relaxation room, where sensory auto-training and acoustics are employed to help students and staff escape stress and overtiredness. The "psycho-hygienics" of the academic process is also under the control of the Valeological Service in order to provide for the "harmonic psychic development" of children by maintaining a positive emotional climate, which helps avoid tiredness, mental pressure, and conflict situations. The overall effect of these efforts is to promote a positive emotional atmosphere, improve clarity of thought, promote good circulation, and increase mental and physical activity.

THE BEGINNING OF AN EDUCATIONAL MOVEMENT

Similar projects were subsequently implemented in Moscow and other regions of Russia. These were principally influenced by one of the initiators of the Volgograd experiment, former president/rector of VSPU, Dr. Stanislav N. Glazachev. After Glazachev moved to Moscow to serve to the Ministry of Russian Education, he initiated a number of research projects on ecological issues in education and soon emerged as the leading expert in ecological education and culture in Russia. His first postgraduate student, Dr. Sergey Kashlev, defended his doctoral theses in 2005 on the approaches for training teachers to have an ecological consciousness and readiness to promote ecological values and knowledge in schoolchildren. Kashlev also developed an ecological school in Minsk. Others were created in the Moscow region and in northern Russia under Glazechev, who has promoted ecological education throughout the former Soviet Union.

Assessments of the ecological schools in Volgograd, Moscow, northern Russia, and Minsk (Byelorussia) suggest that students develop an attitude of deep responsibility toward nature and the society of the school and university leavers. Ecological literacy is evident; students have a thorough understanding of the interactions between nature, people, and society, about contemporary ecological problems and possible solutions, ecologically correct behavior, a strong commitment to nature protection and environment restoration, and a vision of a sustainable society and healthy way of life. Graduates have taken this learning into government; into colleges and universities (85 % of the graduates continue school), and into ecologically-oriented professions and community service activities.

NOTE

1. This case study is based upon field investigation and interviews and the author's long term working relationship with the Ecological Gymnasium

REFERENCES

Arefyeva, V., Glazachev, S., Dmitrichenko, G., & Zvereva, I. (1997). Systemic Approaches to Ecological Educational Organization from the experience of Gymnasium #6, Volgograd. *Science and School, 3*, 18–25.

Bruner, J. (1996). *The Culture of Education*. Cambridge, MA: Harvard University Press.

Edelstein, M. R. (2004). Sustaining Sustainability. In: G. Chase & P. Barlett (Eds), *Strategies for Sustainability: Stories from the Ivory Tower* (pp. 271–292). Boston: MIT Press.

Glazachev, S. (1997). Ecological Culture, Education and the Civilized Choice of Russia. *Science and School, 3*, 4–10.

Orr, D. (1994). *Earth in Mind: On Education, Environment, and the Human Prospect*. Covello, CA: Island Press.

U.S. Environmental Protection Agency, Volgograd Health Risk Assessment Working Group. (September 1997). "Assessing Chronic Health Risks from Stationary Source Air Emissions in Volgograd, Russia: A Case Study in Quantitative Health Risk Assessment." DDP 601.

PART 4: PERSPECTIVES

SECTION 4: INTRODUCTION

What can we learn about cultures of contamination from our comparison of Russia and the United States? Three final chapters offer some perspective.

In Coming Clean after 9/11: The continuing World Trade Center disaster, Michael Edelstein and Catherine McVay Hughes discuss the response by government to the aftermath of the tragic destruction of the twin towers. They provide a detailed analysis of the validity errors in government communication, focusing particularly on communication error and deliberately constrained and distorted communication. Drawing from their own experiences and the literature, they offer analysis of the role of scientists as information providers, the WTC health directory, the Lower Manhattan Development Corporation projects, demolition of remaining buildings and the EPA Technical Advisory Panel. In the end, they find little hope for the quest by residents of areas impacted by contaminants from the WTC disaster to clearly identify and address continuing pollution exposures given the reticence of government to explore the issue and the inattention of a distracted public. If truth was hard to find as the event unfolded, it has not become very much easier over time.

It is interesting to ponder the significance of this worst-case disaster in light of the next mega-scale disaster to occur in the United States. The effects of Katrina, particularly on the city of New Orleans, combine a natural disaster with ample human causes. If the World Trade Center disaster was a portrait in heroic actions to save survivors, Katrina was a much less pretty picture. Here many survivors of the initial disaster were left to languish, and in some cases perish, because of a failed rescue effort.

As I write, New Orleans Mayor Negrin was in New York City apologizing for a comment he made in response to criticism for how slowly things were moving in New Orleans. His comments about New York still having a hole in the ground five years later were taken as an insult by some. But the analogy of New Orleans and New York City may not have been inappropriate. Recovery is a challenge in the face of complex disasters, as the Edelstein and Hughes chapter shows. And recovery is dramatically complicated when the path is strewn, not only by rubble and debris, but also by deception, distortion and deceit, as continuing revelations suggest in the World Trade Center case.[1]

Much as the afermath of Chernobyl, there was extraordinary self-sacrifice as the World Trade Center disaster unfolded. Clearly, sustainable behavior rests on the willingness of people to serve a community of interests rather than merely their own private interests. In *Environmental Altruism: A Comparison of Russia and the United States*, Margaret Gibbs, Tatiana Andrushenko, Natalya Makarevich and Robert Binford offer an interesting comparison between Russian and US students on issues of environmental altruism. As with other triangulations of qualitative and quantitative results in this volume (see Couch and Mercuri and Hastrup, et al.), the chapter offers multiple perspectives. The study illuminates some interesting differences between the countries and suggests that predictive variables useful in the US, such as degree of personal control and feelings of personal responsibility, may not translate into the context of modern Russian experience. Although Russians were much better informed about both domestic and global environmental problems than were US students, despite the latter's media access or perhaps because of it, it was instructive that neither group of students was adequately informed. And, although Americans were more likely to be active on issues, despite their ignorance, the survey revealed a low level of activism for both samples. It was interesting that moving on to graduate work in psychology appeared to indicate greater altruism in the Russian case. The authors discuss a variety of differences between the national cultures that might account for the findings, largely placing constraints on Russians to move toward activism. One meta-implication of the study, however, is that a much greater degree of information and knowledge is needed in both countries to serve as a foundation for informed action. And, expanded opportunities and expectations for activism are required in both contexts, as well.

In the final chapter, *The Parable of the Lions Bridge: Potentials for Mutual Learning Between Cultures of Contamination*, editors Edelstein and Smironova use a parable to contrast the major emergent conclusions from this Russian–US comparative study. They suggest that neither the US approach of regulating contamination nor the Russian approach of adapting to it is a sufficient answer, although elements of each approach are important and offer an area of mutual learning. The real challenge is one of cultural change. In the end, no effort to regulate and limit a contaminating culture will succeed. And no efforts can work to adapt people to live in world beset by progressively worsening contamination. The nature of that cultural change must address the natural parallel of the assimilative capacity. We must find a cultural basis that is regenerative and renewing, not destructive and limiting.

NOTE

1. See, for example, Anthony DePalma's article, E.P.A. whistle-blower says U.S. hid 9/11 dust danger, *New York Times*, August 25, 2006, and Graham Rayman, Scrutiny of post-9/11 statements in wake of suits, illnesses, *Newsday*, August 27, 2006.

Michael R. Edelstein
Editor

COMING CLEAN AFTER 9/11: THE CONTINUING WORLD TRADE CENTER DISASTER

Michael R. Edelstein and Catherine McVay Hughes

INTRODUCTION

The City of New York was suddenly and deliberately attacked on September 11, 2001, killing thousands of people and leaving unbelievable destruction. Thirty-eight buildings and structures were destroyed or damaged, including seven buildings in the World Trade Center site completely leveled. Almost five years later, two very large contaminated buildings, Deutsche Bank at 130 Liberty Street and Fiterman Hall of Borough of Manhattan Community College, have yet to be cleaned up and demolished. Some 30 million square feet of commercial space was lost. Transportation was disrupted, including the loss of the World Trade Center PATH station, the 1/9 subway line and portions of Route 9A and Church Street. Cars were not allowed south of Canal Street for a week. For Americans this was a terrorist attack and a crime. It was a time for mourning losses and responding to disaster. There was the shock that something like this could happen. And there was more. The destruction of the WTC also posed competing environmental, economic and social threats.

Cultures of Contamination: Legacies of Pollution in Russia and the U.S.
Research in Social Problems and Public Policy, Volume 14, 409–446

ISSN: 0196-1152/doi:10.1016/S0196-1152(06)14018-1

THE ENVIRONMENTAL DISASTER

The World Trade Center Disaster was an environmental disaster, releasing multiple contaminants on a distracted city and its occupants. Environmental attorney Kupferman (2003, p. 2) graphically describes the initial environmental exposures to people.

> As the twin towers collapsed, thousands of fleeing workers and area residents were coated with a thick, white dust that quickly began to irritate their skin and lungs. Grey clouds of highly corrosive material containing concrete particles, asbestos, finely crushed windows and fiberglass and heavy metals choked the streets. Gasping survivors gulped down the toxin-laden dust as the entire area – buildings, streets, and ground, interiors and exteriors – was blanketed with fallout ....[later] The Ground Zero workers were surrounded by a foul, corrosive and irritating odor that permeated lower Manhattan, provoking the widely reported "WTC cough."

Beyond fallout from the fallen building, there was smoke from the fires that burned out of control for weeks at Ground Zero, sending a plume of smoke in a generally southeast direction over Lower Manhattan, Brooklyn and Southern Queens. While the contents of these plumes of contamination were never precisely measured, Kupferman approximated their hazardous nature from the normal contents of office buildings.

> Synthetic fabrics, plastics, laminates and building supplies containing formaldehyde; fluorescent lamps containing mercury; the di-electric fluids that encase electrical cables; approximately four pounds of lead from each computer; PCBs from capacitors, electrical cable insulation, and transformers – were clearly elements in highly toxic fires and dust storms. The plume contained, at the very least, toxic lead, asbestos, volatile organic compounds, dioxins, mercury, nickel, vanadium, sulphur, PAHs, PCBs, and furans.

And, additionally, there was lead from secret services munitions, thousands of pounds of hazardous chemicals stored in a U.S. Customs laboratory, fuel stored at the city's command center and 130,000 gallons of PCB-contaminated transformer oil in an electrical substation (Kupferman, 2003, p. 3). Community Board One (2006a) listed the contaminants as including polycyclic aromatic hydrocarbons (PAHs), lead, man-made vitreous fibers (MMVF), mercury, asbestos and other toxic substances. These were deposited on and within residences, places of employment and mixed-use buildings in Lower Manhattan.

THE ECONOMIC DISASTER

Simultaneously, destruction of the World Trade Center was an economic disaster. An initial precipitous drop of 60% occurred in the residential occupancy rates downtown. With more than 100,000 jobs displaced, Manhattan businesses struggled to stay open (CDC, 2006; HUD, July 2003). Such immediate economic impacts dominated environmental concerns.[1]

> Immediately landlords lost rental income; the retail, entertainment and tourism economies of the city shut down; the real estate market took a nosedive. Schools, the financial markets, government agencies, and all but the most necessary places of business closed. Roads, subways, bridges and tunnels connected to this area would remain closed, or with limited access, for months. Layoffs began, and what had been worrisome signs of recession before September 11 intensified .... It became somebody's job to make sure the secondary economic and political tragedy did not eclipse the WTC fire and collapse itself. A "clean" environmental bill of health for what was left of lower Manhattan was the way to get the city and the country back to work. The push was on to shift the public's understandable shock to an uneasy restoration of order, including the difficult task of trying to keep decimated financial markets from fleeing what was now deemed a major target of our enemies and attracting back tourists. While the world mourned, the clock began ticking to re-open the New York Stock Exchange, a block away from the disaster site (Kupferman, 2003, p. 4).[2]

SOCIAL DISASTER: A DESPERATE NEED FOR INFORMATION

There was also a social disaster (Clarke, 2003). As with other complex modern disasters, the destruction of the World Trade Center created a major demand to provide guidance to a desperate public on questions about threat and safety. Given the novelty of their urgent and frightening situation, a victimized community has a thirst for information precisely at the point where government knows the least, due to uncertainty, and when it seeks to maintain strict control over what it does know.

During the brief duration of destruction and downfall, there was a desperate and agonizing need for guidance about protective action – to evacuate or stay with the stricken buildings. But as the acute crisis graduated to a chronic disaster, the issue of safety changed focus. Scientists had never dealt with focused exposures of environmental contamination at such intense levels during short periods of time (Lioy, 2002). Now emergency workers, construction workers, employees in Lower Manhattan businesses and residents needed guidance about their exposure to the many environmental hazards created and released in the disaster. That guidance first came in the form of a public announcement, made on September 13, 2001 by Environmental Protection Agency (EPA) Administrator Christie Whitman.

> EPA is greatly relieved to have learned that there appears to be no significant levels of asbestos dust in the air in New York City. (EPA, September 13, 2001)

Again on September 18, Whitman reiterated:

> I am glad to reassure the people of New York and Washington DC that their air is safe to breath. (EPA, September 18, 2001).

And her reassurances three days later, after Wall Street reopened, were explicit with regard to "... a host of potential contaminants ... either not detectable or are below the Agency's concern levels." She went on to report for groundwater that results failed to detect asbestos, bacterial contamination, PCBs or pesticides (EPA, September 21, 2001).

As we will see, such statements went a long way toward reassuring New Yorkers that it was safe to return to their homes, workers that it was safe to return to the financial district and emergency personnel that they could work on rescue and recovery, fire fighting and cleanup at Ground Zero without worry about environmental exposures.

Whitman's statements also influenced how insurance companies managed the accounts of affected clients, including what level of cleanup to a home and its contents would be compensated and what level and length of reimbursement would be granted for living expenses while the family lived away from home or a business was shut down. Those with less financial flexibility or lacking the ability to obtain financial assistance from relief services, including the Red Cross, Federal Emergency Management Agency (FEMA) or Safe Horizons, may have had no option but to return.

Finally, Whitman's statements served to help rationalize psychological pressures on families to return home. There was a stress point for families when the need to live at home again outweighed any perceived physical harm that could be inflicted by the WTC fires, smoke and dust. And, at a certain point, some residents just wanted to believe that everything was fine and that government agencies were protecting their property and health.

Government efforts to evacuate residents from Lower Manhattan and keep them out at least until fires were extinguished on December 14 were almost certainly a failure. During the initial period of rescue, a moving line of safety was put in place in Manhattan. Initially at 14th Street, the boundary was shortly pulled south to Canal Street and finally ended up at Chambers Street. The Battery Park City apartments were evacuated. Large parts of Lower Manhattan were without power, water and telephone service. It was thought that most residents fled the area below Canal Street (Kupferman, 2003).

However, about one-third of the enrolled residents in the WTC Health Registry living south of Canal Street on 9/11 reported that they never evacuated their homes at all and approximately two-thirds of those who did evacuate returned to their homes quickly, 38% returned during the first week and another 24% the next week. It is likely that, given the participation of the most concerned residents in the registry, actual percentages who remained or returned quickly are probably much higher (NYCDOHMH, 2004c, p. 8).

A survey undertaken in Lower Manhattan by the EPA Office of Inspector General (2003b) suggests inherent problems with post-9/11 communication. The majority of respondents wanted more and more timely information regarding outdoor and indoor air quality. Having said this, they also indicated that they did not believe the information they received. This distrust was evident when they were questioned about government reassurances and their perceptions of possible health risks from breathing the air in Lower Manhattan. For the majority responding, breathing air in Lower Manhattan in the weeks following the WTC collapse – outdoors or indoors – meant possible exposure to short- and long-term health effects (EPA Office of Inspector General, 2003b).

The nexus of our examination rests on the consequences of this guidance and subsequent efforts by government to address the aftermath of this disaster. Of the affected publics, our primary focus is the residential population. Against the tragic victims, the fortunate who escaped from the WTC, the heroic rescuers, emergency personnel and the recovery and removal workers, residents of Lower Manhattan seem to have earned a comparatively low priority in the response to the disaster.

In this chapter, we present a framework for analysis, examine what policy makers and scientists have said to the public since the disaster to define continuing environmental threats and shape protective responses, how these statements were heard by affected publics and to what extent were secondary threats (i.e., harm beyond the immediate destruction of the towers) mitigated effectively as a result?

COMMUNICATION ERROR

We see from early in the aftermath of the WTC disaster the impacts of communication error. Error may occur at multiple points in the regulatory effort to define the consequences of a disaster, chronic or acute. Table 1 summarizes seven potentially interacting types of error, all affecting the validity of communications about safety and precaution, and, more specifically, whether a person at risk from an environmental exposure could gain the information needed to motivate protective action (Edelstein & Makofske,

Table 1. Threats to the Validity of Addressing: "Is it Safe?".

Error of policy	Error due to asking the wrong questions. Confusion over goals leads to selection of the wrong testing methodology and technique
Sampling error	Failure to adequately sample for contaminants with the result that findings do not represent exposure levels in affected areas
Device and method error	Error resulting from the inherent shortcomings of the sampling device(s) used
Measurement error	Error due to improper use of the testing device or uncontrolled confounding situational factors
Analysis error	Laboratory errors in the analysis of the samples collected in test devices
Interpretation error	Incorrect conclusions are drawn from the measurements
Errors of public communication	Conclusions are not adequately shared with the public, thwarting protective actions

Note: Based on Edelstein and Makofske (1998).

1998). These errors range from policy errors that cause the wrong question to be asked and the wrong indices to be measured, sampling errors that lead to measurements in wrong places, device and method errors that involve the use of the wrong devices and techniques, measurement errors that involve improper use of testing devices or collection of data, analysis errors that cause a misanalysis or misreading of the measurements taken, interpretation errors that involve drawing incorrect conclusions from the information collected and, finally, communication errors that either factually misinform, are unclear, are not believable or are viewed as inappropriate. A slipup at any of these seven stages may cause a cascade of errors. The end result is a failure to convey correct information or the creation of wrong impressions, in either case preventing informed choices by those at risk (see Edelstein, 2004a) and those making the decisions.

In the wake of the unbelievable events of September 11, 2001, there were many reasons for errors to be made. The truth was clouded in uncertainty about environmental health dangers, as noted, in a manner never before encountered. There were many competing demands on decision makers that were viewed as superseding and being threatened by any disclosures of hazards. These included criminal investigations, the vague but urgent umbrella goal of "national security" and the aforementioned desire to open Wall Street (which had been closed for September 11, 12, 13 and 14 – the 15th and 16th were a weekend) to establish confidence in the nation's economy and return some normalcy. As environmental information became available, there was a further challenge that a flood of hazard information would outpace the ability to respond. Such "mitigatory gaps" are a major source of stress for individuals and institutions alike. These inherent problems in the situation were further complicated by three core dynamics of contamination events. The first is the risk of environmental stigma due to the disclosure of environmental bad news. Second, there is the fact that victims become disabled and disempowered by their wholesale dependence on government to assure their safety and their potential efforts to offset such dependency through activism. And the third is the tendency of government to violate this trust by distorting the information it provides. The combination of dependency and distrust is particularly damaging (Edelstein, 2004a).

Environmental Stigma

The issue of environmental stigma has in fact weighed heavily on the response to the World Trade Center Disaster from the earliest efforts to quickly reopen the New York Stock Exchange and other financial institutions in the immediate wake of the attack. Property owners, large and small, have faced the fear that excessive publicity about environmental hazards downtown will accentuate the decline in real estate demand and values felt in the post-disaster period. Even owners of residential apartments face the dilemma that their appearing in the media demanding environmental protection might have the effect of depreciating their apartment. And landlords and building owners reeling from the exodus from the neighborhood want to do nothing to discourage people from moving back in. The massive loss of office capacity caused by the destruction of the twin towers and neighboring buildings was matched by the relocation of employers from Lower Manhattan in the wake of the disaster. Once relocated elsewhere, it is difficult enough to lure them back into Lower Manhattan without environmental stigma as a factor. Politicians and others eager for the revitalization of the neighborhood, therefore, echo these concerns. They were also motivated by the need to restore the city's tax-base.

Even as the office rebound has been problematic, Lower Manhattan has recovered as a residential environment. An investigation by the *New York Times* on the first anniversary of 9/11 surveyed more than 7,000 housing units near Ground Zero, from four-unit walk-ups to 1,700-unit complexes. The newspaper documented a dramatic turnover in many of the buildings surpassing 70%. In 27 buildings closest to the fallen towers, the *Times* found an average 50% vacancy rate in the first year, with many buildings totally vacant for an extended period of time. At one year, the vacancy rate in these buildings had dropped to an average of 4% (Winter & Flynn, 2002).

The Lower Manhattan Development Corporation (LMDC) also ran a residential grant program that promoted individuals to remain in, or move to, Lower Manhattan. Among the financial incentives offered was a Two-Year Commitment-Based Grant and grants for businesses and institutions owning or leasing residential property in Lower Manhattan.[3] The programs attracted both existing and new Lower Manhattan residents with 60% of approved applicants classified as low to moderate income individuals and families, representing 20% of approved applicants.[4]

> Since the Residential Grant Program was first announced in February 2002, there has been a tremendous resurgence of interest in moving to Lower Manhattan, which is reflected in the increase in residential occupancy rates. For example, the occupancy rate of Battery Park City was 60 percent after the attack, rose to 74 percent shortly after the announcement of the draft plan, and is now over 95 percent. Over 50 percent of the residents in zone 1, the area closest to the World Trade Center site, are new residents to the area since September 11, 2001. The new and renewed commitment of residents to the area, despite the disaster, provides a strong foundation for the revitalization of Lower Manhattan."[5]

The stigma extends into other areas as well. Restaurateurs and food sellers, for example, encountered a reticence by some customers to buy fresh foods or eat in restaurants downtown after the terrorist attacks. And, with so many fewer people working downtown, there continues severe financial pressure on such service industries generally.

Avoiding environmental stigma emerges as the only substantive justification for EPA's subsequent behavior in distorting the truth about downtown contamination. If the EPA had released all of its concerns regarding pollutants to the public immediately, it almost certainly would have resulted in an immediate and alarming number of departures of the remaining businesses and residents from the City (Hayner, 2005).

Disabling Circumstances

Contamination victims become disabled and disempowered by the situation, becoming dependent on officials and experts to explain their risks and tell them what precautions are required (Edelstein, 2004a). Residents living around the WTC site were captured, in this way, by a confused array of city agencies, including the Department of Health and Mental Hygiene

(NYCDOHMH) and the Department of Environmental Protection (DEP), state agencies, principally the LMDC, and federal agencies, including EPA and FEMA. Such dependency readily results in distrust, particularly where officials fail to recognize the need to restore victims' sense of control over their lives through respect, participation and empowerment.

Enabling Response

One alternative to continuing dependency on a distrusted institutional network for protection and assistance is the propensity for contaminated communities to organize a grassroots response that helps the public gain power, support and information. Early local leadership in downtown Manhattan came from Community Board One, the local branch of City Government. Most Lower Manhattanites returned home within the first few months after the accident and began to ask questions and demand information and assistance. The focal points of concern were the schools. Stuyvesant High School was located adjacent to the loading dock used to transfer World Trade Center debris from truck to barge. The decisions to allow children to return to Stuyvesant and the nearby elementary and middle schools P.S./IS 89 and PS 234 and reopen the Borough of Manhattan Community College raised serious concerns about the safety of these buildings, all of which had been commandeered as emergency quarters after September 11 and now had to be decontaminated.

The safety issues were highlighted when some neighborhood corporate spaces were cleaned to a much higher level than the schools and other public institutions. Parents rightfully questioned why such fibrous materials as couches and carpets were too dangerous to be retained by Shearson Lehman but were safe to remain in school buildings. As a critical body congealed around such issues, the potential for forming a grassroots organization or coalition of action groups ripened.

The leading community organization responding to the World Trade Center disaster is 9/11 Environmental Action, a community-based nonprofit organization formed in April 2002. Membership was drawn from groups responding to September 11 that comprised residents, parents and occupational safety, public and environmental health advocates. Under the very able leadership of Kimberly Flynn, a longtime activist on NYC issues, 9/11 Environmental Action brought key questions to the forefront and engaged community members to participate on the environmental issues of the World Trade Center. Flynn has collaborated closely with other key players, like industrial hygienist David Newman of New York Committee on Occupational Safety and Health (NYCOSH), to advance 9/11 Environmental Action's mission (9/11 Community Action 2006) to:

> end the federal Environmental Protection Agency cover-up, to fight for a comprehensive cleanup, to demand medical monitoring and health care for everyone harmed by WTC contamination and to prevent further exposures to 9/11 pollution from remaining indoor hazards and potential toxic releases from continuing demolitions of heavily contaminated buildings and rebuilding the WTC site.

The organization maintains a comprehensive website, sends frequent news alerts and conducts public outreach and networking efforts aimed at informing and mobilizing concerned individuals and organizations. Networking and organizing is done by Rachel Lidov, whose daughter was a student at Stuyvesant High School during the 9/11 disaster and "cleanup." In support of the community organizing efforts, Jo Polet, a downtown resident, focused on analyzing the scientific and technical information that was required to determine and support the community's position. Of particular success has been the organizations' cooperation with labor in the highly successful World Trade Center Community–Labor Coalition (9/11 Environmental Action, 2006).

Such grassroots "enabling responses" offset an otherwise disabling situation. Disabling is particularly compounded by the conflicting roles played by the government official and victimized resident. While all might unify in the midst of the acute disaster, as 9/11 transformed into a chronic and continuing disaster, government efforts to understand and control the situation contradicted the goals and approaches preferred by prospective victims. Government officials and scientists were concerned with three ambivalent and perhaps contradictory concerns – protecting and maintaining public health, pursuing science and operating within their range of regulatory freedom. The latter constraint arises from the fact that agencies have limited mandates and resources, multiple demands and responsibilities and must act in ways that do not create undue precedent or policy limits and that do not suggest unauthorized commitments. Approaches to the situation also differ. For government, the definition of risk is an abstract question to be answered by experts and in ways that use accepted procedures, maintain agency control over the situation and employ "objective" quantitative analysis, standards, thresholds and comparisons to other risks. The paramount question for officials is "What is the most I can divulge without causing panic, environmental stigma and future liability and consistent with the policy directives I am receiving?"

In contrast, citizens and their advocates take threats personally and absolutely, regardless of thresholds or comparative risks. They expect to participate in key decisions that affect them. They consider how problems fit within the narrative context of their lives and then assess their needs, whether justified by objective criteria or not. They want all precautions to be taken and believe that no exposure is acceptable. Residents of Lower Manhattan ask "Is it safe for me (and by extension my family) to be here in our home and community?" And "What precautions must be taken to make it safe?" They want their lives back.

The two readily talk past each other.

DISTORTED COMMUNICATION

Victims' trust in their dependency situation was further undermined by the systematic distortion of information (see Edelstein, 2004a) in the wake of the WTC tragedy. As a result, victims were denied the objective information necessary to understand what was happening around them and to participate in making crucial decisions about their health and safety. According to the EPA Office of Inspector General (2003a, p. iii),

> After the WTC terrorist attack, people received information from many different sources, and many factors – in addition to government communications – could have influenced their actions. Information is a critical component in helping the public minimize their exposure to potential health hazards. However, evidence gathered through government hearings, news polls, health studies, and our interviews indicated that the public did not receive sufficient air quality information and wanted more information on associated health risks. Also, evidence indicated that government communications were not consistently effective in persuading the public to take recommended precautions.

There were many reasons for this ineffectiveness.

First, in the aftermath of the WTC disaster, government communication was directed by the patchwork quilt of regulation, policy and practice that restricted who the actors were and how they acted. Thurston suggests that the uncertainty after 9/11 was enhanced by the absence of applicable government guidelines. And post-disaster policy was shaped in the absence of basic data that hampered general response to the disaster, as Kupferman (2003, p. 8) noted.

> To be fair, on September 11 no one could comprehend the full severity and repercussions of what had happened. But as the weeks passed, the agencies' evasions became policy. Medical experts were seeing health effects but could not properly diagnose or help patients because they did not have adequate information. Environmental scientists were expecting to learn the components of the fallout in order to make immediate decisions that would affect cleanup, recovery and future systemic planning. And meanwhile, the EPA and OSHA kept saying there was no problem.[6]

In the midst of the resulting government silence, media filled in the gaps. And some media stories amplified the risk, as when stories inappropriately used short-term exposure measurements with long-term standards for a variety of contaminants. The resulting escalation of public fear in Lower Manhattan (Thurston & Chen, 2002) presumably put agencies under additional pressure to quiet such anxiety.

Second, the broader public was not initially prepared to play a critical role in this environmental decision making that followed the disaster. Even well educated members of the public were not equipped to critically screen government communication, such as the exclusive early attention to asbestos exposure. When asbestos was not detected over operative benchmarks, it paved the way for subsequent claims of safety, discussed below.

The public seemed to be mollified at the time by EPA assurances of no long-term adverse health consequences from downtown exposure despite peoples' direct but contrary experience of air pollution-related health problems. Acrid odors and clouds of soot and smoke poured from active fires for months. And many people were directly affected by "World Trade Center cough," a persistent cough that appeared in people who were at Ground Zero and lingered for a month or more. Other widespread symptoms at the time included raw throats, burning eyes, nose bleeds and intermittent asthma attacks and respiratory problems (NYU School of Medicine, 2006). The anomaly of this cough and other symptoms underscored the inadequacy of government decision rules, occurring despite "safe" and unregulated conditions when ambient air fell within "acceptable levels" for average daily exposures to particulate matter (PM). What caused the unhealthy conditions, although overlooked by the agencies, were the hourly peaks of PM often in excess of the standards (Thurston & Chen, 2002; Chen & Thurston, 2002).

Deliberately Constrained Communication

Perhaps the most serious source of communication error was deliberately constrained government communication.

> Examples of constrained communication by government officials include giving false reassurances, withholding information, responding slowly, taking inadequate actions, mishandling testing, arguing among experts, taking actions secretively, drawing arbitrary boundaries, using insensitive or incomprehensible language, denying citizens' perceptions and fears, hiding behind scientific reasoning for political decisions, and appearing to not know what is going on. There are times, of course, when these techniques are used unintentionally. But they are also used strategically at times to distort an issue. (Edelstein, 2004a, p. 175)

In the WTC case, Kupferman (2003, p. 10) illustrates how EPA committed three of the errors under current examination.

> When EPA began posting a "representative sampling" of air-quality monitoring data on its website, on October 3, the postings involved three grades of filtering of information: first, the EPA had tested narrowly as to location and as to matter tested, as noted above; given that the samples posted were selective and so few, and taken mostly outdoors rather than indoors, they did not give an accurate picture of what people were exposed to. Second, the postings were a selection from the total pool of EPA data available – all of which should have been made available for scientific, health-agency, journalistic and public-health communities to examine. Third, the EPA's explication of the selective data it posted was disingenuous and scientifically misleading, and as with the other filtering this minimized the findings of dangerous toxins.

Distorting Safety

As suggested, the actions of the EPA and New York City in declaring the WTC area environmentally safe are a clear example of institutionally constrained information, inviting public backlash and distrust. In all, EPA issued five press releases within 10 days after September 11, 2001, four more through the end of December and another four through the end of May 2002.

> EPA's WTC press releases from September through December 2001 reassured the public about air quality .... EPA's basic overriding message was that the public did not need to be concerned about airborne contaminants caused by the WTC collapse. This reassurance appeared to apply to both indoor and outdoor air.... EPA's overriding message was that there was no significant threat to human health. (EPA Office of Inspector General, 2003a)

EPA's Office of Inspector General released a highly critical report on August 12, 2003 that targeted EPA's actions in Lower Manhattan, including the fact that EPA's early public statements following the collapse of the WTC towers reassured the public regarding the safety of the air outside the Ground Zero area. The Inspector General closely examined, for example, EPA Administrator Whitman's "air is safe to breathe" announcement of September 18, concluding that the agency lacked sufficient data and analyses to make such a blanket statement. The Inspector General learned from EPA Region 2 officials that the truth of the September 18 statement depended on certain qualifications. It only applied to long-term health, not short run problems caused by the polluted air. It applied to the general public, but not to workers at the WTC site. It referred to effects on healthy adults but ignored the fact that some segments of the public are particularly susceptible and vulnerable, including the ill, the aged and the young. It assumed safety in the outdoor air, but was not intended to imply that indoor air was safe. And it focused only on asbestos when many other air pollutants were of concern (EPA Office of Inspector General, 2003a, p. i). Of course, these qualifications were not shared with the public.

The Inspector General also questioned the reassurance by the head of the Occupational Safety and Health Administration (OSHA) that it was safe for New Yorkers to go back to work in New York's Financial District the beginning of the next week after 9/11.

> We could not conclusively determine the extent to which required work practices regarding the control of asbestos were followed at the WTC site during demolition and debris removal. Since asbestos is a known human carcinogen, EPA has established stringent work practices to control emissions of asbestos resulting from demolition and renovation projects. We found that a significant requirement to reduce emissions in emergency demolitions – wetting damaged buildings before demolition and keeping the waste material wet after demolition – was followed. However, work practices applicable to the transport of debris from the site were employed inconsistently. The specific impact on air quality of any variance from EPA's asbestos emergency work practices is unknown. (EPA Office of Inspector General, 2003a, p. ii)

Political Interference

Distortion of information by EPA and OSHA was not fully explained by the understandable confusion about what to measure, what to report and how much information could be shared without muddying the message. Instead, the Inspector General concluded that this was a documented case of political interference into what could be said.

> The White House Council on Environmental Quality (CEQ) influenced, through its early process, the information that EPA communicated to the public through its early press releases when it convinced EPA to add reassuring statements and delete cautionary ones. (EPA Office of Inspector General, 2003a, p. i)

In particular, the CEQ told EPA not to discuss PM (particulate matter) pollution, not to discuss health effects and not to instruct the public that they needed to obtain "professional cleaning" (EPA Office of Inspector General, 2003a, pp. 14–16).

Lying with Statistics

EPA engaged in statistical manipulations to minimize problems. An example was the deliberate minimization of data showing spikes of asbestos at levels demanding action. Instead of responding to a toxic threat, as federal law would dictate, the agency merely averaged its data to wash out these spikes, allowing it to publicly report that there was no risk to public health. In doing so, it made two specious arguments contradicting its own regulations, that short-term exposure to higher levels of asbestos is not dangerous and that long-term exposure is required for health damage to occur (Kupferman, 2003).

Simplifying Cleanup

The cleanup response to the WTC disaster was reminiscent and perhaps modeled after the Reagan administration's ideological response to geologic radon gas, a significant indoor health threat.[7] The experience with radon gas in the 1980s and 1990s demonstrated the use of "respecting privacy" as an excuse for government to relinquish responsibility for cleanup of private spaces to the building owners and occupants (Edelstein & Makofske, 1998). In New York City, this bias materialized as the focus shifted from rescue to recovery and government focused on cleaning up downtown streets not private living areas. Although justified as a means to prevent toxic dusts from being tracked into buildings off the street, Kupferman (2003) argues that the real purpose was to support commercial interests by getting business back to normal. It probably also served to delineate realms of responsibility.

EPA's proclamations of safety combined with New York City DEP inaction served, on one hand, to make landlords the arbiter of cleanup and, on the other, to get landlords off the hook for doing more than superficial cleaning. Meanwhile, NYCDOHMH protocols issued within a week of the disaster clearly implied that a do-it-yourself amateur cleanup was acceptable. The "Fact Sheet for Cleaning Homes Near the World Trade Center for Re-Occupation" included the following recommendation (NYCDOHMH, 2004a, p. 8):

> How can I clean up dust safely? Proper dust cleanup will prevent dust from getting in the air again. The best way to remove dust from surfaces is to use a wet rag or wet mop (wet wiping). Sweeping the dust dry may cause it to get in the air you breathe.

Referring to this same advisory, Kupferman (2003, p. 6) notes that

> Nowhere in the advisory did the health department inform that the "dust" inside these homes might well contain asbestos and myriad other toxic substances; nor did the protocols suggest that professional testing and de-contamination (otherwise known as abatement) should be sought, for resident's protection and as required by law ....

EPA, however, bore legal responsibility for assuring a thorough cleanup, not only after conventional sources of contamination, but also after acts of terrorism, according to the National Contingency Plan, the National Response Plan and Presidential Decision Directive 62. After receiving authorization and funding in May 2002, EPA implemented its residential test and clean program in the summer of 2002. A voluntary program that made residents and residential building owners responsible to choose testing or cleaning, the initiative depended on winning cooperation. Yet, critics claim it was poorly marketed; too few potential participants understood the risk and knew about the need to cleanup. Nor did they have the help, information and resources necessary to assure that a proper cleanup was implemented (Lioy & Gochfeld, 2002). As a result, the program did not adequately characterize and cleanup up WTC contaminants. By the summer of 2003, EPA had monitored for asbestos in only 4,225 and cleaned only 3,425 apartments of the more than 25,000 apartments below Canal Street (EPA, 2005). It is interesting to note that the subsequent downtown resident survey conducted by the EPA Inspector General found that the fact that relatively few eligible respondents had their homes tested for asbestos or professionally cleaned could not be blamed on a lack of awareness of the need for testing and professional cleaning (EPA Office of Inspector General, 2003a, p. ii). Rather, the pronouncement of no hazard helped remove the motivation, as did the sense that the proposed testing was ineffectual and a hassle. They did not trust turning their apartments over to the lowest bidder. Echoing such criticisms, a subsequent law review article concluded that EPA's efforts to alert residents to clean with wet cloths and inform landlords on how to clean buildings constituted "more than a failure to warn: the EPA was not silent to what was going on, but instead misrepresented the situation at hand" (Hayner, 2005, p. 24).

The Inspector General was broadly critical of EPA's 2002 test and cleanup efforts. The agency watchdog criticized EPA for not ensuring that cleanup of the WTC area meets minimum Superfund site cleanup goals, for failing to treat impacted buildings as a system, to include workspaces as well as residential buildings in the cleanup program and for not including all geographic areas impacted by WTC dust in the cleanup effort. The Inspector General expressed concern that, in its focus on testing and cleaning individual apartments rather than addressing buildings as integrated systems, EPA had failed to provide for proper assessment and cleanup of building mechanical ventilation systems. Cleaned apartments could, accordingly, become recontaminated (EPA Office of Inspector General, 2003a). As we shall see, these issues presage concerns later raised by the EPA's planned second round of testing.

A subsequent FEMA program offering HEPA air purifiers, HEPA filters for the machines and HEPA vacuum cleaners to residences for air purification and cleanup also reached only a small percentage of its potential users, and there were difficulties with its implementation. The uncontrolled program provided little guidance on consumer choices and was easily abused by people who were not likely to have actually been exposed to WTC dust.

Contradictory Standards

The EPA Inspector General has not been the only critical voice on these issues. Scientists questioned the early judgments about safety. EPA Whistleblower chemist Cate Jenkins (2002) wrote a memo four months after 9/11 in which she concluded that asbestos in areas affected by the WTC disaster was as serious a health risk as found in Libby, Montana, where EPA declared a Superfund site based on risk to residents.

> The logical question thus arises: Why is EPA leaving people to their own devices in the cleanup of New York City, while intervening to clean homes at taxpayers expense in Libby because of an "imminent and substantial endangerment to public health"?

Limiting Liability

In a scathing but well researched review of government's handling of the post-disaster environmental hazards, attorney Kupferman (2003, p. 2) blames EPA oversights and distortions on the desire of the agency to limit liability.

> ... it would be overly simplistic to attribute the actions of public health and environmental officials to confusion under pressure, at least after the first few days. It has become evident that federal, state and city agency actions and decisions were closely tied to economic and political motivations that placed other goals ahead of public health, and that these decisions led agencies to withhold critical health information from even the communities most at risk. Through its air-quality testing choices and its management of information to the public, the EPA sought to limit the government's responsibility, and liability, even at the cost of exposing people to health risks they had a right to know about and the option to avoid.

Suppression of Exposure Data

Kupferman (2003, pp. 10, 12) reports on government efforts to prevent the public from knowing the hazards surrounding Ground Zero. Freedom of

Information Act requests for government contaminant exposure data were ignored for extended periods citing the criminal investigation underway. When the Environmental Law and Justice Project undertook its own sampling to discover actionable levels of both asbestos and fiberglass, its reports of significant contamination were labeled as alarmist and the laboratories used to do independent tests of the contaminants received government threats to discourage such work. Kupferman cites other suppressed or delayed results that corroborate the unsafe conditions in the wake of 9/11; the results reveal the presence of not only asbestos, but also dioxins, PCBS, PAHS, heavy metals, radionuclides, CO, VOCS and other extremely dangerous substances. The Environmental Law and Justice Project's publicizing of exposure data helped to force EPA to broaden its responsibilities for interior cleanup and other operations.

Taking stock of such information, in February 2006, the judge considering a toxic tort filed by Kupferman and others on behalf of WTC exposure victims,[8] Federal District Judge Deborah A. Batts, termed former EPA Administrator Whitman's actions "conscience-shocking." She elaborated in her ruling,

> No reasonable person would have thought that telling thousands of people that it was safe to return to lower Manhattan, while knowing that such return could pose long-term health risks and other dire consequences, was conduct sanctioned by our laws.

Extending such criticisms to the remediation, a subsequent law review article concluded that EPA's efforts to alert residents to clean with wet cloths and inform landlords on how to clean buildings constituted

> more than a failure to warn: the EPA was not silent to what was going on, but instead misrepresented the situation at hand. (Hayner, 2005, p. 24)

And attorney Susan Mattei (2004), writing on behalf of the Sierra Club, summarized a trail of deceptions surrounding the WTC disaster that included the following: the federal government's failure to obtain and provide health risk information; EPA's misrepresentation of the meaning of the asbestos test results by knowingly mischaracterizing its own technical detection limits as health standards; the CEQ having provided misleading data to US senators; the failure of EPA to find health hazards "because it did not look for them," including PAHs, asbestos, ultra-fine particles and indoor contamination; EPA's failure a dozen times to correct its assurances of safety even after it had more data; FEMA and EPA's assurances to residents that they could just cleanup contaminated WTC dust themselves; and EPA's failure to give warnings for vulnerable populations.

We will now focus on a number of crucial dynamics that characterize the continuing issues affecting Lower Manhattan in the wake of the WTC disaster. These issues include the unique role thrust on scientists who ended up as proxy to an absent EPA, the controversies surrounding the WTC Health Registry, the redevelopment of Lower Manhattan necessitated by the disaster, concerns over demolition of remaining buildings in the WTC area and the plan to sample the extent of continuing contamination developed by EPA in conjunction with the WTC Technical Review Panel.

THE UNIQUE ROLE OF SCIENTISTS

Thurston and Chen, (2002) notes that distrust of government after 9/11 forced people to turn to alternative sources of information, including others in the community, experts from academia, environmental and medical fields and physicians. They had the expertise and possibly the neutrality to fill in the huge gaps created by government communication. Academic scientists in fact came to play an uncommonly dominant role as an alternative source of information.

The National Institute for Environmental Health Science (NIEHS) funded a group of existing grant recipients to quickly shift to WTC research. The effort began to yield results after three years, often generating the need for additional research. At the point of this writing, five years out, the effort has begun to yield invaluable information about the consequences of the disaster.

At one year and even two years out, however, there was little to report based on this work. Researchers made demands on the public to participate as subjects in their research projects but were unable to return information in a timely fashion. Adding to the delay were the norms of science, namely that it is difficult to publish in a peer reviewed journal if the data were previously released. Residents wanting information about whether to move back home or whether it was safe to remain there not only had to wait for a project to be completed, analyzed, written up and reviewed by authors and institutions, but they had to wait for publication as well. This agonizingly slow pace generated a great deal of impatience on the part of the public. The final insult came from the fact that victims might then merely hear about results on the news rather than hearing from the researchers themselves (O'Brien, 2003). As a result of these accumulating frustrations, members of the science community were readily viewed by residents as promoting research to benefit their own careers. Their participation as subjects in this research became just another form of victimization for residents.

But there was one major bright spot evident. Among the universities funded by NIEHS, New York University played a pivotal role beyond their direct WTC research by virtue of their unique effort to share findings and hold meetings with the public. The importance of these meetings was magnified by the fact that agencies and elected officials did not make themselves available to meet with WTC victims from the neighborhood. President Bush, Governor Pataki and Mayor Giuliani had numerous photo sessions with fire fighters, but they avoided meeting with the residential community. In terms of inclusion, NYU's outreach was the only show in town for Lower Manhattanites.

In its proposal to NIEHS, NYU proposed to hire a community coordinator[9] and hold an annual forum for the purpose of sharing data with the community. Three such events were held, in 2001, 2002 and 2003; each attracted substantial community participation. All recipients of NIEHS funding for WTC projects were asked to appear and share their results; many did. The dynamics of these programs are of interest for what they show about the public's need for information. While the attending scientists did highly professional presentations of their planned work and, by the final forum, of preliminary results, there were evident gaps in the presentations affecting public perception and evident need for considerably more research to address the issues of health and safety for residents.

First, as noted above, there was the fact that the research in question was being conducted on the academic/government clock, reflecting lead time to write proposals, the lag until they were funded and the time it then takes to implement a project to the point of getting results and getting it published. Residents of the community were living in the WTC neighborhood while this work was in progress. As a result, they could hardly avoid seeing themselves as guinea pigs, subjected to possible exposure during a period of research. While respecting the work of the scientists, audience members were not getting what they needed – information about whether their neighborhood and homes and schools were safe then and now. Scientists awaiting long-term data could not answer this question professionally.

Second, also as noted, neither EPA nor the relevant City and State agencies held their own public events. The result was that the NYU programs represented the only available occasions for members of the public to learn about health issues, to raise questions and direct comments. People arrived with pent up frustrations and big expectations fed by the lack of other opportunities to learn and question. Consequently, there was intensive participation by the audience during question periods at the NYU programs.

Third, the EPA and other government agencies declined to appear on the programs of the NYU sessions and were not even present in the audience on several occasions; when agency personnel and lawyers were present, they were there only as observers. Given the absence of a public EPA presence, NYU Professor George Thurston unwittingly became the stand in for the absent agencies. He was the only accessible authority able to address issues of safety and willing to take questions. However, Thurston could not speak for the agencies, and it was outside his role to answer many queries. The resulting frustration at times turned to anger on the part of some program attendees. Lacking any other target, emotions were directed at Thurston. In taking responsibility for communicating with the public, Thurston was exposed to challenges that rightfully belonged to government agencies that shirked their duty. Nevertheless, Thurston stood tall while government ducked. He ended up being the lightning rod for displaced concerns that had no appropriate avenue for expression. In the wake of this horrible disaster, government was absent.

By the final session, in fall 2003, the public's frustration boiled over when a spokesperson for the City Department of Health finally appeared. She was there to discuss the WTC Health Registry created for the purpose of tracking the health of those exposed to WTC hazards. This public appearance was necessitated by the public's initial reticence to participate. It quickly became clear that members of the public did not see how the registry would address their health concerns. They did not like how it was structured, how it would work and the fact that it was not attached to any resources to help them if they fell ill. The issues surrounding the Health Registry deserve their own discussion. But it is noteworthy that the NYCDOHMH piggybacked onto the NYU program rather than holding their own event.

Our first case highlights that the failure of government to respond to the public's need for information directly or in a trustworthy fashion created a need for another source of expert information. The shortcomings did nothing to obscure the value of the NYU/NIEHS effort. Based on this model, for future disasters, one can foresee mobilization of the expert community to carry out short- and long-term data collection and analysis aimed at creating public informed consent.

WORLD TRADE CENTER HEALTH REGISTRY

As of April 2006, millions of federal dollars have been allocated by the U.S. Centers for Disease Control and Prevention (CDC) to treat rescue and recovery workers affected by the WTC disaster. For example, in March 2006 CDC announced a plan to allocate nearly US$ 75 million, a large but inadequate sum, to treat those sickened by the toxic air and dust at Ground Zero. It is striking, in contrast, that no federal money has been allocated to screen and treat World Trade Center area residents or those who go to school or work nearby. Downtown residents were upset that they have been completely ignored, again.

In September 2003, the World Trade Center Health Registry (henceforth, "the Registry") was created by the NYCDOHMH to track and evaluate any long-term health effects of 9/11. By the end of 2005, 20% of those registering lived south of Canal Street on 9/11 (14,725 of 71,609 enrollees). This number comprises one-quarter of the 58,562 individuals living below Canal Street according to the 2000 U.S. Census. If we focus more specifically on residents living south of Chambers Street on September 11, 2001, one-third (7,093 of 22,000 residents) enrolled in the registry (NYCDOHMH, 2005a, pp. 7, 11).

Registry Limitations

The Registry only collects data on those that enrolled in their program. It is not known what self-selection bias and omissions are created by this need for the individual to take initiative to register. Furthermore, there are restrictions on eligibility for the Registry (NYCDOHMH, 2006a), limiting participation to people who were in a building, on the street or on the subway south of Chambers Street on 9/11/2001; those involved in rescue, recovery, cleanup or other activities at the WTC site and/or WTC recovery operations on Staten Island any time between 9/11/2001 and 6/30/2002; and students and staff enrolled in schools (pre-K-12) or day care centers south of Canal Street on 9/11/2001. Residents outside of the demarcated boundary were ineligible to enroll. The assumption that health effects from the WTC disaster can be bounded to this degree is questionable. For example, there is already evidence of increased asthma rates among children with pre- and post-9/11 medical records in an area of at least five miles from Ground Zero (Szema et al., 2004).

There are other evident shortcomings of the registry that also deserve mention. First of all, there are the inherent shortcomings of the registry concept. The Registry is simply a data base created from those completing a health registry survey. It is based on self-reported data. Until recently, it offered no medical exam, medical treatment or assistance with problems (now a referral guide is available). It does not address the current or future unmet health needs of residents. Moreover, the validity of the data can be questioned. The Registry is based on self-reported data. The first data came from enrollees who answered a 30-minute telephone survey inquiring where they were on 9/11/2001 and their reported health status. A second survey was launched during the summer of 2006. Finally, the utility of the data can be questioned. The huge delay between information collection and eventual public release of findings means that the results will come too late to influence public policy decisions, such as allocation of funds to screen and treat residents. Residents also question the objectivity of the Registry because of the earlier actions of its administrator, the New York City Department of Health and Mental Hygene, for example, for giving overly simplistic advice to residents after the disaster.

Emerging Evidence

Despite its shortcomings, the Registry has already revealed some interesting data about the health of downtown residents. Most Registry enrollees reported new or worsened respiratory symptoms after 9/11/2001 (NYCDOHMH, 2004b). And, although non-respiratory problems were less prevalent than respiratory problems after 9/11, two new or worsening non-respiratory symptoms [heartburn, indigestion or reflux (23.9%) and severe headaches (21.0%)] had a prevalence exceeding 20% (CDC, 2006).

Two-thirds of the enrolled residents that were south of Chambers Street on 9/11 reported that they were caught in the dust/debris cloud (NYCDOHMH, 2004a, p. 8). We can therefore infer that downtowners share the substantial physical and mental health effects reported by WTC building survivors who were caught in the dust cloud. These survivors were two to five times more likely to report physical or mental health problems after 9/11 (NYCDOHMH, 2006b).

Likewise, it is reasonable to assume that downtown residents share other impacts identified in the overall WTC health literature. On September 8, 2004, the Government Accountability Office (GAO), released a report, "September 11 Health Effects in the Aftermath of the World Trade Center Attack." The report described a broad and continuing health impact, based on the evidence of thousands of people involved in the rescue, recovery and cleanup, as well as those who lived and worked in the WTC area (GAO, 2004).

Other information supports health concerns for downtowners. In April 2006, a former NYPD detective's death was linked to his exposure to the

9/11 toxic cleanup, according to a New Jersey medical examiner (NY1, 2006). The effect of such information on how downtown residents view the future was expressed by Community Board One member John Fratta. "We've become the new victims of 9/1 .... Down the road, it's going to come out that we were all the victims of 9/11" (Kaysen, 2006).

Finally, a Pace University poll taken in summer 2006 reported that residents' perceptions of the environmental cleanup dropped from a positive rating in 2003 (61%) to a decidedly negative rating (32%). The shift in these numbers has been occurring steadily over the intervening two years. What is striking from our perspective is that nearly the same percentage (59%) now believe that they face a long-term health risk from 9/11 (Trichter & Paige, 2006).

Strengthening Health Response

On May 17, 2005, Community Board One unanimously passed a resolution supporting post-9/11 health screening to track and treat current and emerging September 11-related disease for Lower Manhattan residents, office workers, students and staff. In February 2006, it further unanimously supported the appointment of a federal September 11 Health Czar to coordinate the federal government's response to short- and long-term adverse health effects resulting from September 11 (WTC Redevelopment Committee, 2006). Such a post was subsequently created with bipartisan Congressional support.

In April 2006, members of the community met with the 9/11 Health Czar and made the following request (Community Board One, 2006a):

> The Federal government must assume responsibility for this [the short and long term adverse health effects resulting from September 11] and should allocate the necessary federal funds to ensure that all Lower Manhattan residents, workers, students and responders have reasonable access to health and mental health monitoring and treatment.

The World Trade Center Disability Law was signed on June 14, 2005 to provide just such coverage for eligible New York City employees who worked at one of the four specified WTC sites for at least 40 hours between September 11, 2001 and the first anniversary of the disaster or who were disabled by the disaster. Thus, fire, police, sanitation department and other WTC workers have coverage for conditions not evident on their pre-disaster health checkups (NYCDOHMH, 2005b, p. 3).

However, as of May 2006, residents, office workers, students and staff in Lower Manhattan still do not have access to free post-September 11 medical screenings or proper physical medical treatment, with the result that some people may be undiagnosed for illness or may be receiving inadequate medical care. Their sole government assistance is their eligibility to participate in the WTC Health Registry. In addition, there is a very limited program coordinated by the 9/11 Community Health Initiative of the Beyond Ground Zero Network at Belleview Hospital. The World Trade Center Health Impacts Treatment Program was funded by the Red Cross for the uninsured and low income and has a very long waiting list.

REDEVELOPMENT OF LOWER MANHATTAN

Then there is the issue for Lower Manhattan residents of rebuilding their devastated neighborhood. In the wake of the disaster, New York State Governor Pataki and New York City Mayor Giuliani created a joint city and state corporation, the LMDC, to assist in the planning and coordination of the rebuilding and revitalization of Lower Manhattan (i.e., below Houston Street). Despite its promised commitment "*to an open, inclusive, and transparent planning process in which the public has a central role in shaping the future of Lower Manhattan*" (LMDC, 2006a), community trust for the LMDC has been tested on the following issues.

First, as a joint City–State Corporation governed by a Board of Directors jointly appointed by the Governor and Mayor (LMDC, 2006a), LMDC is controlled by political appointees attuned to the election cycle. The complex influences on the agency were evident in the contentious vision for the site. The Governor insisted that a tall office building, the "Freedom Tower," be included in the redesign for the site. Many have viewed the project as more reflecting the Governor's presidential aspirations than local needs. Given post-9/11 fear of working in tall towers, the space will likely be filled by government agencies rather than commercial interests. Moreover, the name "Freedom Tower" makes the project into an inviting new target. Indeed, security concerns for the building and other issues of control over the project delayed construction from commencing until the end of April 2006, stalling redevelopment.

During this interval, in conjunction with other agencies, LMDC prepared a general plan for a 9/11 memorial as well as for the overall redevelopment of the WTC site, including preparation of a Generic Environmental Impact Statement (GEIS). As both authors observed, there were numerous inadequacies in the GEIS. The subsequent hearings, the first public government meetings on the WTC, were contentious as a result (LMDC, 2006b). Many residents felt that their concerns were not being represented by the LMDC.

The promised open and transparent consultative process was abrogated on many occasions in the redevelopment planning process. In one instance, a major project change to drop a planned museum, the International Freedom Center, was made unilaterally by the Governor minutes before a scheduled public presentation to Community Board One. Such surprises led the community board to pass a resolution in October 2005 demanding an open collaborative process and consultation on all substantive changes in the Master Plan (Community Board One, 2005).

The community also mobilized to pressure LMDC to address the continuing threat of air contamination. As a result, limited air monitoring data are collected at various locations and are now made publicly available on the website of the Lower Manhattan Construction Command Center (LMCCC, 2006).

Finally, the matter of demolition of various ghost buildings in the neighborhood, discussed below, remains a major sticking point with the LMDC and is raised at every opportunity.

DEMOLITION OF DAMAGED BUILDINGS

Even as concerns for exposures during the months after 9/11 and the cleanup of residences have continued, the threat of recontamination emerged as a significant concern almost five years after 9/11. The issue involves several larger standing but damaged buildings around the WTC site that have not yet been demolished or abated but which represent serious sinks for WTC dusts. The public fears that careless demolition of these buildings will again recontaminate Lower Manhattan with toxic WTC dusts (Kaysen, 2006).

Very early in EPA's WTC Expert Technical Panel Process, discussed below, the WTC Community–Labor Coalition expressed concern about unmet public health needs including exposures related to on-going demolition and construction. As the community representative to the EPA Panel, the second author asked EPA to become lead environmental agency for such projects. She also asked the agency to take the oversight role and develop a plan to protect against WTC-related contamination from on-going demolition, renovation and construction activities and for the control and minimization of new contamination problems (Hughes, 2006).

The most celebrated of the ghost buildings is the Deutsche Bank building at 130 Liberty Street, a 44-story building acquired by the LMDC on August 31, 2004 for the purposes of removal. The decision by Deutsche Bank to scrap the building had involved a debate between the corporation and its insurers over long-term liabilities from its restoration. The outcome is consistent with the Precautionary Principle embraced by Germany. This presumption of danger rather than safety has been given little credence in the U.S. and was certainly not evident in the planned actions of the LMDC to clean and demolish the building.

The need for precaution during demolition has become a central community objective. Responding to community pressure, in April 2006, EPA denied permits for LMDC's proposed demolition plan for the Deutsche Bank building pending improved safe handling of demolition dusts (EPA, 2006a). EPA has yet to approve the demolition plan submitted by LMDC for Fiterman Hall, at 30 West Broadway, a 15-story building formerly used by Borough of Manhattan Community College viewed as a toxic eyesore by residents (EPA, 2006b). Additional buildings awaiting approved demolition are 133–135 Greenwich, a two-story retail/commercial space, and 21–23 Thames Street, a five-story former residential apartment building (EPA, 2006c). A 12-story building located just south of the WTC site, 130 Cedar Street, awaits EPA approval for a combination of abatement and limited deconstruction/demolition (EPA, 2006d).

Additional attention has gone to buildings not seriously damaged by 9/11 but slated for demolition to make way for community redesign projects as part of the overall reconstruction of the neighborhood. In particular, the Metropolitan Transit Authority plans to take down a two-story building at 189 Broadway and four other nearby buildings for the new Fulton Street Transit Hub. In response to community pressure, the authority agreed to demolish the buildings using EPA guidelines that assume it is contaminated with WTC dust.

Despite this progress, doubts over precaution forced Community Board One to again take a stand on the demolition in April 2006. The impetus was a report in the *New York Daily News* that the two contractors hired by LMDC to demolish 130 Liberty Street had ties to organized crime and a poor safety record (Smith, 2006). In a unanimous resolution regarding the demolition of 130 Liberty Street, the board reiterated its previous demands that demolition activities take precaution for the safety and well-being of local residents and workers and called for work to be done by responsible and qualified contractors, following EPA guidelines and under active EPA supervision (Community Board One, 2006b).

THE EPA WORLD TRADE CENTER EXPERT TECHNICAL PANEL

The potential for continuing WTC-caused environmental hazards has also been a source of on-going community concern. Citing EPA's abysmal record in keeping the public informed about safety, in March 2004, Senator Hillary Clinton of New York forced EPA to convene a scientific advisory panel. The panel was charged with identifying the scope of any remaining environmental hazards to New York City residents and their unmet public health needs. EPA was also to recommend steps to further minimize risks associated with the aftermath of the WTC attacks.

While the panel was initially greeted as a major victory for Lower Manhattan residents, it quickly became apparent that the EPA panel was not the promised scientific discussion of how to best survey and sample for WTC contaminants of concern. There was an apparent political agenda aimed at finding as little as possible. Within the panel's membership, a very clear distinction was evident between the activist (labor and community) perspective and that of most (but not all) of the scientists, whose aims were not applied to the practical task as hand. The public quickly came to define the panel process as a threat to the cleanup of their community rather than a means to that end.

The panel was tasked with identifying the spread of WTC contaminants and pointing to any needed remediation. In considering its approach to this task, the panel was quickly sidetracked by the issue of how to identify WTC dusts from other dusts. The majority of the panel agreed that this question had to be resolved before they could consider how to sample for dusts and subsequently approve a protocol for testing.

The Search for a WTC Signature

From the onset, a major pre-occupation of EPA and its scientific panel was the identification of a "WTC signature," a scientifically valid fingerprint of WTC contamination. The concept had its own internal logic. If the task of the panel was to identify the extent of continuing exposure to WTC contaminants, it had to first identify how far WTC dusts were carried during the disaster and the resulting fire. If this was to be done by environmental sampling, it was necessary to be able to identify the particular characteristics of WTC dust that would allow it to be identified in an environment full of dusts from multiple sources not caused by the WTC disaster. The obsession with finding the signature effectively became a Holy Grail for the panel, distracting it from other considerations. It reflects a fundamental policy error, directing the panel to focus on the wrong questions.

In its early considerations, the panel relied on asbestos as the likely signature. But as multiple sources of ambient asbestos came to light, the WTC marker of choice became the slag wool used as insulation in the WTC building. The presence of slag wool came to be considered the defining characteristic distinguishing WTC from non-WTC contaminants.

Even as EPA embraced slag wool as the long sought marker, however, it was not able to answer a logical question posed by critics. Because slag wool is a relatively heavy substance, it is likely not to have traveled as far from Ground Zero as would lighter contaminants. As a result, a slag wool-based methodology risked false negative findings. It would be warranted in assuming that contaminated samples bearing slag wool were a result of the WTC disaster. But it would be invalid to conclude that the absence of slag wool meant the absence of WTC dust. Furthermore, the reliance on slag wool would introduce a sampling error. By implication, the slag wool samples would be closer to Ground Zero, limiting the scope of the area that the study would identify as contaminated from the WTC disaster. The fact that contaminants had dispersed over a much larger area would be entirely missed. As a result, contamination of these outlying areas would likely not be addressed. Finally, the reliance on a marker using a building material from the twin towers restricted the search to WTC contaminants associated with the initial explosion, fires and collapse of the buildings. It might not represent contaminants resulting from the subsequent months of continuing fire and smoke.

Complex issues of underlying value are exposed in this debate. Thus, while the public's *precautionary* concern was that the use of a slag wool signature would introduce false negative results, a different concern drove agency and scientific thinking. These *cautionary* thinkers were focused on minimizing false positives. EPA did not want to confuse WTC contaminated buildings with those perhaps equally contaminated, but from a different source than the WTC disaster (Edelstein, 2004a). And it did not want to create an even wider liability than it might be given from its sampling, a responsibility to cleanup toxic hazards originating from sources other than the WTC.

On first blush, this appears to be simply a question of scientists concerned with accurately measuring pollution dispersal from the WTC disaster. In tolerating false negatives, the goal of the panel becomes clearly focused on avoiding false positives. This is the general bias of pure science concerned with theory testing. However, the panel exists in a very different applied context. The goal here should be the protection of public health and safety, where a shift to a precautionary stance would be appropriate. The signature focus thus represents a confusion of the pure and applied.

The signature question highlights potential underlying errors of policy and, by extension, communication. The cause is the confused and contradictory mission of the panel. Was its goal to discover a scientifically valid method for contaminant signature? Or was the goal to implement a plan to identify the extent of the contamination from the WTC disaster, determine if risks continue and, if so, to mitigate those risks? On the surface, these are compatible goals. The signature becomes the means for identifying the extent of contamination and thus the area requiring mitigation. In practice, they are at loggerheads. Difficulty agreeing on the first or agreeing on a sufficiently limiting approach serves as an excuse for never doing the second.

But this may not be merely a question of the cautionary norms of pure science. As noted above, the signature issue may have less to do with determining a valid means of identifying dispersal of WTC contaminants than with EPA's desire to limit liability and cost for government cleanup. The greater the area identified (falsely or correctly) as contaminated from the WTC disaster, the greater the potential scope of a remediation required. The agency, and the government it represents, thus has a vested interest in shrinking rather than expanding the scope of identified impact. As a result, precaution is, literally, thrown to the wind (see Edelstein, 2004a).

In either case, the focus on finding a WTC signature complicates achieving the goal held by the community, namely to have a clean, hazard free and safe place to live and work. The public position in the matter is clear. And, as the first author testified to the EPA panel, the signature interlude did not serve the public's goal. A precautionary sampling plan could be implemented to identify contaminated buildings and units and target them for cleanup without concern for the source of the contaminant. It could start at Ground Zero and move outward in concentric circles until a point of insignificant dust was encountered. Even if non-WTC dusts were confused with WTC dusts in doing so, public health would be maximally served (Edelstein, 2004b; Edelstein, 2005).

Sampling for WTC Dust

EPA's final "Test and Clean Program" was released in November 2005. The plan stated that Lower Manhattan residents could have their homes tested for four contaminants (asbestos, MMVF, PAHs and lead) and if high levels were found, then the apartment would be cleaned for free. Elected officials and residents had serious concerns with details of the plan. Consequently, Senator Clinton and Representative Nadler and others called for an investigation by the U.S. Government Accountability Office (GAO) into EPA's efforts to establish an effective testing and cleanup plan.

Community representatives were quick to note that the plan has numerous shortcomings. The program was grossly underfunded, inadequate to meet the need, technically and scientifically flawed and was set to repeat many of the errors identified in the Inspector General's Report on the earlier 2002 program. Some of the crucial shortcomings deserve reiteration.

First, the EPA plan retained obvious errors in geographic sampling. It limits sample collection to the area south of Canal Street and west of Pike and Allen Streets, overlooking the potential for continuing hazards in areas known to have been impacted by the dispersal of dust and smoke after 9/11 that stretch further north and east in Manhattan and encompass parts of Brooklyn. The expanded sampling area had received strong support on the expert panel from scientists living in Brooklyn who personally attested to the fact that WTC dust and smoke reached their neighborhoods. Likewise, the EPA Inspector General's survey indicated that contamination from the collapse of the WTC towers spread into the homes of respondents located beyond the perimeter of the zone designated as eligible for the EPA-led testing and cleaning program (EPA Office of Inspector General, 2003b).

Second, this sampling error was further confounded by limiting the types of units sampled within the chosen boundaries. Workplaces, schools, small businesses and firehouses were excluded, even though these settings offer the potential for sustained human exposure. Thus, EPA could fail to assess the extent of remaining contamination in buildings and areas known to have been impacted. The result would be continued exposures.

Third, the sampling error is compounded by EPA's plan to address individual apartments rather than understanding and addressing buildings as integrated systems. Furthermore, the agency does not intend to provide for proper assessment and cleanup of building mechanical ventilation systems, plenums, closets and other "hot spot" areas known to be reservoirs for contamination. As a result, the sampling plan risks a failed cleanup and exposure of residents to a continual health threat.

Fourth, the planned testing relies on passive air sampling. An air sample would be grabbed from the living unit without disturbing dusts where they had collected. As labor and community members on the panel were quick to note, such passive sampling would not assess how much dust becomes airborne during the activities of daily life. A passive sample might well show little or no airborne WTC contaminant when, in fact, the living space was periodically rich with pollutants after they were disturbed. In short, the use of passive sampling might result in a real hazard being overlooked, another source of false negatives. The alternative, active sampling, applies negative pressure to the living space to stir up the contents. EPA ruled this approach out because it is so disruptive, requiring occupants to leave their homes for a period of several hours, that many people might not volunteer. There was also concern that stirring up contaminants through active sampling might recontaminate the residence, placing occupants at greater risk than if there was no cleanup. The issue of passive sampling again exposed the basic dilemma of EPA's tolerance and even preference for false negatives.

Finally, there are additional errors due to the failure to use appropriate sampling methodologies and protocols for actual dust sampling and a failure to specify properly sensitive detection limits that may cause a measurement error. Communication error is tolerated by the further failure to provide for an independent monitor on behalf of the affected community to raise just such concerns as the above. In sum, then, despite a great deal of scientific rhetoric, EPA and its advisors appear ready to tolerate errors that will impede the identification of the very hazards whose discovery is the very purpose of the study.

Our analysis suggests that EPA's plan was designed to find as little contamination as possible through the principle of exclusion. First, the plan excludes testing and/or cleanup in areas of buildings most likely to be reservoirs of contamination. And the plan is drastically skewed toward using results from cleaner areas to trigger cleanups. Half of the samples taken in an apartment that will be used to determine whether a cleanup is offered for that unit will be taken from "accessible" locations such as table or counter tops likely to have been cleaned repeatedly in the past five years. Whether or not an HVAC unit is cleaned will also be based, in part, on the results of the samples taken from clean, accessible locations in individual apartments.

Mitigatory Gaps

In his testimony to the EPA Technical Panel in the morning of October 5, 2004, the first author warned the panel of the potential for a "mitigatory gap," where the public would learn of their exposure to contaminants, but no remediation, a delayed remediation or a faulty remediation would follow. The day's panel discussions reinforced these concerns. During the final public participation segment, therefore, he asked the Chair of the panel point blank what plans had been made to mitigate apartments found to be contaminated through the sampling plan. Chairman Gilman refused to answer, responding instead that he would speak to the author privately after the meeting, which he did. That brief encounter yielded a quick answer to the question. There were no plans in place or money allocated to actually mitigate residences found through the sampling program to have levels of concern of WTC contaminants. This response was disseminated to activist groups and subsequently dominated public concern during a period when the Chairman abruptly resigned (Edelstein, 2004c).

The End of the Panel

In December 2005 EPA terminated the WTC panel meetings after the 12th technical meeting. By this point, the original split between public-focused committee members, on one side, and scientific and agency-focused members, on the other, had disappeared. Now the panel was largely arrayed against EPA.

Most of the panel members thought that EPA had prematurely ended the process. Many felt that the revised proposed EPA Test and Clean Up Program Plan failed to fully consider the panel's comments, that it was an inappropriate plan in need of reconsideration and that the plan was doomed in any case by its lack of community support. After a 21-month deliberation process, EPA's proposed plan did not receive the endorsement or support of a single panel member at the final December 13, 2005 meeting. Pushed by this uniform dissent, EPA indicated that it would continue to explore the WTC dust signature during the sampling, but not make it part of its decision-making process. The program, according to EPA, would remain voluntary (EPA, 2005, pp. 3–4).

In February 2006, the 50 individuals comprising Community Board One, who live and work around the WTC site and the immediate area, passed a resolution calling for the New York City Council to ask EPA

> to abandon its technically and scientifically flawed 2005 Test and Clean Program, and work with the residents and workers, community and labor organizations and elected officials to design and implement an effective, science based sampling and cleanup program for residences and workplaces in all affected areas. (Community Board One, 2006a)

These same concerns were later reinforced through a resolution introduced by New York City Council Member Alan Gerson, whose district covers the WTC area. His resolution passed unanimously in May 2006.

As of May 2006, no final EPA sampling plan has been shared with the public, and no second round of interior testing and cleaning has taken place.

If EPA's actual goal was to delay and scuttle further testing and cleanup in the Lower Manhattan community, it had succeeded.

CONCLUSION

The story of the EPA Technical Advisory Panel is reminiscent of the Thomas Commission convened to review the health data for Love Canal. As described by Levine (1982), the Thomas Commission was created and managed to distort the facts and minimize the recognized hazards, and thus the required mitigation. The parallels were interesting because the WTC panel was set up to meet in public, with community representation and with public participation. Moreover, its membership shifted against EPA's goals. How could this comparatively open process end up mimicking a closed one? In the end, the panel was only advisory. And while it may have done little to further the urgent goals of testing and cleanup, it served a blatant practice of the use of distorted communication by government.

In its entire scope, this analysis of the dynamics of post-9/11 decision making demonstrates a range of policy and communication errors that unfortunately contribute to the continuing disaster in Lower Manhattan. This disaster is less from the terrorist acts perpetrated on the World Trade Center. Its source, instead, is the dysfunctional and distorting approach taken by EPA and others in the Bush Administration to minimize recognition of persistent problems. By virtue of their approach to addressing the disaster, the repercussions of such policy choices become impacts themselves. These secondary impacts may have more to do with the inability of the community to recover from the disaster than even the issues associated with the original disaster event itself (Edelstein, 2004a; Couch and Mercuri, this volume). Thus, the acute impacts of the events that toppled the World Trade Center towers and damaged many of the surround buildings become protracted and magnified by the chronic and pervasive consequences.

The Katrina disaster adds one final postscript to this paper. Media coverage and public revelations after Katrina thoroughly exposed the U.S. government's failures in addressing the disaster. In response, massive national mobilizations were under way. Yet, as the long task of envisioning how to recreate New Orleans proceeds, one might recall Kupferman (2003, p. 3) words that, in the wake of 9/11,

> ... the nation began an outpouring of affection for the city that had perhaps not been seen since the Giants won the pennant in '51. Whatever our grief was locally – and it was great – we in New York sensed that we were not alone.

It is doubtful that any New Yorker could think those thoughts today. Disasters receive protracted attention and its victims support for only so long. And then attention is captured by whatever unfortunate event is next in queue. Assistance and support, testing and remediation, are a challenge in the short run. Over time, in no small part, it is the victims who are left with the carnage – physical, social and psychological.

NOTES

1. Kupferman's Environmental Law and Justice Project took the lead in filing a lawsuit on behalf of downtown residents and workers against the federal government for failing to warn and protect people downtown.

2. The New York stock is actually several blocks away; American Stock Exchange is one block away from Ground Zero.

3. LMDC's web site, discontinued link.

4. An award winning website to market Lower Manhattan was funded by LMDC. A successor site can be found at www.lowermanhattan.info

5. http://www.renewnyc.com/FundingInitiatives/HUD%20QR%202003-07-30%20 submitted%20to%20HUD.htm

6. Given the overload of contaminants, some residents interpreted the non-detect results to mean that sampling devices had become clogged, skewing the results.

7. The radon threat message, in contrast, was extremely blatant. As a naturally occurring and therefore blameless hazard, EPA was not under the same pressure to mince words about radon as it was with the WTC. In fact, the ability to blame nature for a major hazard was an attractive feature of geologic radon, just as the ability to blame terrorists became the overriding motif for the WTC disaster.

8. The judge's comments occurred in the context of a decision allowing the class action law suit to proceed that seeks to hold Whitman and others responsible for failing to warn people of hazardous substances in the air following the collapse of the WTC and then failing to carry out an adequate cleanup of building interiors.

9. The second author served in this role. The section is primarily based on the first author's observations.

REFERENCES

9/11 Environmental Action. (2006). http://www.911ea.org/Who_We_Are.htm, visited on June 3, 2006.

CDC (Centers for Disease Control). (April 7, 2006). Surveillance for World Trade Center health effects among survivors of collapsed and damaged buildings. *Morbidity and Mortality Weekly Report* 55(SS2), pp. 1–24, at http://www.cdc.gov/mmwr/PDF/ss/ss5502.pdf

Chen, L. C., & Thurston, G. (2002). World Trade Center cough explained. *The Lancet Supplement*, *360*(December), S37–S38.

Clarke, L. (Ed.). (2003). Terrorism and disaster: New threats, new ideas. In: *Research in social problems and public policy* (Vol. 11). New York: Elsevier.

Community Board One, Manhattan (2005). Meeting of October 2005, retrieved December 24, 2006 from http://www.cb1.org/pdf/cb1_minutes_oct05.pdf

Community Board One, Manhattan (2006a). *WTC redevelopment on EPA's 2005 final test and clean program*. Resolution of February 21, 2006 at http://www.cb1.org/

Community Board One, Manhattan (2006b). Resolution of April 2006 at http://cb1.org/

Edelstein, M. R. (2004a). *Contaminated communities: Coping with the impacts of residential toxic exposure* (2nd ed.). Boulder, CO: Westview Press.

Edelstein, M. R. (2004b). Comments to the EPA World Trade Center Technical Advisory Committee. Borough of Manhattan Community College, July 26.

Edelstein, M. R. (2004c). Comments to the EPA World Trade Center Technical Advisory Committee. Borough of Manhattan Community College, October 5.

Edelstein, M. R. (2005). Comments to the EPA Technical Advisory Panel. July 16.

Edelstein, M. R., & Makofske, W. A. (1998). *Radon's deadly daughters: Science, environmental policy, and the politics of risk*. Latham, MD: Rowman Littlefield.

EPA. (September 13, 2001). Press release, attached to EPA Office of Inspector General (2003) as Appendix H, p. 87.

EPA. (September 18, 2001). Press release, attached to EPA Office of Inspector General (2003) as Appendix H, p. 77.

EPA. (September 21, 2001). *NYC monitoring efforts continue to show safe drinking water and air*. Press release found at http://www.epa.gov/wtc/stories/headline2_092101.htm

EPA. (2005). WTC Panel summary report at http://www.epa.gov/wtc/panel/pdfs/12-13-05_WTC_Summary_Report_2-2.pdf

EPA. (2006a). *Demolition of 130 Liberty*. At http://www.epa.gov/wtc/demolish_deconstruct/130liberty.htm

EPA. (2006b). *Demolition of 30 West Broadway*. At http://www.epa.gov/wtc/demolish_deconstruct/30wbroadway.htm

EPA. (2006c). *Demolition of 133–135 Greenwich*. At http://www.epa.gov/wtc/demolish_deconstruct/133_135greenwich.htm

EPA. (2006d). *Demolition of 130 Cedar*. At http://www.epa.gov/wtc/demolish_deconstruct/130_cedar.htm

EPA Office of Inspector General. (2003a). *Evaluation report: EPA's response to the World Trade Center Collapse: challenges, successes, and areas for improvement*. Report 2003-P-00012. August 21.

EPA Office of Inspector General. (2003b). *Survey of air quality information related to the World Trade Center collapse*. Report No. 2003-P-00014. September 26. online at http://www.epa.gov/oig/reports/2003/WTC-survey_20030926.pdf

Government Accountability Office (GAO). (September 8, 2004). *September 11 health effects in the aftermath of the World Trade Center attack*.

Hayner, E. K. (2005). Failure to warn and the federal Tort Claims Act: An overview of potential claims of Lower Manhattan Residents in the wake of the World Trade Center disaster. *Albany Law Environmental Outlook Journal*, *10*, 115–183.

HUD (Housing and Urban Development, July 2003). Retrieved from http://www.renewnyc.com/FundingInitiatives/HUD%20QR%202003-07-30%20submitted%20to%20HUD.htm

Hughes, C. M. (2006). *Community report to the EPA Expert Panel*. At http://www.epa.gov/wtc/panel/pdfs/hughes-20040726.pdf

Jenkins, C. (January 11, 2002). *Preliminary assessment (memorandum)*. USEPA, retrieved June 3, 2006 from http://www.911digitalarchive.org/objects/104.pdf

Kaysen, R. (April 21–27, 2006). More EPA concerns. In: *Downtown Express* (p. 1). http://www.downtownexpress.com/de_154/moreepaconcerrns.html

Kupferman, J. R. (2003). The public health fallout from September 11: Official deception and long-term damage. In: C. Brown (Ed.), *Lost liberties: Ashcroft and the assault on personal freedom*. NY: The New Press.

Levine, A. (1982). *Love Canal: Science, politics and people*. Lexington, MA: DC Heath, Lexington Books.

Lioy, P. J., Weisel, C. P., Millette, J. R., Eisenreich, S., Vallero, D., Offenberg, J., Buckley, B., Turpin, B., Zhong, M., Cohen, M. D., Prophete, C., Yang, I., Stiles, R., Chee, G., Johnson, W., Porcja, R., Alimokhtari, S., Hale, R. C., Weschler, C., & Lung, C. C. (2002). Characterization of the dust/smoke aerosol that settled east of the World Trade Center (WTC) in Lower Manhattan after the collapse of the WTC 11 September 2001. *Environmental Health Perspectives, 110*, 703.

Lioy, P., & Gochfeld, M. (2002). Lessons learned on environmental, occupational and residential exposures from the attack on the WTC. *American Journal of Industrial Medicine, 42*, 560–563.

LMCCC. (2006). *Air monitoring for Lower Manhattan*, at http://www.lowermanhattan.info/health/airquality.asp#LMCCC

LMDC. (2006a). *About us*, at http://www.renewnyc.com/AboutUs/index.asp

LMDC. (2006b). *Planning design and development*, at http://www.renewnyc.com/plan_des_dev/default.asp

Mattei, S. (August 2004). *Pollution and deception at Ground Zero: How the Bush administration's reckless disregard of 9/11 toxic hazards poses long term threats for New York City and the nation*. New York: Sierra Club.

NY1 News. (April 28, 2006). *Former NYPD detective's death linked to 9/11 cleanup*, at http://www.ny1.com/ny1/content/index.jsp?stid = 1&aid = 58553

NYCDOHMH. (2004a). *World Trade Center report for 2004*, at http://www.nyc.gov/html/doh/downloads/pdf/wtc/wtc-report2004-0430.pdf

NYCDOHMH. (2004b). At http://www.nyc.gov/html/doh/html/press_archive04/pr151-1122.shtml

NYCDOHMH. (April 2004c). *World Trade Center health registry data snapshot: Understanding the health impact of 9/11* (Vol. 1, No. 3, p. 8).

NYCDOHMH. (2005a). *World Trade Center report for 2005*, at http://www.nyc.gov/html/doh/downloads/pdf/wtc/wtc-report2005-0131.pdf

NYCDOHMH. (2005b). *World Trade Center report for 2005*, at http://www.nyc.gov/html/doh/downloads/pdf/wtc/wtc-report200507.pdf

NYCDOHMH. (2006a). *About the registry*, at http://www.nyc.gov/html/doh/html/wtc/about.html

NYCDOHMH. (April 7, 2006b). *Substantial physical and mental health effects reported by WTC building survivors*, at http://www.nyc.gov/html/doh/html/pr2006/pr021-06.shtml

NYU School of Medicine. (2006). *The World Trade Center pollutants information*. Department of Environmental Medicine, at http://www.med.nyu.edu/environmental/research/wtc

O'Brien, E. (August 12, 2003). Moms question W.T.C. baby study findings. In: *Downtown Express* (p. 18), retrieved from http://www.downtownexpress.com/DE_16/momsquestionwtc.html

Smith, G. B. (April 16, 2006). Mob link eyed in bank demolition. In: *New York Daily News*, at http://www.nydailynews.com/news/local/v-pfriendly/story/409209p-346370c.html

Szema, A. M., Khedkar, M., Maloney, P. F., Takach, P. A., Nickels, M. S., Patel, H., Modugno, F., Tso, A. Y., & Lin, D. H. (2004). Clinical deterioration in pediatric asthmatic patients after September 11, 2001. *Journal of Allergy and Clinical Immunology*, *113*(3), 420–426.

Thurston, G., & Chen, L. C. (2002). Risk communication in the aftermath of the World Trade Center disaster. *American Journal of Industrial Medicine*, *42*(6), 543–544.

Trichter, J., & Paige, C. (July 8, 2006). *The rebuilding of Lower Manhattan: If the rebuilding effort's progress to date is a victory in disguise, then Lower Manhattan resident's say: "It's a very effective disguise indeed."* A pace poll survey research study, New York, NY.

Winter, G., & Flynn, K. (2002). Living Space: About the survey. *The New York Times*. Retrieved from http://www.nytimes.com/packages/html/nyregion/20020911_911_RESID/met_911_RESID_020911_03.html and http://www.nytimes.com/packages/html/nyregion/20020911_911_RESID/met_911_RESID_020911_01.html

WTC Redevelopment Committee. (February 21, 2006). *Community Board One resolution*, at http://www.epa.gov/wtc/panel/pdfs/hughes-20040726.pdf

ENVIRONMENTAL ALTRUISM: A COMPARISON OF RUSSIA AND THE UNITED STATES

Margaret Gibbs, Tatyana J. Andrushchenko, Natalya Makarevich and Robert Binford

INTRODUCTION

The chapter compares the factors involved in environmental altruism, the willingness to involve oneself in working for environmental and ecological goals, in Russia and the United States. The lead authors developed the idea for the research from their participation in the grant from the Trust for Mutual Understanding, examining how Russia and the United States deal with contamination in their communities. We were the only clinical psychologists in the group that toured in Russia, and we both taught young people at universities, sometimes in courses that dealt specifically about the environment, but more usually in courses that helped prepare them for their roles as psychologists. Because of our clinical perspective, we had several discussions about the role of the individual in dealing with the environment. The role of the individual is particularly critical to environmental issues in Russia. Although Russia has even more strict environmental laws than the United States, its level of enforcement has been minimal. At the same time, environmental activism is a relatively new phenomenon there and it faces cultural and social barriers. While activism plays a major role in the United

Cultures of Contamination: Legacies of Pollution in Russia and the U.S.
Research in Social Problems and Public Policy, Volume 14, 447–463

ISSN: 0196-1152/doi:10.1016/S0196-1152(06)14019-3

States to keep the system functioning, efforts to deal with contamination in the United States generally occur on a political and legal level in the context of creating and enforcing laws. While the U.S. activists are hardly satisfied with the system's functioning, a greater level of creativity and commitment is needed from Russian environmental activists to help the nation even reach this level of routine enforcement of environmental laws.

In Andrushchenko's opinion, environmental change in Russia depends on educating the populace about ecological culture and the dangers of contamination, while empowering them to feel they can change the environment. To that end, she suggested that we conduct a survey in our two countries to see what kinds of contamination students are aware of, the dangers that they associate with contamination, and what might motivate them to take action to change the environment. The suggestion meshed with Gibbs' existing research on environmental altruism.

METHODOLOGY

The two lead authors brought different goals to the research. Andrushchenko expressed a pragmatic interest in finding ways to change the culture of contamination using an "emic" approach to qualitatively identify levels of awareness and knowledge possessed by students within each culture. In contrast, Gibbs brought an "etic" approach to research aimed at empirically comparing the two cultures on common variables that might be used to predict environmental altruism.

Recognizing the differences in our goals and approaches, we opted for a simple survey that could meet both our needs. A short survey was also the best strategy for minimizing the peril of cross-cultural research, where the meaning of questions is distorted by cultural differences. The survey asked students to list the issues of environmental contamination they were aware of in their own country, in other countries, and in the immediate region surrounding them. It asked for an assessment of the hazards caused by exposure to contamination. Students were then asked a series of attitudinal questions that might be predictive of environmental altruism; finally, they were asked to check off environmental actions that they had taken in the past.

The first draft of the survey was written in Russian, and translated into English. Then the authors met and discussed the items to make sure they had the same meaning in both languages. The clarified English version was then revised and sent to Russia, where it was back translated into Russian.

Some wording differences were necessary. For instance, in terms of political orientation, the "liberal/independent/conservative" breakdown routinely used in the United States was not appropriate in Russia, where we instead used "left wing/neutral/right wing." The question about ethnicity was omitted in Russia because the categories of ethnicity appropriate for the United States were not relevant there.

A list of environmental actions was developed by Andrushchenko and Gibbs with the help of students. It included behaviors like writing a paper about an environmental topic for a class, taking a class on environmental issues, recycling, signing an environmental petition, attending an environmental demonstration, and contributing money to an environmental cause. These behaviors obviously varied in personal and financial cost to the individual. Accordingly, rather than simply summing the total of number of behaviors, we asked graduate students at Fairleigh Dickinson University (FDU) to weigh each behavior on a Likert scale according to its perceived personal cost. The sum of these weights comprised the score for environmental action, which served as the dependent variable for our statistical analyses. A limitation of the study is that we do not know if Russians would give the same ratings to these behaviors.

CONCEPTUALIZING ENVIRONMENTAL ALTRUISM

In selecting a construct that might explain and predict proenvironmental behavior, altruism was preferred to the competing theory of cost/benefit analyses (e.g., Ajzen, 1991). Given the high costs of environmental action in Russia, weighing costs and benefits might rationalize a strategy of inaction, unless long-range consequences were considered. In contrast, a model based on the development of altruistic behavior is of interest to psychologists and has a wider relevance. Environmental altruism has been linked to three prerequisites (Schwartz, 1977):

- awareness of harmful consequences to others;
- taking self-responsibility to effect change; and
- concluding that personal costs are not prohibitive.

Actions to improve the environment have been found to correlate (albeit weakly) with personality characteristics related to self-control and an internal locus of control (Pettus & Giles, 1987). Although the locus of control finding has been replicated a number of times (e.g., Allen & Ferrand, 1999), these variables failed to predict individual opposition to the expansion of a landfill known to cause pollution (Gibbs, Ross, Weinman, & Watson, 2001). In that instance, opposition to the landfill was predicted instead by the social and political context, that is having parents and friends who had been involved in civic causes and having more liberal political views (see also Olli, Grendstad, & Wollebaek, 2001; Villacorta, Koestner, & Lekes, 2003).

Given the literature on altruism, we expected environmental activism to be influenced by these factors:

- awareness of local, national, and global contamination;
- the severity of the perceived consequences;
- feelings of personal responsibility to try to change the situation; and
- a lower perceived personal cost of trying to influence the environment.

We additionally hypothesized that proenvironmental behavior would be correlated with several other variables:

- internal locus of control, including belief that powerful groups can block one's efforts (Levenson, 1983);
- more liberal political views, and
- having parents and friends who have been involved in civic causes.

We also included other demographic variables like age, gender, and academic major, all of which have been found to be related to environmental behaviors in previous research.

Andrushchenko alternatively conceptualized altruism in the context of a psycho-pedagogical approach to solving the ecological crisis based upon instilling an ecological consciousness through curricular change in secondary education. Ecological consciousness (Deryabo & Yasvin, 1996) is defined around four constructs:

- individual and collective beliefs and values about interaction within the system of human beings in nature and within nature itself,
- subjective attitudes of a person to the world of nature,
- appropriate strategies and technologies for people's interaction with nature, and
- life values and ethical character issues that dictate individuals' ecologically oriented behavior.

Ecological consciousness is accordingly conceptualized as a combination of ecological values, beliefs and attitudes about nature with appropriate strategies and technologies.

In contrast to the self-centered assumptions of western psychology evident in our first approach, the form of the desired ecological consciousness is eco-centric (Yasvin, 2000). Its understanding of the world is characterized by ecological expediency, harmony between individual and nature, the subjective (as opposed to objectified) perception of nature, and a balance between utilitarian and authentic interactions with nature.

Andrushchenko's focus here was specifically with professional education. With training, psychologists develop their helping behaviors, knowledge, and attitudes. Therefore, she reasoned that an educational program focused on the preparation of skilled and motivated helping professionals would also foster increased altruism toward the environment.

The Russian sample

Andrushchenko and Natalya Makarevich collected surveys from psychology and social work students enrolled at Volgograd State Pedagogical University. With a population of more than one million people, Volgograd is a heavily-industrialized city stretched out along the Volga River in southwest Russia. The Volgograd region is recognized for the severity of its ecological situation, including pollution of the air, water, soils, and destruction of fauna and flora, particularly forests. Volgogradskaya oblast has one of the highest rates of atmospheric pollution in Russia. It has two critical areas, the highly-industrialized southern district, with its chemical and petroleum plants, and the northern district, home of the deadly triangle formed by the aluminum, steel, and tractor factories. These enterprises pump into the atmosphere vast quantities of sulphurous compounds, nitric oxides, metal oxides, hydrocarbons, and other harmful substances. Volgograd also faces serious water pollution problems resulting from household and industrial sewage and agriculture. Water treatment systems are far from adequate. Finally, Volgograd lacks any way to address its mounting garbage problem. Large fields of waste are dumped outside the city lines. Given these well-known problems, we expected Russian participants to show a high level of awareness of contamination of their local environment.

Table 1 shows the composition of the Russian and the U.S. samples. The 100 Russian participants were evenly split between first and second year psychology majors, on one hand, and students in their fifth year and graduate students, on the other. This sample constituted almost 60% of the total population of psychology students at Volgograd State Pedagogical University (VSPU). Reflecting the population demographics, the sample was almost entirely (93%) female. In both Russia and the U.S., psychology majors are more likely to be female. The mean age for the sample was 19.5, ranging

Table 1. Characteristics of the Two Samples.

	Russian Sample	U.S. Sample
Gender	male 7 female 93 total 100	male 22 female 90 total 112
Age	X = 19.5, range 17–28	X = 24, range 18–54
Year in school	undergraduate 51 senior & graduate 49	undergraduate 92 graduate 22
Psychology major (%)	100	71
Political affiliation	65% neutral, 35% other categories	36% liberal, 45% independent, 13% conservative

between means of 18 and 21 for the two respective groupings. Nearly all those sampled identified themselves as politically neutral, rather than left or right wing.

The United States Sample

Robert Binford, Mia Hernandez, and Doron Marbe gathered data from a sample of undergraduate and graduate students enrolled in psychology classes at Fairleigh Dickinson University in Teaneck, New Jersey. The survey was administered in psychology classes, usually for extra credit.

Teaneck is located just west of Manhattan and most students commute from within this region. Bergen County is home to more than 900,000 people. The entire metropolitan region has long been in violation of the Clean Air Act for tropospheric ozone. Traffic congestion is a major cause of air pollution. Dumping of toxic wastes and garbage are well documented in the region. The county ranks 13th of 21 in New Jersey, with 190,000 pounds of toxins released to the air per year. While total environmental releases are just above average for the U.S. counties, Bergen County is at the 60th percentile for cancer and noncancer air and water risks and at the 70th percentile for air releases of developmental and reproductive air pollutants (Environmental Defense Fund, 2006). Northern New Jersey has long suffered the nickname "cancer alley." Numerous Superfund and Brownfields Sites surround the metropolitan area (see Hoffman, this volume). We expected that students would show a relatively high level of awareness of problems of contamination, although not necessarily local pollution.

Of the 114 U.S. students in the sample, 82% were undergraduates,[1] 80% were female, and the mean age was 24, with a range of 18–54. Politically, students mostly ranked themselves as moderate (45% independent, 36% liberal, and 13% conservative). While more than two-thirds of the sample was comprised of Psychology majors, the fact that Psychology classes are taken by nonmajors prevented matching the more academically segregated Russian sample.

RESULTS

We will first present the descriptive data for the two samples, comparing them, with emphasis on the Russian sample. Table 2 shows the responses written by Russian students about issues of contamination in their country.

As predicted, Russian students were highly aware of issues of contamination in their country and in local areas. On average, Russian students mentioned 3.18 different contamination issues in their country. This was in contrast to the U.S. mean of 1.67, indicating that Russian students in this sample were significantly more aware of contamination issues in their own country than the U.S. sample was of the issues in the U.S. Responses varied qualitatively as well as quantitatively. United States students were likely to write in "air and water pollution" when asked about their awareness of environmental contamination in their country. In contrast, 92% of the Russian students wrote about radiation, 57% of them mentioning Chernobyl. And 78% of the Russians also wrote about water pollution, mostly citing oil spills, waste and problems in the Caspian Sea and Lake Baikal. Half also cited atmospheric pollution. Thus Russians revealed a much higher degree of specificity and accuracy in their knowledge of environmental issues. The Russian ecological crises appeared to be an integral part of the daily lives of Volgograd students.

In contrast to their responses on domestic pollution, Russian students evidenced much less awareness of contamination in other countries (see Table 3). The average number of responses (1.53) was half that seen above. Furthermore, responses on problems abroad were much less specific and tended to focus on general issues of pollution such as water contamination. This may reflect the fact that international events can be relatively inaccessible for Russian students because of the way the media operates in Russia.

The U.S. students revealed little knowledge of contamination issues in other countries. They gave fewer average responses (1.27) than did Russian students, and fewer than they gave for domestic problems, although not significantly. It would appear that the U.S. students also receive little international news about environmental issues outside of major catastrophes

Table 2. Awareness of Situations of Environmental Contamination in Russia.

		Freshmen Students		Graduate Students	
		Amount of Statements	%	Amount of Statements	%
I	Radiation Pollution (Contamination)	43	30.28	49	27.84
1	Chernobel	30		27	
2	Radioactive waste	7		13	
3	Atom Explosions	3		7	
4	Higher radiation level	3		2	
II	Water Pollution	37	26.06	41	23.30
1	Baikal	10		10	
2	Oil Spill	8		10	
3	Wastage	13		15	
4	General Water Pollution	6		4	
5	Intermittent Caspian Sea	0		2	
III	Atmospheric Pollution	25	17.61	25	14.20
1	Fires	9		7	
2	Exhaust	7		5	
3	Pollutant Emissions	8		12	
4	Air Changes	3		2	
IV	Soil Pollution	7	4.93	23	13.07
1	Chemicals	1	0.70	6	3.41
2	Garbage	6	4.23	17	9.66
V	Disturbance natural system	3	2.11	6	3.41
VI	Combined Pollution	27	19.01	27	15.34
1	Industrial Pollution	15		19	
2	Transport	0		3	
3	Military operations	2		2	
4	Oil escape	8		4	
VII	Other	0	0	3	1.70
	Sum Total – 318	142		176	
	Average (Statements for one student)	2,78		3,59	

Table 3. Comparison of Russian and U.S. Samples on Awareness of Environmental Contamination.

Question	RF Sample (n = 100)		U.S. Sample (n = 112)		*t*
	Mean	SD	Mean	SD	
Awareness of contamination in own country (Number of statements)	3.18	1.36	1.68	1.02	8.77*
Awareness of contamination in other countries (Number of statements)	1.53	1.12	1.27	1.04	1.67
Awareness of contamination in own community (Number of statements)	1.78	1.34	0.83	0.73	6.46*
Consequences (Number of statements)	3.44	1.53	1.18	0.88	11.96*
Seriousness of consequences (7 point scale, 1 high)	1.53	1.05	2.40	1.61	4.46*
Extent of personal responsibility (7 point scale, 1 high)	4.45	1.36	4.53	1.56	0.47
Extent of personal cost (7 point scale, 1 high)	2.61	1.48	2.72	1.42	0.30
Own actions impact world? (7 point scale, 1 high)			4.28	1.61	
Can actions succeed? (7 point scale, 1 high)			4.72	1.27	
Can powerful groups block action? (7 point scale, 1 high)			5.32	1.45	

*$p < 0.05$.

like Chernobyl. Given ready access to media, this lack of knowledge defies a simple explanation.

In their response to questions about environmental contamination near their own community, the Russian participants focused on soil pollution from garbage, atmospheric pollution, and water pollution, all realistic concerns in Volgograd. The average number of responses provided was 1.78. In the U.S., the average number of responses about contamination near one's own community was only .83, a significantly smaller number.

Another question on the survey asked participants what they saw as the consequences of environmental contamination. Russian students listed an average of 3.44 consequences, as compared to 1.18 consequences from the U.S. sample, indicating a significant difference. The focus was on health consequences (79%), on animals (46%), and on plant life (39%). Responses again were more specific and thoughtful than within the U.S. sample. Russian students talked about such specific issues as global climate change, the greenhouse effect, and disturbing the ozone layer while the U.S. students were more likely to make general statements such as "bad for your health."

The next question on the survey asked students to rate the severity of these consequences on a seven-point Likert scale with 1 indicating most serious. Russian students rated the consequences as significantly more serious than the U.S. students (1.53 vs. 2.37).

Personal responsibility to try to change the environment was rated on a seven-point scale, where 7 indicated "very much" and 1 "not at all." Here, similar ratings were found for Russian (4.45) and the U.S. (4.53) participants placing both in the middle of the scale showing some, but modest, personal responsibility. Similarly, a rating of difficulty in changing the environment, where a rating of "1" indicated "very hard," found nearly identical ratings from Russian (2.61) and the U.S. (2.72) students, indicating a belief that change would in fact be difficult. This item was used in determining "personal cost" for our regression analyses.

When we consider the dependent measure of weighted number of pro-environmental actions taken, the reported actions of Russian students averaged 6.16 compared to the rating of 9.78 for the U.S. students. This measure suggests that the U.S. students have personally done more to change the environment than Russian students. While the Russian students reported having "written a paper on environmental issues," "taken a class on environmental issues," and "voted for a candidate or issue based on environmental concerns," for the most part, they had not signed petitions, participated in environmental demonstrations, or been involved with recycling. While opportunities for such activities exist in Russia, they are less widespread and traditional.

Predictors of Environmental Action for both samples, it was expected that the number of environmental actions would also correlate with a series of independent variables:

- a higher level of awareness of local and national environmental contamination;
- a higher level of perceived dangerousness of these problems;
- a sense of personal responsibility to change the environment;
- a lower perceived cost in trying to change the environment;
- a more internal locus of control;
- having family involved in environmental action;
- and more politically liberal views.

Table 4 presents the correlations between variables and environmental actions in the U.S. sample. For the most part, our hypotheses were

Table 4. Pearson Correlations between Variables and Environmental Action in the U.S. Sample.

Variable	Pearson Correlation
Awareness of contamination in own country	0.217*
Awareness of contamination in other countries	0.210*
Awareness of contamination in own area	0.042
Number of consequences listed for contamination	0.197*
Seriousness of contamination	−0.134
Personal responsibility	0.229*
Personal cost	0.091
Own actions impact world	−0.183*
Actions succeed	0.265**
Powerful others can block	0.200*
Age	0.029

*$p<0.05$.
**$p< 0.01$. $N = 110$.

confirmed for the U.S. sample. Higher levels of awareness of environmental problems on a national and international scale (but not in one's own community) successfully correlated with higher environmental action. The more consequences the participant listed, though not the seriousness attached to them, the greater number of environmental actions were reported.

Sense of personal responsibility to try to change the environment also proved to be a significant predictor of environmental action, but not the cost, measured by the perceived difficulty of success. Locus of control strongly correlated with environmental behavior in the U.S. sample. Personal efficacy, as measured by the question, "How much do you feel your actions can have an impact on the world?" correlated with environmental behavior. A similar question on total efficacy, "To what extent do you believe that efforts to preserve or improve the environment can succeed?" also correlated with environmental behavior.

Results for the question, "To what extent do you believe that powerful groups can block efforts to preserve or improve the environment?" ran contrary to the direction of our hypothesis. We expected that belief that powerful others would block one's efforts would discourage environmental involvement (Levenson, 1973). In fact, we found that those who believed powerful groups could block efforts were more likely to be involved in environmental behavior. It would appear that the sense that powerful groups are blocking efforts to improve the environment acts motivates

Table 5. Awareness of Contamination and its Consequences as Predictors of Environmental Action in the U.S. Sample.

Effect	Coef.	Std. Error	Std. Coef.	*t*	*p*
Constant	8.13	2.08	0.00	3.90	0.00
Awareness in own country	0.64	0.83	0.09	0.77	0.44
Awareness in other countries	0.91	0.79	0.12	1.15	0.25
Awareness in own area	−0.22	0.97	−0.02	−0.27	0.82
Consequences listed	0.71	0.86	0.08	0.86	0.41
Seriousness of consequences	−0.47	0.45	−0.10	−1.051	0.30

	Sum of Squares	df	Mean Square	*F*	*p*
Regression	425.17	5	85.13	1.58	0.17
Residual	5664.11	105	53.94		

individuals by angering them and by making clear the importance of their taking responsibility to try to change the situation.

None of our demographic variables predicted environmental action, neither age, gender, major, political affiliation, nor having family and friends involved in social causes. Political affiliation may have failed to predict activism because the sample did not provide a wide variety of political views. Similarly, most of the majors were psychology students, and the sample had an imbalance of females to males.

Because of the number of variables, we divided the multiple regression in two parts. Table 5 lists the regression for variables relating to awareness of environmental contamination and the seriousness of the consequences. In spite of the significance of the individual correlations, the multiple regression coefficient was not significant.

Table 6 lists the multiple regression for the variables relating to responsibility, cost, and locus of control. This regression was significant at the 0.003 level. Variables contributing most to the relationship were the sense of personal responsibility for the environment and the two measures of personal efficacy; the sense that one's actions can impact the world and that one's actions are successful.

In summary, correlations showed most of our variables correlated with environmental action for the U.S. students. Those variables that were supported in the regression analyses were locus of control and a sense of personal responsibility.

Within the Russian sample, a highly different picture emerged. None of the items in the survey predicted environmental behavior. Looking at the

Table 6. Locus of Control and Other Variables Predicting Environmental Action in the U.S. Sample.

Effect	Coef.	Std. Error	Std. Coef.	*T*	*p*
Constant	1.05	4.16	0.00	0.25	0.80
Personal Responsibility	0.97	0.45	0.20	2.14	0.03
Personal Cost	−0.38	0.50	−0.07	−0.75	0.46
Own Actions Impact World	−0.74	0.42	−0.16	−1.74	0.08
Can Succeed	1.08	0.57	0.18	1.88	0.06
Powerful Groups Impede	0.64	0.49	0.12	1.31	0.19
	Sum of Squares	df	Mean Square	*F*	*p*
Regression	947.05	5	189.41	3.83	0.003
Residual	5236.82	106	49.40		

year in school, Andrushchenko's main variable of concern, graduate students and undergraduates in their final year carried out more total actions, 87, as compared to 79 for first and second year students. This difference, however, was not significant. There was a slight tendency for graduate students to have been more likely to carry out two or more actions as contrasted with none or one actions. Fifty percent of the graduate students carried out two or more actions, while 34% of the first and second year students did ($X_2 = 2.63$, $p < 0.15$). In an analysis of the actions, Andrushchenko felt that those actions taken by graduates were likely to be more active. Table 7 provides a listing of the actions taken by both groups of students. She considered passive behaviors to be the first four listed in the table, and active behaviors to be the remaining ones. While the difference between groups does not yield a significant chi-square, the differences in behavior seem illustrative of something deeper that is going on in the groups. Andrushchenko feels that they help to demonstrate that education, in particular professional education in psychology, can play a critical role in improving Russia's environmental situation by motivating students to become more aware and more active.

DISCUSSION

Russian students were more aware than the U.S. students of environmental contamination in their own country, providing both more and more specific examples and offering more and more serious consequences. However, when

Table 7. Environmental Action Carried out by the Two Samples of Russian Students.

	Junior (%)		Senior (%)	
Contributed money to foundations	4	2.41	2	1.20
Participated in meetings	4	2.41	3	1.81
Signed a petition	0	0.00	10	6.02
Attended special lectures	18	10.84	13	7.83
Involved in the programs on waste usage	3	1.81	1	0.60
Became a member of a green organization	2	1.20	2	1.20
Wrote protest letters	0	0.00	1	0.60
Was the leader of a civic organization	2	1.20	0	0.00
Prepared a report	18	10.84	21	12.65
Cleaned a territory	5	3.01	4	2.41
"Greened" a territory	1	0.60	5	3.01
Didn't litter	9	5.42	12	7.23
Participated in ecological activities	2	1.20	3	1.81
Conducted a project	1	0.60	0	0.00
Collected signatures	0	0.00	2	1.20
Didn't allow others to litter	2	1.20	3	1.81
Uninformed	8	4.82	5	3.01
Total	79		87	

it came to acting to combat contamination, they did less than the U.S. students.

Do methodological issues account for these differences? Perhaps of most concern was that environmental behaviors were weighted by a U.S. sample. It may be considerably more difficult to carry out these same environmental actions in the Russian context. Nevertheless, most Russian students were not especially active, with the primary activities being those related to university course work. Any mode of comparison would likely indicate that American students were more active.

The finding that Russian students are highly aware of contamination issues is not surprising given the extent of environmental damage in their country. As noted, Volgograd is recognized as a highly-polluted city, and much of Russia as a whole has been devastated by past disregard for the environment.

It is hard to explain why the U.S. students were not more aware of the significant contamination issues in their own country. Although environmental contamination is certainly a critical issue in the U.S., and is constantly in the news, it may be that students do not see the immediacy for their own lives. Perhaps the U.S. respondents are complacent because there is so much open media attention to the problem and because government agencies and the many environmental organizations give the impression that somebody else is taking care of the problem. In this regard, it is interesting that the U.S. students who believe that powerful groups are able to block efforts to preserve or improve the environment are the ones most likely to take environmental action.

The Schwartz model predicted fairly well for the U.S. sample, with personal efficacy and responsibility, but not personal cost, predictive of environmental action. Awareness of contamination issues and seeing the consequences as serious were correlated with environmental action, though they did not emerge as significant in the multiple regression. Do our findings suggest that the Schwartz model is culture specific, or emic, relevant only to the U.S. and western European countries in which it has been primarily studied?

Perhaps, but it is interesting to examine the Russian findings from within that model. It is possible that level of awareness did not predict environmental action because of a ceiling effect, that is, the Russian level of awareness is so high for the whole sample. Personal cost failed to be predictive for the U.S. sample, but it may be highly relevant to the Russian situation. The personal costs of activism are very different for Russians than in the U.S. Russians lack a history of public participation, along with the repertoire of norms and skills that go along with it. Moreover, many Russian citizens who engaged in past environmental protests had difficulties with or were pursued by the authorities. While the political situation changed with Perestroika, it remains dangerous to speak out and to act. For instance, one member of our group in the Trust for Mutual Understanding has a member of the security services always present at her public presentations. When members of our team traveled to the Urals to conduct a course for regional activists, a suspected government agent dogged their every step. And even the act of conducting research can generate considerable trouble, as some participants in this volume discovered. It may be that the cost of environmental action is simply too high for most students to take actions outside those within the classroom or the voting booth.

While the Krishi (see Tsepilova, this volume) and Chelyabinsk cases (see Mironova et al., this volume) illustrate the power of public environmental movements in Russia, the high cost of Russian protest may translate into fewer everyday opportunities for environmental action. The U.S. citizens are spoiled by the ready availability of institutionalized opportunities for environmental participation (recycling opportunities, causes to donate money to, and other environmental infrastructure) and the ease with which local protest groups can form, comparatively lacking in Russia.

Locus of control may itself vary between the two countries, as suggested by Edelstein and Smirnova (this volume). Suggested is a greater tendency for the U.S. citizens to think of themselves as self-directed and Russians as other directed. Likewise, the U.S. citizens are more likely to assume efficacy and Russians more likely to see the system as unmovable. Americans are unlikely to accept the situation as "given," Russians to accept their lot. Nevertheless, our samples from both countries rated themselves equally responsible to make changes in the environment. But what they mean by this may differ. Russian students may see themselves as having to act in more creative ways outside the system in order to make these changes, while the U.S. students, as noted, probably have more opportunities to act within the system.

The importance of professional education, particularly psychological education, in Russia has been suggested by this research. It is possible, of course, that other self-selection factors influenced the Russian findings, since students entering a graduate psychology program may themselves differ from those in the undergraduate program. However, at least half of the higher level education sample were advanced undergraduates. It is interesting to note that it is probably the education, not the age of the students that is of greatest influence. Szagun and Pavlov (1995) found that the level of environmental awareness decreased with age in a Russian sample. Neither age nor year in school was a factor within the U.S. sample, suggesting that education does not play the key role that it does in Russia in increasing awareness and activism around the environment.

Finally, in evaluating the success of our project, it would appear that the mixture of both etic and emic and quantitative and qualitative approaches is valuable to cross-cultural research. The qualitative and emic aspects of the work clarify and add to our understanding of the etic and quantitative.

NOTE

1. While VSPU seniors are fifth year students whose age and experience is similar to graduate students, U.S. undergraduates are quite different from those pursuing advanced degrees. There were not enough graduate students available to directly match the Russian sample.

REFERENCES

Ajzen, I. (1991). The theory of planned behavior. *Organizational behavior and human decision processes, 50*, 179–211.

Allen, J. B., & Ferrand, J. L. (1999). Environmental locus of control, sympathy and pro-environmental behavior: A test of Geller's actively caring hypothesis. *Environment and Behavior, 31*, 338–353.

Deryabo, C. D., & Yasvin, V. A. (1996). *Ecological pedagogy and psychology*. Rostov-on-Don: Phoenix.

Environmental Defense Fund. (2006). Bergen County, New Jersey. www.scorecard.org

Gibbs, M. S., Ross, H., Weinman, E., & Watson, H. (April 2001). Pro-environmental behavior: Awareness of environmental risk, perceived control and demographic predictors. *Paper presented at the Annual Convention of the Eastern Psychological Association*, Washington, D.C.

Levenson, H. (1973). Multidimensional locus of control in psychiatric patients. *Journal of Consulting and Clinical Psychology, 41*, 397–404.

Olli, E., Grendstad, G., & Wollebaek, D. (2001). Correlates of environmental behaviors: Bringing back social context. *Environment and Behavior, 33*, 181–208.

Pettus, A. M., & Giles, M. B. (1987). Personality characteristics and environmental attitudes. *Population and Environment, 9*, 127–136.

Schwartz, S. H. (1977). Normative influences on altruism. In: Berkowitz, L. (Ed.). *Advances in experimental social psychology* (10, pp. 221–279). San Diego: Academic Press.

Szagun, G., & Pavlov, V. (1995). Environmental awareness: A comparative study of German and Russian adolescents. *Youth and Society, 27*, 93–112.

Villacorta, M., Koestner, R., & Lekes, N. (2003). Further validation of the motivation toward the environment scale. *Environment and Behavior, 35*, 486–505.

Yasvin, V. A. (2000). *The psychology of attitudes toward nature*. Moscow: Smysl.

THE PARABLE OF THE LIONS BRIDGE: POTENTIALS FOR MUTUAL LEARNING BETWEEN CULTURES OF CONTAMINATION

Michael R. Edelstein and Lyudmila V. Smirnova

THE PARABLE OF THE LIONS BRIDGE

Four cast iron lions guard the charming Lions footbridge crossing St. Petersburg's Griboedov Canal. The first author strolled across the bridge on an August evening in 1998 with a Russian friend, Polina. The experience presents a parable for concluding this volume:

> As we crossed the bridge in the dark, we barely avoided stepping into a gaping hole in the deck half way across. A person could easily fall through. I reacted as a typical American, immediately taking responsibility for doing something about the problem, looking for some board to place over the hole or a barrier to warn pedestrians or someone to report the hazard to who would address it promptly. Polina indicated that no Russian would make such a fuss, and she could think of no one to report the hazard to who would respond. As we stood discussing this problem before two of the guardian lions, a group of drunken soldiers began to cross the bridge from the far side, arm in arm, singing loudly. Sure enough, one slipped through the hole and, but for his comrades holding his arms, he would have plunged into the canal. Surely they will report it, I said, but Polina was doubtful. She saw little chance for some protective action to fix the bridge or even to warn passers-by to beware. We went on our way mindful of the problem left behind. Perhaps a month after my return to the U.S., I received an email from Polina

Cultures of Contamination: Legacies of Pollution in Russia and the U.S.
Research in Social Problems and Public Policy, Volume 14, 465–492

ISSN: 0196-1152/doi:10.1016/S0196-1152(06)14020-X

that she had gone to the bridge and, to her surprise, it was fixed. "Perhaps," she wrote, "there is hope for Russia after all."

Given this experience, it was ironic to read the opening to Oleg Glazechev's chapter (Into Thin Air, this volume). Glazechev recounts a remarkably similar story (holes in bridges must be no rarity in Russia). Sharing Polina's assumption that the bridge will not be fixed, he concludes that the only way to protect children in a land whose bridges sport holes is to make sure they are good swimmers.

The Lions Bridge parable suggests two very different paradigms. The first is outwardly focused on changing polluting conditions and thus avoiding exposure. The second is inwardly focused on learning to adapt to environmental conditions. The parable presents an interesting framework for concluding this comparative study.

A COMPARISON OF RISK IN THE U.S. AND RUSSIA

Appreciation of contaminated communities in the U.S. dates from the late 1970s with the discovery of buried hazardous wastes at Love Canal in Niagara Falls, New York and, in Europe, with the dioxin contamination due to an explosion in 1976 at a pharmaceutical plant in Seveso, Italy. Such awareness in Russia awaited glasnost and Perestroika in the later 1980s into the 1990s.

Reflecting on such events, Beck (1986) posits that the world has entered into a "Risk Society," a post-modern era characterized by the realization that it is no longer nature that we must fear; rather it is human-caused hazards that have become our principle sources of danger. Accordingly, questions of risk insinuate themselves into all facets of life. Is the air we breathe safe? The water we drink? The food we eat? People are forced to make constant risk choices and tradeoffs in the face of uncertainty. As a result, whereas moderns accepted risk as a cost of the good life (i.e., "a tradeoff"), post-moderns have become averse to risk. It follows that, in the Risk Society, a constant state of vigilance must be maintained against human-caused dangers.

There is little doubt that environmental and other risks are ubiquitous; the question that arises is whether vigilance against such risks is also widespread. While the Risk Society formulation is highly descriptive of Northern Europeans, who have entered into a precautionary society as a result, the first author has long questioned whether it applies to the U.S. under normal circumstances (Edelstein, 2000). Work in this volume further calls into question the applicability of risk society theory to Russia as well.

Cultural Immunity in the U.S.

American culture grants its adherents a "cultural immunity" to environmental threat (Edelstein, 2000, 2004). The immunity involved does nothing to prevent toxic exposures, even shielding abuses from attention and action. Rather, it affords freedom from worrying about environmental threats. It is a form of widespread and shared denial.

There is nothing mysterious about the form taken by this cultural immunity. It simply reflects the influence of modern life through experiences of employment, consumerism, isolation within the nuclear family, extreme time pressure, media seduction, media and information overload, mindless recreation, and the disconnection from community, polity, and nature. Most people are concerned with home, family, work, entertainment, and their immediate social circle – but not environmental risk – unless a particular hazard intrudes into their lives in a visible and believable fashion. The dominant social paradigm of Western society assures that individual values include such elements as individualism and the beliefs that humans are above nature, that risks are justified by progress, and that technology and growth can solve all problems (Milbrath, 1984; Olsen, Lodwick, & Dunlap, 1992). And, of course, our society can be characterized by the synthetic, rather than natural, elements upon which it depends – oil, chlorine, etc. – which are highly polluting and toxic. While aware that others have become toxic victims, Americans generally see themselves as immune from catastrophes viewed as potentially happening to some, but not to themselves. Given their lack of ecological literacy, they may not link abstract environmental problems to their own personal vulnerability, failing to see even alarming environmental threats to the biosphere as personally threatening. One consequence of cultural immunity is that exposure to contamination, should it occur, catches victims by surprise, often despite many signs of trouble that had been ignored.

Russian Risk Acceptance

The cloak of cultural immunity does not seem to fit Russians. Rather, it has been observed that Russians pathologically accept risk as an inescapable condition. And, this observation has been confirmed during the transition from communism to capitalism. Paradoxically, environmental risk has been magnified during transition due to a simultaneous shift to the new dangers of high modernization and a slide back to a pre-modern absence of institutional and social mechanisms for controlling these risks (Bulatova, this volume; Yanitsky, 2000, 2003). Such risk acceptance was evidenced in Sokol (Kuliasova & Kuliasov, this volume), where respondents are reconciled to the fact that ecological pollution is a given of life, a condition beyond change or, at least, beyond their own influence. There is no expectation that conditions will improve. The choices presented are to ignore the situation or take personally protective actions.

The perception of a hopeless situation is not the only explanation for "the blind eye turned by Russians to ... pollution" (Bolatova, this volume). Various cultural predilections and practices may also make it easier to choose risk acceptance, much as American culture favors denial. What are the bases for Russian risk acceptance?

Control

A core issue is the apparent difference between Russians and Americans with regard to perceived personal control (see Gibbs & Andruschenko, this volume). While the presumption of control underlies Western psychology (Edelstein, 2004), Russians are characterized by psychological passivity (Zykova, this volume). In contrast to Americans, Russians have developed a cultural view based upon the ready acceptance of fate, no matter how cruel, and for low expectations for personal control over the course of life. One of the Russian participants in the 2002 Urals Conference, organized in conjunction with our Trust for Mutual Understanding project, summed it up succinctly:

> In our life, we live on a one year plan. We are not planning for a long time.

There are numerous possible contributing causes for a lack of control and passivity, contributing to Russian risk acceptance.

Fate and Fatalism

Beneath the atheism of the Communist period, many Russians held a deep and fervent belief in Russian orthodoxy plainly evident now with religious freedom. This tradition harkens back to the medieval belief in fate and the "other worldly" look to heaven for salvation from worldly sin (see Arendt, 1958). Countless amulets are worn to ward off everything from bullets in battle to the evil eye. This same faith in fate and fatalism is also evident in the rural Tatar and Bashkir victims of the Mayak disasters, described as

> ... devout Muslims, whose faith pointed to the workings of divine will, not the incompetent Ministry of Nuclear Energy, for the cause of suffering. Many Muslim women in the region saw the high mortality and disease rates as punishment for their sins on earth. Rather than questioning or reacting to the unfolding disaster, local villagers referred to the heightened illness simply as "River Disease" and took no action whatsoever. (Mironova, Tysiachniouk, & Reisman, this volume)

It is not just that one is not in charge of their fate, but there is a presumption that the worst may as well happen and there is nothing to be done to change it. A similar analysis was made by a Bashkir informant, who described the impact of her Muslim religion on environmental victims:

> They look at life with fate. They don't need to take an active position to change it. They continue to be passive.

Powerlessness

One reason that Russians tolerate extreme levels of pollution that Westerns would not is a fundamental perceived inability to affect change. There is a sense that individuals can do little to change the situation in a way that would reduce their risks. Experts in places such as Dzerzhinsk make dire assessments of calamity while people just accept the pollution that surrounds them as a given (Bulatova, this volume). This sense of powerlessness comes through even in everyday Russian speech, which is conveyed by the frequent exclamations of the words "zhal" and "gospady" ("it is a pity" and "oh God").

Environmental injustice is another factor, as discussed for the Tartar and Bashkir minorities as a result of Mayak releases (Edelstein & Tysiachniouk, this volume). It has been noted that the Mayak facility was built in an area inhabited largely by minority Bashkir and Tartar people who represented an available labor force for the original construction. These minority groups were subsequently the principal victims of the various chronic and acute disasters unleashed by Mayak. And, in the case of Muslumovo and other areas still inhabited within the "footprints" of the various disasters, these national minorities are in the local majority (see Edelstein & Tysiachniouk, this volume; Mironova et al., this volume). Their disproportionate but innocent exposures; their lack of power, information or warnings; and their disadvantaged treatment subsequently all speak to issues of systematic environmental injustice resulting, if not from bias, then from disregard.

A lawsuit won in 2004 offers additional insights into the complex issues of environmental injustice involved at the Mayak site (McDonald, 2004).

According to the records of the case, in 1957, when the Kyshtym disaster occurred, some 1,500 local Tartar and Bashkir people including children were forced to become liquidators of the accident, working in the hot zones. Apparently, most of the children involved died shortly thereafter from radiation-related diseases. These liquidators were told that they were responding to an oil spill; they were never told about the radiation. Soldiers came to a village, Karabolka,12 miles from Mayak, a few days after a huge blast was felt and a dark cloud had hovered overhead. The soldiers sealed off the Muslim area of the village of 2,900 from the Russian area and put ethnic Tartar and Bashkir people to work. Children were released from school on two occasions for this purpose. They later found that the Russian area had been evacuated and the people resettled. The conscripted people dug up contaminated vegetables for burial, buried dead cattle, filled in wells poisoned by the release, dismantled wood houses, and cleaned bricks. One child laborer who survived recalled:

> Our hands were bleeding. Everybody was vomiting. My vomit was very green. The doctor looked at it and said I had eaten too many peas, and he sent me back to work. But of course I hadn't eaten any peas at all.

This girl of 11 years had borrowed a bright dress that day for what she thought was to be a school outing. In the recent court action, three residents of Karabolka won reparations of a monthly stipend of $8 a month and a yearly visit to a spa. Other lawsuits filed address the legacy of water contamination from Mayak in the Techa basin.

Forbearance

Russians have been repeatedly called upon to survive under extreme conditions. This forbearance is evident in the broad population, as well as in the populations victimized by specific events. The meta-cultural message of the battle of Stalingrad and the siege of Leningrad in World War II is to "grin and bear it." The Russian people have borne so many hardships that they have culturally adapted to dismissing risks and bad conditions. The result is a level of patience not readily found in the U.S., expressed by the words "nyet problem" (i.e., no problem). Surrounded by hazards, people become desensitized to risk (see Bulatova, this volume). As master organizer Natalia Mironova commented at the Urals Conference:

> When I talk about how the U.S. organizes, in Russia it is entirely different. In Russia, if the community faces risk, the perceived risk is pushed away from your head if you cannot solve the situation. You delete it from your head.

Avos'

Another factor contributing to powerlessness found in our discussion of Chernobyl is the concept of avos', "it won't touch me" (Smirnova & Edelstein, this volume). Seemingly contradicting fatalism, "avos'" was evident both among soldiers sent to liquidate the fires and the populations receiving radiation exposures due to Chernobyl. Zykova (this volume) found respondents to estimate lower levels of radiation danger for themselves than for others living in the contaminated areas, professionals confronting the hazards and even their children, relatives, and neighbors. In both the cases of the liquidators and the residents, avos' amounted to a denial of danger and a rationalization for not taking protective action.

Folk Remedy

Russians also call upon long-standing social and behavioral adaptations to cope with risk. In several instances in this volume, we encountered discussions of the use of vodka as a general health tonic. Different from the positive issues of social drinking and the negative issues of alcoholism and over consumption that are also part of vodka's impact on Russian society, vodka plays an important folk medicinal role as a means of counteracting poisons that may have been introduced into the body. It is a cure that invokes deep confidence.

A second folk custom is the banya, the Russian hot steam sauna. In combination with the long-standing practice of medicinally shocking stresses from the body with extremes of hot and cold, the banya serves as a way to sweat out any toxins from the body. After a period of steaming, the banya user routinely jumps naked into a lake or river, a pool, a snow bank or some other source of freezing water. It is telling that not only were these practices used for the liquidators at Chernobyl, but they were also accepted as remedies by those exposed to radiation there (see Edelstein & Tysiachniouk, this volume; Smirnova & Edelstein, this volume).

Both drinking vodka and doing banya are social customs, often shared with friends and kin. They exemplify another element of Russian coping process, namely the heavy reliance on family and friends to provide a supportive social context allowing people to overcome their trials and tribulations.

Environmental Connection

Likewise, Russians tend to have a very high dependence upon nature for providing or at least aiding subsistence. Many Russians own or share access to a dacha, a country house (often a shack) with an intensive garden plot.

Food grown at the dacha is preserved for the entire year. Many Russians also travel periodically into the woods to harvest berries, fruits, and mushrooms. Some hunt. Cumulatively, this contact with nature and use of local nature as a source of sustenance occurs to a much greater extent than for U.S. residents. Russians also swim in lakes, hike, and enjoy nature to an even higher degree than Americans, many of whom are totally acclimated to a human-built environment. This point was stressed by one of the participants in the Ural Mountain conference:

> We have a pagan history. In nature we are sure somebody will protect us. During war, forests and nature and wilderness are always friendly in our perception. It is historic. We swim in lakes. Americans swim in swimming pools. Americans do not collect mushrooms. Russians have lived for a 10.000 years. The idea of environmental risk does not fit with the idea of a safe environment. There are deep psychological issues for why we push risk out of our heads.

This informant, a Tartar woman from Muslumovo, went on to explain how this affected behavior in a hypercontaminated environment:

> In Muslumovo, we keep fishing. We cannot accept the aggression of the wilderness. Our society and psychology is invested there.

Nadezdha Kutepova cautioned, however, that Russians enjoy such an expanse of nature that they also take it for granted.

> I have studied the attitudes of the public to the environment. Americans come form small countries and they value territory. The Slavic nation has been spread across nature. We were sheep herders. Who thinks about environmental protection here? There is a joke, "cut the tree but leave the sticks."

There is another dimension to the connection between Russians and environment, namely attachment to place. Americans certainly evidence place connections, but as a nation, their mobility (on the order of moving every five years on average) has no direct Russian counterpart. Russians lived in the same villages for centuries except for displacement by war, relocation due to disaster, and the punitive resettlements of the Stalin era. There is, therefore, an inbuilt resistance to change of place and related lifestyle that is very deeply ingrained, as seen with residents that remained behind or returned to areas affected by Chernobyl (Zykova, this volume). The same is true for urban Russians. To this day, Russians are assigned to a residence, where they "belong." To be without a permanent residence or living away from ones residence can result in arrest. It is not only citizens of closed cities (Kutepova & Tsepilova, this volume) who have restricted mobility. Theirs is just restrictive with greater totality.

Given this fixed mentality, people are more concerned about the risk of relocation than of staying put. This stands in sharp contrast to the reaction of Americans, who are likely to want to get out of a situation that they perceive to be dangerous. One Russian participant in the Urals conference suggested that this resistance to change is so strong that it erases whatever fears are introduced by the situation:

> Radiophobia does not exist. People just go back to their previous lifestyle. They have no fear at all. People just live there.

Another participant described the results of a survey done by Greenpeace in Muslumovo:

> People do not want change. Some still swim in the Techa.

And, Olga Tsepilova recounted her observations from her longitudinal study, suggesting that the health issues so central to Americans are offset by place attachment for Russians:

> In Kirishi, I asked people if they wanted to change their home. Do you translocate if you have a good house and a good place to go to? Eighty-five percent still did not want to go. "It is a small town where I have lived my whole life. My relatives are here. My dacha. The river." There is no value placed on health. People do not want to be resettled.

Communism

The Communist period additionally spawned cultural traits that might be contributing to Russians' acquiescence.

First, Communism introduced the double-edged sword of a collective society. On one hand, the role of the individual was strongly sublimated to the group. People learned to act for the good of the whole, not for their own private self-interests. One of the Ural conference participants stressed this point:

> People in the U.S. are individuals. We are a communal people. We have a communal mentality. There are pluses and minuses to both systems. In Russia people listen to the local environmental leaders.

At the same time, she stressed that people do not act in their own interests, something she saw as a deficit:

> I seek to educate people about the benefits of individualism.

Under communism, there was often extreme loyalty to the work group and sacrifice to help the group meet its quota and goals. That is precisely the kind of lesson one wishes individualistic Americans could easily grasp in making different consumer and environmental decisions. On the other hand, the group was not necessarily supportive of the individual, setting rules for behavior and enforcing its rules with punishment ranging from shunning to reporting misbehaviors to the secret police, possibly leading to torture, imprisonment, and death. As a result, one did not disclose one's inner or true thoughts in public. It was too dangerous. One reserved disclosure for family and close friends at home. The lack of public discourse, public spaces designed for discourse, and the norms and protections for free speech later emerged as limitations to grass roots action (see Bulatova, this volume).

Heroism is a second factor. Russia has a long tradition of celebrating its heroes in elaborate war memorials. For example, in the former Stalingrad (Volgograd), the immense statue of Mother Russia wields her sword over the vanquished Germans. The visitor confronts acres of exhibits that recall the horror of war and emphasize what brave Russians forbore. Mironova et al. (this volume) discuss at length the Soviet mentality of heroism. This selflessness and sacrifice on behalf of the motherland was established and reinforced by propaganda. And it was remembered and rewarded. Yet, except for the Chernobyl liquidators (see Smirnova & Edelstein, this volume), Russia has hardly recognized or addressed the needs of its toxic and radioactive victims. To do so is to expose tragic failures of government policy. Of a similar nature is the cultural tradition of ignoring worker and environmental risks (Bulatova, this volume).

Repression represents a third characteristic of Soviet communism. Kutepova begins her chapter in this volume with a description of the "Black Raven" (a symbol historically used in songs and poems as a metaphor for "threat, death, or danger"), the secret police vehicle that came to collect those who fell out of favor (Kutepova & Tsepilova, this volume). Mironova et al. (this volume) argue that such memories of Stalin's period of repression continue to stifle open criticism. Tolerating adverse environmental conditions and wrongdoing is a rational response when the whistleblowers, activists, and even those asking the wrong questions are more likely to be punished than the perpetrators.

Fourth, communism relied upon propaganda and distortion; its information is not trustworthy. With a free press and open public discourse forbidden, information has remained under state control. The resulting projection of some messages (propaganda) combined with the suppression of others (secrecy) was enforced by secret police and state security (see Kutepova & Tsepilova and Mironova et al., this volume), often with brutality. A result, for example, was that radioactive releases from Mayak were covered up to the extent that doctors were expressly forbidden to make diagnoses or attribute illness and mortality to radiation. Those releasing state secrets as environmental activists are liable for prosecution.

Fifth, bureaucratic rationality characterized communist practice. Bolatova's analysis of Dzerzhinsk (this volume) cites the organization of Russian society to meet quotas as a partial reason for why worker health and environment came to be ignored. Communist society was organized around 5-year plans and managers were required to meet quotas set forth by the plan. Workers were exhorted to sacrifice for the plan; social pressure was used to reinforce it. The resulting species of bureaucrat, called the apparatchik, was a bean-counting quota meeter. We encountered classic apparatchik behavior when the head of environmental affairs for the oblast of Volgograd repeatedly claimed that all laws were followed; therefore, there were no contamination problems, and therefore, no toxic victims in the city. Illustrated is a deeply ingrained official inability to admit problems (see Bolatova, this volume). No one looks for or admits pollution problems; thus there is little experience responding to them to report.

Sixth, Russia evidenced a strong technocratic influence even beyond that found in the U.S. Engineers and applied scientists had enormous influence; there were stronger institutional capabilities to get things done than there were to evaluate and decide what it was best to do. Technocracy was further supported by the nature of Soviet education, with its emphasis on the production of outstanding specialist scientists, but lack of context for interdisciplinary generalism. There was an excess of know how and a shortage of wisdom.

Two consequences of technocracy and overspecialization are relevant. First, while Russians are well educated, there is a major gap in applied environmental skill areas needed to supply government, nongovernmental organizations, and the business sector with policy-trained individuals capable of playing a guiding role in environmental decision making. This is evident, for example, in OVOS (Ozenka Vozdeistvia Orkuzaiushaya Sreda), Russia's statute for "assessment of environmental impacts." Because most scientists belong to government institutions, there is also a paucity of scientists available to serve as expert witnesses about projects during a "public expertise," or a public instituted project review. And there is a risk associated with being seen as an obstructionist scientist. A second consequence is that the knowledge and free thought problem-solving skills needed to play the broad role of citizen in a democratic society were historically lacking. The lack of cultural models and support for democratic action has been a noticeable limiting factor in efforts to implement more open forms of discussion.

Environmental Stigma

Fear of environmental stigma serves as a more recent constraint on responses to contamination threats in Russia. As the new economy invites tourism and boosterism, those who question environmental safety are denigrated. Pollution is tolerated for fear jobs will disappear (Kulyasova & Kulyasov, this volume).

ENVIRONMENTAL TURBULENCE

An exception to the denial or numb acceptance of contaminating events occurs when people come to see themselves as victims of toxic exposure and find that their normal lives are replaced by "environmental turbulence" (Edelstein, 2004). In such instances, the costs predicted by the "Risk Society" become evident. Toxic victimization challenges whatever was secure and stable about the prior life. Uncertainty and worry replace any previous ability to cognitively separate oneself from or passively accept environmental threats. Often, those victimized become vigilant, even hypervigilant.

We have seen evidence in this volume of environmental turbulence in both Russians and Americans exposed to toxic hazards. The belief in one's exposure to environmental contamination challenges three realms of the person (Edelstein, 2004): (1) their ability to cope with daily life, (2) their lifestyle or normal pattern of activities and behaviors that constitute daily life, and (3) their lifescape, or underlying assumptions about everyday life. Of course, detailed impacts to these three realms will vary from instance to instance and certainly from culture to culture.

Coping skills are tested as people accept frightening bad news while being forced to make adjustments in the way they live and view the world. Tension often results as relationships are strained by adjustments, fear and worry. Of particular frustration to Americans is the sense of helplessness that occurs as their lives are captured by unexpected events, they become dependent upon others to make crucial decisions, they fear for their health under uncertainty and insufficient information and they are unable to protect themselves or dependents. Toxic victimization is also often marked by feelings of anger and blame arising from the fact that exposures are human caused. Such events are stressful, at the least, and may contribute to psychopathology and social conflict, at the worst. One reason for conflict is that those not perceiving exposure to contaminants do not understand the changed realities of toxic victims, rendering them as outsiders, now seen as different and often inferior (Edelstein, 2004; Erikson, 1994). There is ample reason to believe that Russians go though parallel but perhaps different challenges in coping with the situation that reflect cultural differences, such as those noted above (see Edelstein & Tysiachniouk, this volume).

A second realm of impact is lifestyle (Edelstein, 2004). Daily activities, including work and home life, may be forced to change and familiar and important places may become off limits. Muslumovo residents were restricted from the Techa River running right through their village (Edelstein & Tysiachniouk, this volume; Mironova et al., this volume). Many people were relocated as the result of the Kyshtym 57, Chernobyl and Love Canal disasters (Levine, this volume; Mironova et al., this volume; Zykova, this volume). Subsistence activities that evidence dependence on nature may have to be abandoned (Edelstein & Tysiachniouk, this volume; Hastrup, Thomas, & Edelstein, this volume).

Five lifescape changes also occur after contamination, inverting core life assumptions central to "normal" thinking and causing a loss of the cultural immunity that previously served as a protective shield. First is health. For Americans, their presumption of good health (Weinstein, 1989) is inverted, forcing a reassessment of past and current health problems and a fear of future health problems such as cancer (see Hastrup et al., this volume).

This American presumption of health drew a response form participants in the Urals conference. One activist observed,

> Russia has some NIMBY, but in Russia organizations think about the country fist and then about the self. American women are more concerned about health. They take care of their health like it was a car. Russians do not care about their health until they are seriously sick. It seems that we are all healthy, and there are no sick people.

Nadezhda Kutepova then commented about a frequent Russian gripe in encountering Americans, their rhetorical rather than sincere use of the question "How are you?"

> Americans say "I am fine." Russians tell you their problems.

While Russians do not begin with the same level of health optimism, it is evident that they can become concerned about dread outcomes after exposure (see Edelstein & Tysiachniouk, this volume; Zykova, this volume). Thus, a participant in the Urals conference spoke about people continuing to live in the contaminated areas along the Techa River.

> Radiophiobia is not present here, but there is some victim's behavior. People are pessimistic. They think about themselves as victims. Victims of disease take on the sick role. There is an accepted scenario – learned helplessness, fate and depression. People are sick along the Techa. They did not know it is from the environment. Now they are convinced to be radiophobic. The worry itself can cause disease.

But here again there is a relative difference between Americans and Russians who learn they have been exposed to contaminants. Natalia Mironova noted at the Urals conference,

> Americans are optimistic even if they know.

Second, the sense of perceived control over their lives that Americans assume is also inverted by contamination, as victims realize their inability to prevent exposure and address subsequent challenges. This fundamentally disabling situation invites those commonly affected to organize a grassroots response that enables them to regain considerable control, at least in the short run. We have already noted that, lacking such control, for Russians, contamination may be just another in a history of social threats beyond their influence. One can ignore such conditions or can take whatever individually protective precautions one selects (Kulyasova & Kulyasov, this volume). When it does occur, grassroots environmental activism in Russia is a powerful force (see Mironova et al., Tsepilova, and Edelstein and Tyschiachniouk, this volume).

Third, American's former trust of the surrounding environment is inverted as well, leaving suspect whether conditions are safe or not. There seems to be little difference, in this regard, between the situation of Appalachians and Ural residents faced with contaminated environments, other than that rendered by the economic ability to forgo subsistence sources of food (Edestein & Tysiachniouk, this volume; Hastrup et al, this volume; Zykova, this volume).

Fourth, victims' homes likewise are redefined from places of security to places of danger; both the psychological and financial value of the home becomes inverted. In the U.S., a resulting dilemma ensues about whether toxic victims should remain or leave their homes; they are literally forced into a tradeoff between economics and perceived safety. In Russia, where mobility has long been severely curtailed, experience with moving large populations has occurred during forced relocation after nuclear disaster. However, such displacements result in some residents returning to their homes, despite the danger, either as an indication of attachment to place or the fatalistic sense that, given past exposures, future exposures make little difference. Meanwhile, many of those not relocated may seek to be moved (see Edelstein & Tysiachniouk, this volume). The act of being left behind with hazards appears to be a powerful motivator for grassroots action. At Love Canal, the principle goals of activists were to force resettlement of those left behind (Levine, this volume). This was precisely the goal of the Muslumovo activist who served as a primary informant (for Edelstein

& Tysiachniouk, this volume). Although grassroots action is not always successful in the U.S., it often is. Residents of Muslumovo remain behind 50 years after their environment became contaminated.

Finally, there is contradiction of social trust and the belief that good people avoid harm. Fingers are pointed at those who caused the pollution, those who failed to stop or control it and those who fail to help the victims. The result is an abrupt shift for many Americans, who tend deep down to trust their government and others to protect them and to help them when they are in need. Considerable anger and disappointment is felt over failed protections and poor post-disaster response, reflecting a failed belief that bad things won't be allowed to happen in the first place and that victims will be assisted to the level of restoring their previous lives. Victimization is treated as an inappropriate injustice that demands rectification.

The issues of trust may be different for Russians. Many Russians hold a historic distrust for their government, and they lack the assumptions of protection and the expectations for relief, as well as the expectation that injustices will not be allowed to occur. Such distrust was evident in Sokol, directed to the plant owners and government authorities. It was a primary reason for people assuming that they could not change the conditions and, instead, selecting personal strategies of action (see Kulyasova & Kulyasov, this volume). Neither is there trust for the press or its official sources. However, there is an inclination to trust experts (Zykova, this volume).

Despite the overall tendency to feel distrust, however, Russians are not inclined to protest. Exceptions occurred during Perestroika, when there was a short-term suspension of constraints and an invitation to participate. In fact, those who experienced successful activism during that period are the one group who remain ready to protest (Tsepilova and Mironova et al., this volume).

Another issue bearing on trust of government is that extreme environmental disasters have created issues of dependency for many victims now reliant upon receiving government support because of their victimization. One of the participants in the Ural conference noted the inherent contradiction that such disabling implies.

> People have government assistance. It is a contradiction because they do not trust the state.

While neither Americans nor Russians fit expectations for being constantly vigilant over the threat of environmental harm, known toxic victimization in both countries causes environmental turbulence and attendant consequences. In both societies, those who accept that they have become environmental victims and who are sufficiently informed enter into membership of the Risk Society.

RESPONDING TO CONTAMINATION: PARADIGMS OF EXTERNAL AND INTERNAL THREAT CONTROL

An interesting difference between Russians and Americans emerges in the response to environmental threats, once recognized. In this sense, we might suggest that Americans are externally focused on the ecology and biosphere. In contrast, Russians appear to be internally focused on the "endoecology" or inner ecology of the human organism.

Control of external threats

Americans tend to view issues of contamination from the psychological perspective of personal and social control (Edelstein, 2004). Accordingly, we are likely to view contamination as an unacceptable condition. The emphasis in our response is upon changing the contaminating conditions rather than adjusting to them. Our regulatory ideology is of "pollution control." We measure and regulate "point sources." Our pollution laws require use of "best available technology."

These strategies are external. The idea that people tolerate hazardous external conditions and, instead, modify their own behaviors and physiology as an adaptive step is anathema to the American way of thinking. Placing themselves above and outside of nature and assuming a high degree of personal control readily extended by technology, Americans are activists, targeting the threat rather than the consequence.[1] Their preferred remedies involve technical rather than social engineering.

At the same time as Americans believe in "tech fixes," however, there are frequent clashes between technologies perceived to be hazardous and activism in defense of personal property, lifestyle and health. Americans consistently engage in local, collective defense of place termed by detractors who view it as a form of selfishness as the "Not-In-My-Backyard" or NIMBY syndrome. Collectively, NIMBY has amounted to a societal rejection of hazardous and stigmatized technologies, contradicting the dominant paradigm's acceptance of risk (Edelstein, 2004). While the labeling of facility opponents in the U.S. is not dissimilar in form from the dismissal of Russian opponents as "chemophobic" (see Filatov, this volume), the difference is that the process of siting a hazardous facility offers the public notice, information and access to the environmental review, the right to comment and the right to challenge in court not available in comparable type or degree in Russia.[2] It is not just that Americans fear the proposed hazard, they participate within formal processes that guarantee their right to be heard.

In a conversation held on June 5, 2002 in Volgograd, Boris Filatov recounted a Russian instance of NIMBY, the social uprising that blocked the use of the nation's first chemical weapons disposal facility.

> The "Chapiavsk Syndrome" came from a revolution of the population against the planned Samara site so it never happened.

Officials learned from this situation and subsequently succeeded in opening a radioactive disposal site shortly before I spoke to Filatov.

> In Saratov, they had the prior experience to work with the population. The facility opened two months ago.

Efforts to import the "control paradigm" based on the U.S. regulatory approach to Russia occurred in the 1990s. Funded by USAID (United States Agency for International Development), the U.S. Environmental Protection Agency (EPA), undertook a series of collaborative environmental projects. In Volgograd they conducted a pilot study for air quality improvement techniques applicable throughout the former Soviet Union (EPA, 1997; Filatov et al., 1998; Wolff, Ozkaynak, & Filatov, 1997).Working with a Volgograd Citizen's Environment Task Force set up to oversee the projects, a model quantitative health risk assessment was conducted by a joint Russian/U.S. task force, the Volgograd Health Risk Assessment Working Group. A quantitative health risk assessment of current, stationary-source, industrial air emissions identified a toxic triangle formed around the second author's native neighborhood in the city of one million formed by the Red October steel plant, the aluminum plant and the silica brick plant. Fugitive plant emissions from the triangle (those escaping by by-passing "the stack") were found to be predominantly responsible for an estimated 2,700 additional deaths per year in Volgograd due to particulate matter (PM10). Another 13 annual deaths were estimated to occur due to carcinogenic emissions.

Mitigations were planned and piloted. In conjunction with a proposed "Russian Air Management Program," low-cost pollution prevention and control measures in several factories led to an 8% reduction in emissions by stationary-sources. It was further estimated that another 30–35% reduction was possible. Low cost air pollution reduction measures were particularly expected to work at the brick enterprise. Furthermore, an inexpensive and simple way was proposed to halve fugitive emissions and increase production from between 400 and 600 electric-arc furnaces in Russia using a U.S. technology for making better furnace lids.

The collaboration also proposed improved methodology and upgrading of the Russian air quality dispersion model, new equipment and control strategies and a program of pollution control training. Steps requiring little capitalization or funded by the U.S. were undertaken. However, the irony was that, in the 1990s, the Russian economy did not support even modest investments to reduce environmental risk. It remains to be seen if such steps will be taken in richer times. The "control paradigm" does not work if there is no way to exercise control or if priorities in a lean time are elsewhere.

Filatov co-authored a key risk report as part of the EPA project. He explained during the first author's June 5, 2002 visit that he had made a sustained effort to import the American practice of risk assessment to Russia. But the effort was largely futile because of Russian reticence to address causal relationships, both in assessing the potential consequences of known instances of environmental contamination and also in looking backward from symptoms to potential causes.

> Russians will assess the situation and identify specific locations that are above pollution levels but then not try to look for the source of pollution.

Of course, if the source is not identified and studied, no remedy will be forthcoming. Filatov also confirmed this comment made by a local environmental official about inherent corruption in environmental protection.

> Money from the environmental protection fund is not spent on environmental protection. Officials are frequent travelers.

Filatov's version of the corruption, however, had the funds coming back to the agency, not necessarily into the pocket of the official.

> The Russian approach is to just note the fact that pollution has taken place and then administer fines. There is a conflict of interest in not cleaning up contaminated industrial sites because the fines can continue to be assessed. Part of these monies go to the environmental agency for its budget.

As a result not only is there no incentive to see the problem solved, but there is a decided vested interest to see that it continues. Russian corporations renewing their "environmental passports" pay these fines as the cost of doing business. The fines are much less than the cost of solving the problem. The result is a perpetuation rather than solution of problems.

Endoecology

In the course of this work, we have encountered in Russia an alternative perspective on environmental control. Endoecology involves attention to the internal balance of human health and the means for restoring it in areas of ecological distress. Our first encounter with endoecology occurred across the Volga River from Volgograd in the highly polluted industrial satellite city of Volzhsky.

In late January 2000, the first author visited the Volzhsky Maternity Hospital. The hospital takes high risk late term pregnancies (past 33 weeks) from the entire Volgograd oblast. Although not equipped to test expectant mothers for chemical or radioactive exposures, a marked increase in bacterial infections and bone disease prompted a search for ways to boost the immune systems of mothers and children.

On the advice of a Moscow physician, the doctors focused their effort on the lymph system as a pathway for cleansing the body. A video demonstrated their use of very long needles to make special injections directly into the lymphic system. Blood was also passed through lasers for detoxification. These approaches were combined with traditional and holistic health care to address the health of the whole family: mother, father and prior children. There is also an effort to assure that parents have a healthy spirit as well as body. The doctors described their focus on a safe delivery as that of a fire brigade approaching a fire.

Outcome data that was shared showed a dramatic drop in incidence of health problems and shorter hospital stays after the new program was implemented. The earlier the mother began prophylactic treatments before birth, the less hospitalization and medicine has been needed. Because the maternity hospital deals with high risk pregnancies, it is hard to compare their statistics to other hospitals in the region.

Since there is no control over the healthfulness of environment in which their patients are forced to live, these doctors consciously refocused on the inner ecology or "endoecology," in order to compensate for pollution. In their view, endoecology is more important than outer ecology because it is under the patient's control.

Within days, a visit to the Ecological Gymnasium revealed another instance of endoecological practice (Smirnova, this volume). Much later we met Oleg Glazechev, a physiologist who has undertaken an active program for developing endoecological methods (this volume). His father was the second author's rector, as well as her mentor and collaborator during the creation of the gymnasium. Her reunion with the father led to our meeting the son.

In many ways, endoecology is a logical extension of the prophylactic Russian traditions of the baths (banya) and of the sanitorium, a health spa where one would vacation for rest and renewal and possibly cure. It fits well with the teachings of Ivanov, an early 20th century health guru in Russia who taught means of stress control and who has many followers in Russia today. Yet, a brief search for additional literature on the internet yielded only a few examples, including the use of selenium to improve the ability of kindergarten children to counteract the effects of xenobiotics (arsenic and mercury) in their local environment (Andreeva et al., 1996) and some obscure references to Ukrainian programs directed at children in areas exposed to Chernobyl. All references fit the Russian context, offering strategies for helping people adapt to a polluted world while presuming that pollution is a long term reality not subject to mitigation. The paucity of data suggests that it is premature to conclude that Russians widely employ endoecological approaches. Rather, it may be safer to say that palliative coping strategies appear to be more characteristic of Russian responses to contamination than the active responses that more characterize the U.S.[3]

Nevertheless, our own observations reveal that endoecology seems to be diffusing into mainstream practice in Russia. For example, one of the Volgograd environmental officials we met with claimed that practices found at the Ecological Gymnasium had spread to his local neighborhood kindergarten. Apparently both the salt cave for asthma and the stress reduction room had been replicated there and preventative injections were used to build the children's immune systems. Both officials made reference to the prophylactic summer "eco camps" attended by children from local orphanages for stays of twenty days at a time.

Even those Russians unfamiliar with the term and concept of endoecology, therefore, tended to pursue an adaptive and prophylactic strategy. For example, an activist at the Ural Conference from the highly contaminated city of Zlatoust who had never heard of endoecology, per se, described an effort to create a prophylactic system there. The approach was based, interestingly, on local longitudinal health surveys developed in the U.S. by Linda Price King, a former colleague of Lois Gibbs at the Center for Environmental Justice and Health. King left the center to specialize in measuring community health impacts. The activist explained.

> Geptel is a chemical industry. Disease in the youth has gone up 15 times what is expected. There is twice the expected birth deaths. A third die at working age. More are dying than are born. There was no data on the cause. A volunteer organization was formed with doctors and nurses and teachers. They used a survey developed by Linda

> Price King adapted for Russian use. They concentrated on the worst hit sections of the city and found a lot of blood disease and cancer.

Having proven the symptoms, the Zlatoust activists now concentrated on a prophylactic response.

The creative energy placed on adaptive strategies has led to interesting outcomes not likely in the U.S., with its focus on changing the conditions rather than learning to live with them. Yet, many environmental exposures in the U.S. have already proven long term and are likely to continue. Without abandoning a focus on controlling pollution sources, the contaminated legacy that already exists may well warrant applications of endoecological practice to issues such as skyrocketing rates of asthma, cancers, attention disorders and autism and to helping children (and adults) better adapt to living in hyper-contaminated areas.

The first author was openly incredulous when the doctors in Volzhsky espoused a belief that, if enough people practiced internal harmony and balance, their collective endoecological health would bring health to the external ecology, as well. They felt that endoecology should become the focus of U.S. environmental groups for this reason. On reflection, while the idea that making people's inner ecology healthy can cure a contaminated environment steps outside a Western cosmology if taken literally, it may be true that people who learn to carefully safeguard their physical and psychological health are likely to act to safeguard the outer ecology as well. If it serves to motivate protective actions, perhaps endoecology is indeed a path toward curing the external ecology.

THE LIONS' BRIDGE AND THE NEW POSTMODERNITY

The legacy of pollution created during the modern period is certain to endure even as the unsustainable nature of this paradigm forces its abandonment. And, now with the addition of the meta-contamination issue of global climate change added, we have left ourselves without easy choices and no options that would maintain the status quo. The issues raised by the parable of the Lion's Bridge have become all the more urgent.

Given the respective propensities for cultural immunity in the U.S. and denial in Russia, we can reasonably ask, "Must all recognize that they are toxic victims to appreciate the problem?" And how do we respond? Is a sustainable future served, on one hand, by bridge inspections and periodic repair or, on the other hand, by our learning to accept and dodge the hazard or, if that fails, to survive it? In reality, both strategies are important. But we really need to rethink our bridge and the conditions that cause it to fail – to choose what I have called "a third path" (Edelstein, 2004). Let's explore the options.

Repair the Bridge

In the U.S., our lengthy experience in trying to repair the bridge has revealed serious limitations in the strategy, at least as practiced. Employing a regulatory approach, we have created a system of inspections resulting in sporadic and limited technological fixes guided by an ever changing state-of-the-art. Despite some successes in moderating some types of pollution, the approach has hardly stopped contamination from occurring.

The endpoint of regulation is the protection of the environment. Regulation seeks to minimize harm to ecological systems beyond their assimilative capacity. As a result, its goal is not to prevent all pollution, but only that beyond what it is assumed that nature can handle. However, the incremental and fragmented approach employed to measure and control pollution is in no way pegged to ecological limits, which are rarely understood until seriously tested. And, of course, regulation aimed at point sources – tail pipes and smoke stacks and discharge points – does not address the problems of widespread ambient, non-point exposures. Furthermore, assuming that every instance of pollution occurs in a vacuum, it fails to look at cumulative and long-term issues. Within a Culture of Contamination, the regulatory approach is merely the Dutchman's thumb thrust hastily into the dike to plug a leak. The result has not only been a local and regional ecological harm but also a loss of global health and systemic balance.

Even the effort to focus on specific contaminated sites is of limited success. Despite the Superfund program's goal to effect timely cleanup for selected sites, many sites languish unremediated and nearby communities remain at risk. Cleanup levels are compromised due to costs and other tradeoffs. And, besides, Superfund has not been reauthorized, limping along underfunded, understaffed, and underutilized. Given the many faults of the Superfund approach, Brownfields redevelopment has emerged as a market driven alternative. While effectual, Brownfields involve serious compromises when long-term health and environmental integrity are considered (see Hoffman, this volume).

The limitations of the government regulatory approach have led some in the West to advocate a non-governmental voluntary approach (self-regulation) or a mixed public–private solution based upon incentives for private actors to do the right thing. Indeed, pollution trading and carbon credits may help reduce aggregate problems at the cost of perpetuating localized or regional problems. And a change toward Industrial Ecology and the greening of business is beginning to take shape. Nevertheless, it remains unrealistic to expect an unregulated corporate sector to initiate change on its own outside the context of regulatory control.

Finally, it should be noted that American technological fixes do not require people to take responsibility for their actions, and thus will always be undermined by a culture that considers it to be responsible to drive SUVs in a time of air pollution, oil shortages, and global warming.

A Russian effort to maintain and enforce strict laws and to invest in pollution reduction would be a major step indeed. And, in both Russian and the U.S., there is a need to address the current ability of government, media, and corporations to control and distort information, a common problem that we have documented in this volume despite the respective differences between the nation's overall ability to function in an open and democratic manner. Nevertheless, we cannot rely on a regulatory approach by itself.

Learn to Swim

Is endoecology, a viable option in the American context? The best argument for considering prophylactic approaches is the limits of the regulatory system in the U.S. to stop pollution and to mitigate contamination. The failure is evidenced, for example, in the explosion of asthma cases, particularly among children and minorities. As is usual in the U.S., we treat the symptoms, in this case with inhalers. But, short of preparing people for climbing mountains or high altitude sports, we do not engage in the kinds of adaptive preparation that has occurred, at least at the experimental scale, in Russia. Findings by Glazechev (this volume) raise the possibility that we can enhance the lung performance of children as a means to help offset urban air pollution and other asthma causes.

But should we do this? A major roadblock to Americans even considering such approaches is their belief as a matter of right that the environment should be made safe rather than the individual forced to adapt. Is it moral, in this context, to adapt the child to a sick situation rather than creating a safe environment? This question can be countered by the fact that children are currently exposed to polluted environments and, hollow rhetoric aside, will continue to be. Thus, is it moral to not enhance children's capacities to offset polluted settings?

In fact, there are obvious limits to the effectiveness of an endoecological approach. While a prophylactic approach increases certain tolerances, capacities, and resistances of exposure victims, the nature of toxic exposure offers diverse threats to multiple systems. To be truly effective and protective, an endoecological approach would have to enhance human capacity to survive multiple and synergistic threats. Hypoxia training is not likely to help offset environmentally caused cancer. The Russian practice of adapting to contamination inappropriately places too much demand on the ability of the organism to out evolve existing and future dangers.

Despite its holistic nature, its focus on stress reduction and its efforts to improve general immune system functioning, it is infeasible that even the most advanced system of endoecological training and treatment could offset worst case environmental conditions. It is unlikely that the evolved superhuman envisioned in an old magazine cartoon will ever be created – able to breath polluted air, drink polluted water, eat polluted foods, and touch polluted materials without harm. And if a human with those capacities were achievable, would we want to be that person?

Methods for enhancing the individual's general resistance to unhealthy environments deserve consideration for general application. But not as a statement that one has given up on a future where people live in healthy environments.

Despite advantages to both "solutions" to the Lion's Bridge parable, neither the regulatory nor endoecological approaches can cope with a world production system that generates hazards by its very essence. At its core, our technocracies are about solving problems by creating bigger ones, meeting objectives by maximizing resource and energy use and destructive levels and forms of throughput and pollution. Their nature is maximally entropic. To face spending billions on environmental cleanup when we merely cause new forms of contamination makes no sense. One cannot get ahead of the game. Likewise, the prospect of modifying humans to be adaptive to a world that presents ever greater adaptive challenges is also self-defeating. The point is that Cultures of Contamination cannot be modified to be safe. And they represent impossible adaptive challenges. No use or combination of these approaches can change things if we continue to operate within the current frame.

A Third Path

It is a clear conclusion from this comparative study that both the U.S. and Russia have engaged in widespread abuse of the earth, with adverse ramifications for people and nature alike. Despite significant differences, both are contaminating cultures. The differences in response to contamination invite a potential for mutual learning – both for addressing the existing negative environmental legacy and in charting a different future course that recognizes that the future prospects of their respective peoples depends on the health of the planet they share.

The Parable of the Lion's Bridge demands a third path, in the sense of an alternative to the two choices of attempting to fix problems as they arise or else adapt to them. This third path requires a total reworking of human systems to employ renewable, regenerative and non-polluting and natural rather than synthetic approaches that maintain and enhance nature's capabilities. A sustainable world system might preserve future qualities now being sacrificed by the Culture of Contamination. The goal will be healthy individuals living in a healthy world. This is indeed a goal worthy of future collaboration.

There are both different and similar challenges in the U.S. and Russia for achieving sustainability. In the U.S., people must accede to change even when many are not yet convinced of the bankruptcy of our current approach. The challenge is to educate a populace that has tuned out learning, on one hand, without the complexity of our problems causing people to give up, on the other hand. Given the American tendency for short term thinking, the risk is that a sense of hopelessness could accentuate an already substantial thread of "live for today" thinking. And change will require active participation, a vibrant democratic process and an exploring and informing media. Americans have arisen to the challenge of major social changes repeatedly in their relatively short history. All of the avenues for such change will be required.

Russians have engaged in even shorter term thinking, are historically disempowered for various reasons discussed, and lack an open, free, and democratic society able to be led from the grass roots. But Russians bring to the current situation a significant advantage. They have already weathered the break up of the old order. They find themselves in the chaotic center of transitioning to a new set of rules, as yet undefined. Ironically, being pushed back toward pre-modernity and membership in the developing rather than developed world offers a greater potential for crafting an alternative there. The slate has been wiped relatively clean. If a political will can be focused on sustainable ends, the potential for employing green approaches for new projects from the ground level up offers Russia (and the entire third world) a chance to bypass the high entropy (polluting and resource depleting) practices that are so deeply ingrained in the U.S. (Sachs & Agrawal, 2006). Given the failure of the culture of contamination, there is a comparative advantage to starting de novo.

This third path must be guided by sustainable innovation, characterized by several principles. It is freed of the assumption that the old order must be blindly maintained. The focus is to anticipate the consequences of future actions through a precautionary lens. And it engages widespread participation to serve the critical function of preventing narrow interest groups from exploiting the commons (Botkin et al., 1979). Unlike the Culture of Contamination shared by Russia and the U.S., sustainable innovation is oriented toward prevention rather than solving problems once created. It is precautionary; avoiding risk taking that then must somehow be managed. In avoiding or severely minimizing the creation of new threats, it also turns an innovative eye toward the task of resolving the legacy of known environmental threats and the need to manage response to new unfolding or emergent disasters, some human caused.

We must also engage in sustainable innovation regarding how humans deal with the earth, engaging in critical redefinition of the self and its relationship to the environment. Ironically, the concept of "the environment" is an integral component of the Culture of Contamination. Because the abstract notion of "environment" is viewed as a realm that is somehow separate from the person, harm to the environment does not translate in our minds as harm to ourselves. In contrast, our surround or "ambient" envelops our bodies so tightly as to be seen as a second skin. Anything that threatens the ambient threatens us. Therefore, we are vigilant about the fate of our ambient even if we are indifferent to the future of the environment (Von Uexküll, 1984).

The Russian practice of prophylactic endoecology approaches, at its broadest, a technique for taking an ambient perspective to the earth. During the first author's visit with the Volzhsky maternity doctors, he was openly incredulous when all three physicians espoused a belief that, if enough people practiced endoecology, they could collectively repair the earth. On reflection, however, it became clear that people who become balanced in their own physical and psychological health are likely to act to safeguard the outer ecology as well. The stake in protecting the earth comes from the redefinition of self as encased by, rather than separated from, nature.

Within a third path, the externally focused "ecological" approach can work hand in hand with an "endoecological" approach. The approach avoids the creation of a cascade of new problems caused by the nature of society itself. Instead, the goal will be the restoration of both the outer and the inner ecological worlds under conditions in which there is a chance to establish the right conditions for a sustainable future. In short, the goal will be healthy individuals living in a healthy world. This is indeed a goal worthy of future collaboration.

To return, then, to the magnificent Lion's Bridge. There is value in a program of official bridge regulation, inspection, and repair. And there is value in the population knowing how to swim, in case they plunge through the floorboards into the canal below. But perhaps the greatest value comes from learning how to build bridges more sustainably, creating a system of contributory care that draws its effectiveness from a community of many witnesses and many actors that mutually care for and protects its commons and citizens. Hopefully, this volume has itself helped to build bridges toward his end. Cultures of Contamination need to be replaced by Cultures of Caring.

NOTES

1. Of course, in reality, there are exceptions. For example, key areas of the country have been allowed to remain out of compliance with the Clean Air Act for decades while increasing numbers of people carry asthma inhalers.

2. Russians do not have comparable opportunities for public involvement. Environmental litigation remains in its early stages. And environmental review under OVOS, the environmental impact assessment process, is strongly biased in favor of government experts. For a public review to occur, a petition for a Public Expertise must be filed and accepted and suitable experts must be found to file the review. While the process has been used to successfully stop projects on rare occasions, it has not functioned to open up the review process. Recent changes to U.S. laws, such as the most recent energy act, limit the role of citizen opponents in siting decisions. These represent a step backwards.

3. It is instructive that the one instance of endoecological practice evident in the west is the use of hypoxia training (Glazachev, this volume) as a training tool for mountain climbers to prepare them to cope with thin air environments. While the technique may be very similar to that seen in Glazachev's application, the context could not be more different. The mountain climbers are athletes preparing to enter abnormal environments voluntarily and for a short duration. Russian children are just growing up in their hometowns.

REFERENCES

Andreeva, S. R., Volosnikova, T. V., Ganeev, A. A., Dovoretskaja, T. I., Ivanenko, A. A., Ivanenko, N. B., Nosova, E. B., Popov, S. V., Filippova, S. O., & Yakovleva, E. M. (1996). Endoecology, Physical Culture and State of Health of Preschool Children. *Physical Culture*. Accessed on line December 26, 2007 at http://lib.sportedu.ru/Press/FKVOT/2003N3/Index.htm

Arendt, H. (1958). *The human condition*. Garden City, N.J.: Doubleday.

Beck, U. (1986). *Risk Society: Towards a new modernity*. London: Sage.

Botkin, J., Elmandrja, M., & Malitza, M. (1979). *No limits to learning: Bridging the human gap*. New York, NY: Pergamon.

Edelstein, M. R. (2000). Outsiders just don't understand: Personalization of risk and the boundary between modernity and postmodernity. In: M. Cohen (Ed.), *Risk in the modern age: Social theory, science and environmental decision-making* (pp. 123–142). New York: St. Martin's Press.

Edelstein, M. R. (2004). *Contaminated communities: Coping with residential toxic exposure*. Boulder, CO: Westview Press.

Erikson, K. (1994). *A new species of trouble*. N.Y.: Norton.

Filatov, B., Kolodi, T., Wolff, S., Ozkaynak, H., Larson, B., Shaposhnikov, D., Avaliani, S., & Vishnavetskaya, L. (1998). Assessing chronic health risks from stationary source air emissions in Volgograd, Russia. In: I. Linkov & R. Wilson (Eds). *Air pollution in the Ural Mountains* (pp. 231–238). Netherlands: Kluwer.

McDonald, M. (April 7, 2004). *Russia finally acknowledges '57 nuclear disaster*. Seattle Times, downloaded on August 7, 2006 from http://seattletimes.nwsource.com/html/nationworld/2001897370_nukeaccident07.html

Milbrath, L. (1984). *Environmentalists: Vanguard for a new society*. Albany: SUNY Press.

Olsen, M., Lodwick, D., & Dunlap, R. (1992). *Viewing the world ecologically*. Boulder, CO: Westview Press.

Sachs, W., & Agrawal, H. (Eds). (2006). Fairness in a fragile world: A memo on sustainable development. In: T. Schoyer & T. Golodik (Eds), *Creating a sustainable world: Past experiences, future struggles* (pp. 31–58). N.Y.: Apex Press.

U.S. Environmental Protection Agency (U.S. EPA). Volgograd Health Risk Assessment Working Group (1997). Assessing chronoic health risks from stationary source air emissions in Volgograd, Russia: A Case Study in Quantitative Health Risk Assessment. DDP 601.

Von Uexküll, T. (1984). *Ambient and environment: Or which is the correct perspective on nature?* Paper presented at the International Association for the Study of People and Their Physical Surroundings, West Berlin, West Germany, July 25–29.

Weinstein, N. D. (1989). Optimistic biases about personal risks. *Science, 246*, 1232–1233.

Wolff, S., Ozkaynak, H., & Filatov, B. (September 1997). *Assessing chronic health risks from stationary source air emissions in Volgograd, Russia: A case study in quanitative health risk assessment*. The Volgograd Health Risk Assessment Working Group. Cooperative Agreement #118–0003-A-00–5311–00.

Yanitsky, O. (2000). *Russian greens in a risk society*. Helsinki: Kikimora.

Yanitsky, O. (2003). The sociology of risk: Key ideas. *Mir Rossii, 11*(1), 3–35.

ABOUT THE AUTHORS

Tatyana J. Andrushchenko is dean of the School of Psychology and Social Work, head of the Department of Social Work and professor of psychology at Volgograd State Pedagogical University. Dr. Andrushenko earned her Ph.D. in Psychology from the Russian Federation Education Academy, Institute of Psychology. Her primary work is in child development, counselling methods, interpersonal communication, and the evaluation of social psychological services. She has participated in exchange programs in Denmark, the Netherlands, and the US. tandr@vspu.ru

Robert Binford is completing his Ph.D. in Clinical Psychology from Fairleigh Dickinson University. He is currently finishing his internship in the California prison system.

Alla Bolotova is a research associate of the Centre for Independent Social Research in St. Petersburg. She holds a master degree in sociology from the European University at St. Petersburg and a master degree in environmental studies from the State University at St. Petersburg. She is currently completing her Ph.D. thesis titled "The Environmental Movement in Russia: the Politics of Everyday Life." Her interests include conceptualizations of nature, nature and everyday life practices, environmental history, civil society, and social movements. Ms Bolotova has participated in a number of international research projects, mainly dealing with analysis of social aspects of environmental problems in post-Soviet Russia. She has published a number of articles on the Russian environmental movement and the conceptualization of nature in the Soviet Union. xnu@indepsocres.spb.ru

Nikolay G. Britanov, an M.D., is head of Occupational Hygiene Laboratory of Research Institute of Hygiene, Toxicology and Occupational Pathology (RIHTOP) at Federal Medical and Biological Agency, Volgograd, Russia.

Stephen R. Couch is professor of sociology and science, technology and society, Pennsylvania State University, Schuylkill Campus. An accomplished musician and award-winning teacher and researcher, he earned his doctorate in sociology from Binghamton University. As an environmental sociologist, Dr. Couch has co-authored and co-edited four books, published more than thirty research articles, and presented his work throughout the United States, Europe, Mexico, and Japan. His work addresses technological hazards; current research interests include community breakdown and recovery from environmental contamination, the relationship of lay and scientific knowledge concerning environmental risks, the collective construction of meaning by groups victimized by modern technology, and disasters and popular culture. He has served as an expert witness in toxic tort litigation and consulted for government and industry. He has served as editor of Social Problems Forum, held positions of leadership in various sociological organizations and headed Penn State's Master of Arts in Community Psychology and Social Change and Center for Environment and Community. src@psu.edu

Michael R. Edelstein is professor of environmental psychology, convener of environmental studies and co-director of the Institute for Environmental Studies at Ramapo College of New Jersey. He holds a Ph.D. in Social Psychology from SUNY Buffalo. Dr. Edelstein has worked, spoken, and published widely on social impact assessment, transitioning to renewable energy, and sustainability on college campuses, but is best known for his thirty years of work on the psychological and social impacts of environmental contamination. Dr. Edelstein's *Contaminated Communities: Coping with Residential Toxic Exposure*, published in a second edition in 2004, is considered a classic in its field. Dr. Edelstein has served as an expert witness in major toxic tort cases and administrative proceedings, consulted to state and federal agencies, and has been awarded grants by the Fund for Improvement of Post-Secondary Education of the U.S. Department of Education, the Geraldine R. Dodge Foundation, and the Trust for Mutual Understanding. His work on local sustainability involves his twenty-five year long presidency of Orange Environment, a non-profit environmental organization based in Orange County, New York. In 2005, he ran unsuccessfully for Orange County Executive. medelste@ramapo.edu

Boris N. Filatov, an M.D., Ph.D. and Professor, is director of Research Institute of Hygiene, Toxicology and Occupational Pathology (RIHTOP) at Federal Medical and Biological Agency, Volgograd, Russia. He holds both a medical degree and a Ph.D. Originally concerned with monitoring workers' health and conducting research in a former weapons production facility, he has supervised the post-Cold War transformation of that facility into the current institute focused upon environmental health and remediation of chemical waste sites. He is the leading Russian expert on dioxin contamination. He is also globally recognized for his expertise in chemical terrorism. He served as the team leader of the Volgograd Health Risk Assessment Working Group that brought together Western and Russian health experts to assess the local environmental quality. Filatov@rihtop.ru

Margaret Gibbs, Ph.D., is a clinical psychologist who received her degree from the Department of Social Relations at Harvard University. She has been a professor of psychology for many years at Fairleigh Dickinson University, where she chaired the Psychology Department, headed the graduate program in psychology and served as provost. Her research publications are in the areas of disaster, community psychology and women's issues. gibbs1@optonline.net

Oleg S. Glazachev is professor in the Department of Normal Physiology, IM. Sechenov Moscow Medical Academy and heads the Department of Social Physiology, P.K. Anokhin Research Institute of Normal Physiology of the Russian Academy of Medical Science, the International Institute of Social Physiology, and the International Academy of Sciences, Russian section. He holds a degree in pediatrics from the Volgograd Medical Institute and a doctorate in Medicine from the P.K. Anokhin Scientific Research Institute of Normal Physiology. He previously headed a state program "Children of Chernobyl," Dr. Glazachev has authored more than 160 professional publications and co-authored six monographs and registered three inventions. His work addresses health and adaptation for people subjected to situations of psychoemotional tension, valeology, adaptation of children and teenagers living in unfavorable ecological conditions, and rehabilitation technologies employing hypoxia stimulation, contrast thermo-influence, and audiovisual stimulation. glazachev1@mtu-net.ru

William K. Hallman is a professor in the Departments of Human Ecology and Psychology and the director of the Food Policy Institute at Rutgers University. Dr. Hallman received his Ph.D. in experimental psychology from the University of South Carolina. He was awarded the dissertation of the year by the Community Psychology Division of the American Psychological Association for his work on public perceptions of risk, attributions of responsibility, and psychological distress in communities surrounding the hazardous waste landfill in Pinewood South Carolina. Dr. Hallman is an internationally recognized expert in risk perception and risk communication. His current research includes studies of unexplained illnesses in veterans of the Gulf War and the health consequences for New York City firefighters of responding to the World Trade Center disaster on 9/11. As director of the Food Policy Institute he is currently engaged in research efforts that encompass issues of hunger and food insecurity in the United States, preferences for organic and locally grown produce, and the threat of food allergens. He also heads projects examining public perceptions of genetically modified foods, American's mental models of bacteria and viruses, and the health, economic, social, and psychological consequences of intentional and unintentional food contamination. hallman@aesop.rutgers.edu

Janice L. Hastrup has a Ph.D. in clinical psychology and an M.S. in epidemiological research. She is currently on the faculty of the Department of Psychology at the University at Buffalo (SUNY) and is a licensed psychologist in New York State. She currently serves on the board of directors of the Love Canal Medical Fund and on the Western New York Hispanics and Friends Civic Association, Children and Families Task Force. She has consulted on several environmental hazard investigations and has collaborated on behavioral research on cancer prevention at Roswell Park Cancer Institute. Her major area of research involves health psychology, including health communications within families. She has served on several NIH and CDC study sections, and currently holds a grant from the National Cancer Institute to study beliefs about gene–behavior interaction in the causation of disease. jhastrup@buffalo.edu

Madelyn Hoffman, served as director of the Grass Roots Environmental Organization of New Jersey from 1983 through 1998. She is now the director of New Jersey Peace Action. During that time, she worked with more than 200 grass roots organizations on issues related to the cleanup of toxic chemical pollution at the thousands of contaminated waste sites in New Jersey. In 1999, Ms. Hoffman participated in an exchange program between the United States and the Ukraine, Belorus, and Moldova with other women involved in grassroots environmental activism. Ms. Hoffman was the Green Party candidate for governor of New Jersey in 1997 and Ralph Nader's vice-presidential running mate for New Jersey in 1996. njtoxics@aol.com

Catherine McVay Hughes graduated as a civil engineer from Princeton University, working on major engineering projects around the globe before taking an MBA from the Wharton School of Business and shifting her career to investment banking. She subsequently left Wall Street to work on affordable energy and lead poisoning prevention for the New York Public Interest Research Group, where she co-authored "Get the Lead Out," a widely read guidebook on lead poisoning prevention and mitigation. She left after a decade to use the Internet to catalyze grassroots knowledge-sharing networks for the long-term management of chronic diseases. Her company, AsktheMoms.com, LLC launched a web site for "families helping families manage asthma" (www.asthmamoms.com) shortly before the terror attacks that destroyed the World Trade Center and forced her family from their Lower Manhattan home one block from "Ground Zero." After 9/11, Hughes redirected her web site to extensive coverage of air quality issues for downtown residents and workers. Hughes moved her family back to Lower Manhattan in 2002 to be part of its rebirth. Beyond her work at Asthma-Moms.Com, she has served as Lower Manhattan Community Outreach Liaison for New York University School of Medicine, Department of Environmental Medicine; member of Community Board One and chair of its World Trade Center Committee; and from 2003 to 2005, Community Liaison to the U.S. Environmental Protection Agency World Trade Center Expert Technical Review Panel. cmch@verizon.net

Valentina V. Klauchek, a Ph.D., is deputy director for science of Research Institute of Hygiene, Toxicology and Occupational Pathology (RIHTOP) at Federal Medical and Biological Agency, Volgograd, Russia.

Sergei V. Klauchek, an M.D. and Ph.D., is dean of General Practitioners Institute, Physiology Department, Volgograd State Medical University; and project director, lead researcher of Research Institute of Hygiene, Toxicology and Occupational Pathology (RIHTOP) at Federal Medical and Biological Agency, Volgograd, Russia. He has written extensively about psychological issues relating to environmental risk and siting of hazardous waste facilities.

Ivan Kulyasov is a scientific researcher in environmental sociology for the Centre for Independent Social Research in St. Petersburg. He holds a masters in medicine. His fourty publications and four co-edited books have been published in Russian and English. He has collaborated on international projects with scientists from the USA and Finland. The focus of his research is the comparative analysis of mechanisms and practices of ecological modernization in Russia, the transformation of the forest sector and natural resource management in Russia, the Russian environmental movement, and ecovillage and sustainable communities development. Beyond scientific writing, Mr. Kulyasov frequently writes about his work in newspapers, popular environmental brochures and electronic lists. ivan-kulyasov@yandex.ru

Antonina Kulyasova is a researcher in environmental sociology at the Centre for Independent Social Research in St. Petersburg. A Ph. D. in economics from St. Petersburg University, she is an Alumna of the IREX program "Contemporary Issues" and held a 2002 fellowship in residence at Dickinson College. Dr. Kulyasova has more than 40 publications in print and has edited four books. She has collaborated in international projects with scientists from the USA, Finland, and Sweden, serving as coordinator of several research projects supported by international funds and scientific organizations. Her work focuses upon social and environmental issues of sustainable development, development of intersectional partnerships, sustainable management of watersheds, natural resource governance, restructuring of state environmental bodies, the transformation of the pulp and paper industry, localization of the global FSC certification process in Russian forest communities, and public participation in FSC certified forest governance. ku@indepsocres.spb.ru

Nadezhda Kutepova is a native of the ZATO Ozersk, Chelyabinskaya oblast. A graduate in sociology she is now studying law at the Ural State Law Academy. Ms. Kutepova is a seasoned activist focused on issues of life in closed nuclear cities. In 2001 she established the Democratic Information Resource Center to work on such goals as developing a civil community there. In 2003 she initiated litigation over violations of human rights in Ozersk (over the right to a birth place, to reunite families, to relocate and to choose one's place of residence). Subsequently, she was prosecuted by the Russian FSB for her work in human rights protection and for initiating the first research project on the history of ZATO. The prosecution was stopped after the Russian human rights protection movement and ecological community intervened on her behalf. Her continuing work is to promote human rights in Ozersk, seeking to reform the legal system there to conform to the Russian constitution and international human rights. She organized a campaign to protect the rights of widows of the liquidators of the 1957 nuclear accident at PO MAYAK (her own father was a liquidator). Her work has been published in the magazine "Ecology and Law" and on the Bellona Foundation web site (www.bellona.no). In 2006, she joined the working group of the Federal Agency of Atomic Energy on solving the problems of the population living in the zone of nuclear influence PO MAYAK. She is president of the non-profit organization "The Planet of Hopes," a representative of the international ecological group "Eco-protection" and leader of Community Service on Human Rights ZATO Ozersk.

Adeline G. Levine Ph.D. is professor emeritus from the Sociology Department at SUNY Buffalo, where she had a long and fruitful career working on issues related to women's rights and the environment as a scholar and activist. She is best known for her path-breaking study of the Love Canal toxic disaster in Niagara Falls in 1978 and 1979. Her 1982 book, *Love Canal: Science, Politics and People* (Boston: Lexington Books) became a classic in the social sciences and influenced toxic activism.

Natalya Makarevich (now Natalya Agaffonova) is a research student at Volgograd State Pedagogical University in psychology and social work. Her interests include ecologic psychology, the child's inner world and problems of self-consciousness. airahir@yandex.ru

Nicholas D. Martyniak has worked for the NJ Department of Environmental Protection, the National Park Service, and a number of community environmental groups across the New York metropolitan area. He is currently a Ph.D. candidate in urban planning and policy development at Rutgers University with his dissertation research focusing on the motivations for participation by grassroots communities in the environmental justice movement. Other research interests include risk analysis, environmental planning and environmental economics. He has been an instructor at Rutgers University for the last five years and a research assistant at the Food Policy Institute, Rutgers University for the last two years, under the direction of Dr. William K. Hallman, researching consumer responses to agroterrorism. nicdm@rci.rutgers.edu

Anne E. Mercuri is Program Director of Pathways, the Adult Behavioral Health Outpatient Services, at The Pottsville Hospital and Warne Clinic in Pottsville, PA. She also serves as an adjunct professor at Penn State University's Schuylkill Campus, where she earned an M.A. in community psychology and social change with an environmental concentration. Ms. Mercuri co-authored several articles in disaster studies addressing the Three Mile Island nuclear disaster, contaminated communities and the effects of disasters on children and adolescents. She served as assistant editor of Social Problems Forum for six years. aem141@psu.edu

Natalia Mironova is best known for her role as chair, since 1989, of the Movement for Nuclear Safety and leader and organizer of the Russian ecological and anti-nuclear movement. Originally trained as thermo-electric engineer at the Chelyabinsky Polytechnic Institute, she recently defended her Ph.D. in the sociology of governance from the Russian President's Academy of Government Service. In 1990 she was named to the State ecological expertise for the Southern-Ural Atom Station and the State commission investigating the ecological situation in Chelyabinskaya oblast. From 1990 to 1993, Dr. Mironova served as an elected deputy of the Regional Peoples' Deputies. In 1991 she organized the Chelyabinsk city referendum against the spent nuclear fuel import. That same year, she became chair of the Committee for Radiation Safety for Chelyabinskaya oblast. She has been an elected member of the Supreme Ecological Committee of the State Duma of the Russian Federation since 1996. In 2004 she joined the board of the directors of the Chelyabinsk Regional Institute of Public Politics and Law. In 2005 she was elected a member of the organizing committee of the political party "Green Russia" and the chair of its Chelyabinsk regional department. Dr. Mironova has developed numerous projects involving the interaction and cooperation of non-governmental and state organizations related to nuclear non-proliferation and the safety of nuclear arsenals, handling of nuclear wastes, and assistance of ecological victims. She has authored eight inventions and twenty publications in the field of thermo-dynamics and seventy analytical works in nuclear policy, sociology of governance, and civil society published in Russia and abroad. natalia_mironova@rambler.ru

Jonathan Reisman graduated with a B.A. in mathematics and a minor in philosophy from New York University in 2001. He subsequently spent several years in Russia studying international environmental organizations and their role in post-Soviet Russian society. In partnership with Maria Tysiachniouk, he conducted a Russia-wide in-depth study of the logging industry and the role that small and large environmental NGOs are playing in the push toward sustainability. He also participated in water quality monitoring exercises in St. Petersburg and Vologda. He is now a student at Robert Wood Johnson Medical School in New Jersey. sped 76@aol.com

Lyudmila V. Smirnova is associate professor of education at Mount Saint Mary College in Newburgh, New York. Previously she served as professor of education and dean of the School of Foreign Languages at Volgograd State Pedagogical University, where she also earned her Ph.D. Author of more than 70 articles and monographs, Dr. Smirnova has studied innovative educational approaches around the world. Certified in both American and International Montessori Instruction, she brought this long banned approach to southern Russia during Perestroika. Among her other projects in educational innovation was collaboration in the Ecological Gymnasium project in Volgograd. Her most recent interests are in advancing techniques for effective on-line teaching and developing a theory of "Pedagogical Recidivism," which addresses resistance to pedagogical innovation among teachers and students. smirnova@msmc.edu

Sherilyn N. Thomas holds a Ph.D. in clinical psychology and currently is a clinical assistant professor and director of the Psychological Services Center at the University at Buffalo. Prior to this position, she served as the project director for a large federally funded study examining the effects of motivational interviewing in reducing HIV-risk behaviors among heavy-drinking college students. Dr. Thomas is a certified trainer of the motivational interviewing technique. She is also a licensed psychologist in New York State and operates a part-time private practice, specializing in anxiety disorders treatment. sfthomas@buffalo.edu

Olga Tsepilova is a senior researcher at the Sociological Institute, Russian Academy of Sciences, St. Petersburg, where she earned her doctorate in sociology. She is the author of more than 60 publications in Environmental Sociology. In her 23 years at the Academy of Sciences, she has specialized in the study of social-environmental movements, environmental conflicts in new investment policies, the formation and development of Russian environmental policy, Russian environmental safety and, beginning with her involvement in the Trust project in 2002, the social environmental context of nuclear cities' development. She currently heads the Academic Project "Nuclear cities in Russia: Sociological Analysis of the Prospects for Reducing Ecological Risks." Dr. Tsepilova is a member of the board of directors of the St. Petersburg Sociologists Association and executive officer of the political party "Green Russia." tsepilova@mail.ru

Maria Tysiachniouk holds a Master of Science in Environmental Studies from Bard College, NY, a Ph.D. in biology from the Russian Academy of Sciences and a Certificate in Nonprofit Studies from Johns Hopkins University. She is currently earning a Ph.D. in environmental sociology from Wageningen University in the Netherlands. She has taught at Herzen Pedagogical University in St. Petersburg, St. Petersburg State University, Johns Hopkins University, Dickinson College, PA, Ramapo College of New

Jersey, Towson University, and offered short courses at several European universities. Dr. Tysiachniouk has written more than a hundred publications on topics related to the third sector and has had fieldwork experience in Russia, China, the United States, South Africa, and Brazil. She is currently chairing the Environmental Sociology group at the Center for Independent Social Research, St. Petersburg, Russia and doing intense field research on global forest governance as well as investigating the role of non-profit organizations in social transformation. She has completed several projects related to the study of contaminated communities and the role of NGOs in coping with contamination. tysiachn@hotmail.com

Abraham H. Wandersman is professor of psychology at the University of South Carolina-Columbia. He received his Ph.D from Cornell University in social and environmental psychology and social organization and change. Dr. Wandersman performs research, results-based accountability, and program evaluation on citizen participation in community organizations and coalitions and on interagency collaboration. He has directed major studies for the Centers for Disease Control and the RAND Corporation. He is the author of many books and articles, most recently on the concept of Empowerment Evaluation. He has won numerous awards, including the Myrdal Award for Evaluation Practice, the presidential prize from the American Evaluation Association and the Distinguished Theory and Research Contributions Award by the Society for Community Research and Action. He has served as interim co-director of the Institute for Families in Society at the University of South Carolina. Dr. Wandersman serves or has served on a number of advisory committees for prevention organizations addressing AIDS and HIV, mental health, substance abuse, and caregiving. WanderAH@gwm.sc.edu

Irina A. Zykova is the principal science researcher at the Research Institute of Radiation Hygiene in St. Petersburg, Russia, where she has served since 1973. She earned her medical degree from the Leningrad Medical Institute, now St. Petersburg Medical University, in 1969, and her Doctor of Medical Sciences in 2001. Dr. Zykova has studied the psychological and social consequences of the Chernobyl Accident on the Russian Population for the past twenty years. She conducted field interviews in the radioactively contaminated territories after the Chernobyl accident in June, July 1986 and again from 1988 through 2005. She pioneered and validated the methodology for radioanxiety monitoring. She has written and spoken widely about perceived risk and nuclear power, media coverage of nuclear issues, and the psychological consequences of radiation exposure. zia_2004@mail.ru

Printed and bound by CPI Group (UK) Ltd, Croydon, CR0 4YY
16/09/2026
14963402-0007